Mobile Robotics: Design, Methods and Applications

Mobile Robotics: Design, Methods and Applications

Edited by
Andy Evans

Willford Press
www.willfordpress.com

Published by Willford Press,
118-35 Queens Blvd., Suite 400,
Forest Hills, NY 11375, USA

ISBN: 978-1-64728-532-6

Cataloging-in-Publication Data

Mobile robotics : design, methods and applications / edited by Andy Evans.
p. cm.
Includes bibliographical references and index.
ISBN 978-1-64728-532-6
1. Mobile robots. 2. Mobile robots--Design and construction. 3. Robots. 4. Robotics. I. Evans, Andy.
TJ211.415 .M63 2023
629.892--dc23

For information on all Willford Press publications visit our website at www.willfordpress.com

WILLFORD PRESS

Contents

Preface

Every book is a source of knowledge and this one is no exception. The idea that led to the conceptualization of this book was the fact that the world is advancing rapidly; which makes it crucial to document the progress in every field. I am aware that a lot of data is already available, yet, there is a lot more to learn. Hence, I accepted the responsibility of editing this book and contributing my knowledge to the community.

Mobile robots are automatic machines capable of moving without human intervention and performing certain actions. Mobile robotics is an emerging discipline of robotics and information technology that deals with the development of mobile robots. It is solution-oriented discipline, which involves the application of concepts from information technology, engineering science, artificial intelligence and cognitive sciences. Designing a mobile robot requires designing the different mechanical parts and developing a control system for the robot. Mobile robots are used in medical facilities to move patients, assist with surgical procedures, gather and transport supplies, and disinfect rooms. In agriculture, farmers use mobile robots for maintaining, measuring and harvesting fresh produce. This book traces the progress of the field of mobile robotics and highlights some of its key design concepts, methods and applications. It strives to provide a fair idea about this discipline and help develop a better understanding of the latest advances within this field. As this field is emerging at a rapid pace, the contents of this book will help the readers understand the modern concepts and applications of the subject.

While editing this book, I had multiple visions for it. Then I finally narrowed down to make every chapter a sole standing text explaining a particular topic, so that they can be used independently. However, the umbrella subject sinews them into a common theme. This makes the book a unique platform of knowledge.

I would like to give the major credit of this book to the experts from every corner of the world, who took the time to share their expertise with us. Also, I owe the completion of this book to the never-ending support of my family, who supported me throughout the project.

Editor

A Taxonomy for Mobile Robots: Types, Applications, Capabilities, Implementations, Requirements and Challenges

Uwe Jahn [1,*], Daniel Heß [1], Merlin Stampa [1], Andreas Sutorma [2], Christof Röhrig [1], Peter Schulz [3] and Carsten Wolff [1]

1 IDiAL Institute, Dortmund University of Applied Science and Arts, Otto-Hahn-Str. 23, 44227 Dortmund, Germany; daniel.hess@fh-dortmund.de (D.H.); merlin.stampa@fh-dortmund.de (M.S.); roehrig@ieee.org (C.R.); carsten.wolff@fh-dortmund.de (C.W.)
2 Faculty of Information Technology, Dortmund University of Applied Science and Arts, Sonnenstraße 96, 44139 Dortmund, Germany; andreas.sutorma@fh-dortmund.de
3 Faculty of Engineering and Computer Science, Hamburg University of Applied Sciences, 20099 Hamburg, Germany; peter.schulz@haw-hamburg.de
* Correspondence: uwe.jahn@fh-dortmund.de

Abstract: Mobile robotics is a widespread field of research, whose differentiation from general robotics is often based only on the ability to move. However, mobile robots need unique capabilities, such as the function of navigation. Also, there are limiting factors, such as the typically limited energy, which must be considered when developing a mobile robot. This article deals with the definition of an archetypal robot, which is represented in the form of a taxonomy. Types and fields of application are defined. A systematic literature review is carried out for the definition of typical capabilities and implementations, where reference systems, textbooks, and literature references are considered.

Keywords: robotics; mobile robot; taxonomy; requirements; challenges; capabilities; implementations; literature review; survey

1. Introduction

Robotics has become a popular research field over recent decades. The digital library IEEE Xplore (*IEEE Xplore* [1] is a relevant source for systematic literature searches in the field of computer science and engineering. All publications that have been published in connection with the IEEE are searched. The library includes conference, journal, magazine, books, courses and standards publications.) achieves about 240,000 search hits when searching for *"robot OR robotics"* in the metadata. Approximately 77,000 search hits were found in the search for mobile robots (*"mobile AND (robot OR robotics)"*).

Figure 1 shows that the popularity of robotics has been increasing almost constantly over the last ten years. The relevance of mobile robotics in scientific publications, on the other hand, remained almost constant between 2009 and 2016, and has been increasing rapidly every year since then. On the one hand, this shows that the research area *mobile robotics* is more relevant than ever before and on the other hand that there are more and more solutions and research results in this area in the form of publications.

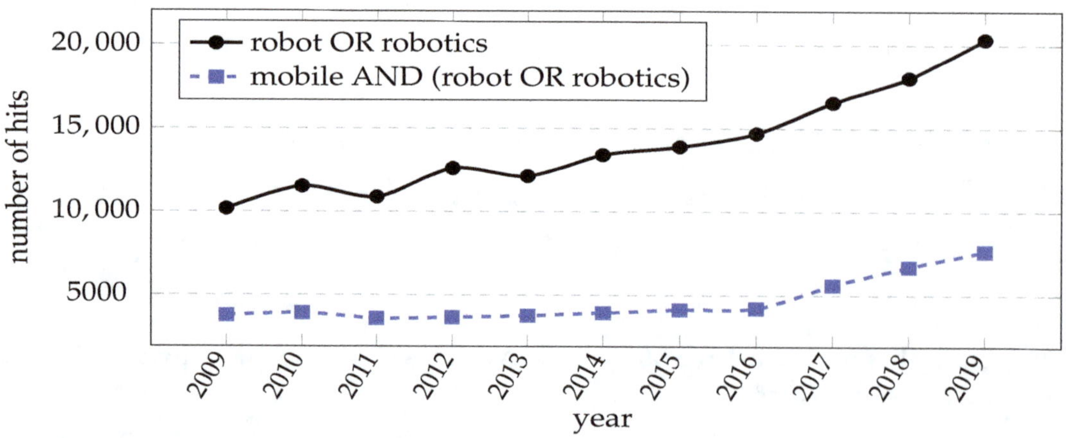

Figure 1. Development of search hits for *robotics* and *mobile robotics* (All search queries have been executed in IEEE Xplore in August 2020.)

The group of mobile robots that can move through its environment, is a carrier for many research topics within robotics, e.g., navigation or autonomy. Often researchers reference to the mobile robot. This article discussed how a typical mobile robot could be described by answering the following questions: What are the capabilities of a typical mobile robot? What technologies are topically used to implement its capabilities? What requirements and challenges result from these capabilities and implementations?

This research's primary goal is to define a taxonomy of mobile robots that describes the classes, applications, capabilities, and implementations of a typical mobile robot. The definition of a taxonomy was initially invented with biology. A taxonomy describes a method to classify objects. A taxonomy is often used to put individual cases into a unified and more generic context to handle those cases with standard methods. With the creation of a taxonomy, structures are established in which an object or structure can be sorted. This research uses taxonomy to define an archetypal mobile robot, whereas the indication "archetypal" means a generally valid definition of a device, here the archetypal definition of a mobile robot.

The taxonomy is mostly based on a systematic literature review, which is shown in Section 2, including the state of the art of taxonomies for mobile robots in Section 2.2. In Section 3 types of mobile robots, in Section 4 applications, in Section 5 capabilities, and in Section 6 implementations are analysed. The derivation of the taxonomy is shown in Section 7 before this article concludes with Section 8.

2. Literature Review

The basis of the definition of a mobile robot is created by running a systematic literature review [2,3]. The first step in the literature review is the definition of search terms. Those search terms are created by reviewing references of three types of literature. Those references will be described briefly in the next sections and will be compared and collected in Tables 3 and 4. First, books from the field of robotics will be examined in Section 2.1. Then, a selection of survey papers will be reviewed, which outlines the state of the art of (mobile) robotics (see Section 2.2). Last, other search terms will be searched by examining some reference systems from mobile robotics (see Section 2.3). The relevance of the selected references are shown in Table 1, which shows the number of citations with the exception of the *reference systems* where the number of search hits of references using this systems are listed. The quantification of the relevance is done by evaluating the search hits and number of citations using search engines for scientific references, namely *Google Scholar* (*Google Scholar* [4] runs its search queries independent from the publisher.) and *IEEE Xplore*. The selection of the survey papers and publications is not only done by pure number of citations and search hits. To minimize the chance to miss new or non popular research terms within mobile robotics, a mixture of old to new and few- to often-cited references are chosen. Reference systems that are known to the authors were selected, as well as randomly selected systems from search results. The quantification of each term will be done later by analyzing the relevance of each term in the literature (see Sections 5 and 6).

Table 1. Relevance of the selected references

Type	Reference: Title [source]	Reference: Year	Number of Citations/Search Hits: Google Scholar	Number of Citations/Search Hits: IEEE Xplore
Books see Section 2.1	Artificial Intelligence [5]	2009	35,011	N/A
	Probabilistic Robotics [6]	2005	10,226	N/A
	Springer Handbook of Robotics [7]	2016	3974	N/A
	Intro. to Auton. Mobile Robots [8]	2011	3665	N/A
	Robotics, Vision and Control [9]	2016	1545	N/A
	Embedded Robotics [10]	2008	568	N/A
Surveys see Section 2.2	Modular Reconfigurable Robots [11]	2007	813	N/A
	Challenges of Science Robotics [12]	2018	303	N/A
	Multi-Robot Taxonomy [13]	2016	126	N/A
	Challenges for Mars Expl. [14]	2000	86	N/A
	Robot Teleop, Taxonomy [15]	2015	33	N/A
	Review of Mobile Robots [16]	2019	20	N/A
	ANR Requirements [17]	2018	3	N/A
	Challenges of Mobile Robots [18]	2020	1	N/A
Reference systems see Section 2.3	Spot [19]	2019	945	70
	Khepera IV [20]	2015	187	46
	TurtleBot3 [21]	2017	180	35
	AMiRo [22]	2018	105	15
	WolfBot [23]	2014	80	10
	e-puck2 [24]	2018	75	8
	OmniMan [25]	2020	32	1
	ArEduBot [26]	2011	24	1
	Savvy [27]	2017	6	2
	Arduino Robot [28]	2018	5	1
	Omnidirectional Mobile Robot [29]	2016	4	0

All search queries have been executed in July and August 2020.

2.1. Books

Artificial Intelligence: A Modern Approach [5] by Russell and Norvig is one of the most popular books in the field of Artificial Intelligence (AI) and robotics. The book was quoted more than 35,000 times since its release in 2009 (see Table 1) and is often used as a textbook in study programs. The book's focus is on AI and its applications, e.g., in robotics. Detailed backgrounds are given on problem-solving, planning, reasoning, learning, and perception with or due to AI. Although the chapter on robotics is comparatively short, an overview of the hardware, software architecture and robot applications is given.

Probabilistic Robotics [6] by Thrun et al. is another indispensable book for research in the field of robotics, especially if the mathematical background of its algorithms and methods are required. In the beginning, the books explain the mathematical theories, e.g., for the Gaussian filter. This filter and many other algorithms and filters are explained in chapters such as localization, mapping, or planning and control in a high level of detail. Some practical examples are also described. The popularity of

Probabilistic Robotics is also recognizable with the high relevance in the scientific literature with more than 10.050 quotes.

Springer Handbook of Robotics [7] is, with about 2.200 pages, a good overview of robotics at a deep level. In the second version from 2016, the authors describe the current state of robotic systems' art. The book shows all basic robotics, including the design of robots, sensors and perception, manipulation and interfaces, moving in the environment, robots at work and robots and humans. Siciliano and Khatib claims to address topics with high topicality. In summary, both editions of this book have been quoted about 4000 times.

Introduction to Autonomous Mobile Robots [8] focuses, in comparison to the three already presented books, on mobile robots. Siegwart et al. start with an overview of classes of robots, following a detailed explanation of mobile robots' kinematic. Perception, localization, planning, and navigation are also shown. Helpful are the numerous examples from mobile robots, e.g., the introduction to sensors typically used with mobile robots.

Robotics, Vision and Control: Fundamental Algorithms in MATLAB [9] simplifies the work with robots if one has the intention to use Mathworks MATLAB [30] for the development of a robot. Corke shows many algorithms and functions which should be ready to be used in MATLAB. Examples are printed with MATLAB code, which helps new developers with entering the complex field of robotics. A big part of the book focuses on Computer Vision (CV) and how its applications are well supported with MATLAB. Modeling reactive systems, such as mobile robots, with MATLAB Simulink [31], is also described with practical examples.

Embedded Robotics [10] by Bräunl shows the combination of embedded systems and mobile robots. At first, the field of embedded systems is explained. Then, the design of mobile robots using embedded systems is described. The last chapter shows the applications of mobile robots. In *Embedded Robots*, the focus is always on hardware-near programming, whereas the other presented books mainly focus on algorithms or mathematical backgrounds. Here, Bräunl explains embedded development with robots up to the bit- and logic-level.

2.2. *Surveys*

In [11], the challenges and chances of robots are presented. The paper from 2007 was quoted about 800 times and is not explicit for mobile robots, but shows some trends in the research field of robots which can be adapted to mobile robotics. The focus of the publication lies on modular and reconfigurable systems. According to Yim et al., modular and reconfigurable systems are flexible to use and future-proof for future use-cases of mobile robots. The authors also describe the challenge of creating robust robots, both on the hardware and software level. Also, robots should be able to repair themselves independently, e.g., if used in unreachable terrain, e.g., on Mars. Algorithms to calculate the optimal configuration of robots and efficient and scaleable communications between robots in a big group are also defined as necessary preparations for future robots by Yim et al.

In [12], trends for robotics in education and research are shown. With (1) new materials, (2) bio-hybrid and bio-inspired robots, (3) energy, (4) swarm robotic, (5) navigation and exploration, (6) AI for robots, (7) brain-computer interfaces, (8) social interaction, (9) medical robots and (10) ethnic and safety ten trends are defined. The publication from 2018 shows a wide range of high-level trends, which can be adapted for mobile robotics. Despite the comparatively short availability of about two years, the publication has already been cited more than 300 times.

Ref. [13] gives an overview of target detection and tracking in multi-robot systems as a taxonomy. Even though the survey does not examine mobile robotics from a higher perspective, some aspects can be unified and adopted. Robin and Lacroix describes the necessity of decentralized algorithms to perform target detection and tracking with mobile robots. Besides, it is claimed that the target system needs to be monitored. This monitoring should be done directly on the target system and not only in simulations.

The requirements on the hardware within the development of mobile robots to explore Mars is shown in [14]. Huntsberger et al. describe the challenge on the mobile robot's hardware, without explaining the software's challenges. The focus is on the development of a suitable energy supply for the mobile robot. Also, the necessity of the robot's capability to recognize faulty parts on its hardware is described. The robot must also be able to repair those broken parts on its own.

In [15], a taxonomy for guidelines for the development of teleoperations with robots is shown. Furthermore, one focus of the publication is on integrating (human) users with the robots by giving usability guidelines. Adamides et al. define eight categories in the presented taxonomy. (1) Platform architecture and scalability, (2) error prevention and recovery, (3) visual design, (4) information presentation, (5) robot state awareness, (6) interaction effectiveness and efficiency, (7) robot environment/ surroundings awareness, and (8) cognitive factors.

Rubio et al. published a survey in 2019 on concepts, methods, frameworks, and applications of mobile robots [16]. Besides the classification into Wheeled Mobile Robots (WMR), Unmanned Arial Vehicle (UAV), and Unmanned Underwater Vehicle (UUV), the application field *navigation* is described in detail, e.g., the authors differentiate between localization and path planning. Besides, typical sensors for mobile robots are described, divided into two categories: (1) internal sensors, which measure internal data such as the velocity of the motors or the battery state, (2) passive and active sensors that measure the environment the robot.

In [17], concepts and requirements on Autonomous Networked Robots (ANR) are shown, and architectural patterns are compared. Therefore, centralized architectures are, according to Chukwuemeka and Habib, inferior to hybrid architectures of layered and decentralised approaches, especially in terms of reliability, robustness, and scalability. On the contrary, centralized architectures are more efficient than decentralized structures. The authors explain that the use of mobile robots in ANR are suitable for exploration and observation use-cases, where single robots are distributed into zones, which must not leave due to communication issues of each robot to the other mobile robots in the ANR system.

Alatise and Hancke show in [18] current challenges on autonomous mobile robots and its sensor data fusion. Mobile robots need to handle navigation in each application field. According to Alatise and Hancke other challenges are path planning, collision avoidance, and localization. Sensor data fusion is vital to gain reliable data from the sensors. Faulty data sets can be detected and corrected with sensor data fusion methods. Also, the authors described the state of the art methods for some popular sensors.

2.3. Reference Systems

As state of the art in mobile robotics the company *Boston Dynamics* stands out, e.g., with the robots *Atlas* or *Spot* [19]. *Spot* is the first of the *Boston Dynamics* mobile robots which is commercially available [32]. *Spot's* movement is highly inspired by natural role models, namely dogs. *Spot* has four legs, which helps the robot move robustly, e.g., to move in difficult terrain or climb stairs or past obstacles. The hardware can be individualized, e.g., it is possible to attach a manipulator or a moving camera on the robot's back. The robot can be controlled via a game-pad, like a remote controller, whereby the kinematics are calculated internally on the robot. Developers and users are provided with a high-level Application Programming Interface (API) and Software Development Kit (SDK) to implement their software (and hardware) for their use-cases. According to *Boston Dynamics*, *Spot* can be used in many scenarios, from transportation to observation of critical infrastructures.

Another popular robot is the *Khepera IV* mini robot, whose first version has been published in 1991. The current version is evaluated in detail by Soares et al. in [20]. The mobile robot processing unit runs an ARM Cortex-A8 CPU, which processes various sensors, e.g., infrared and ultrasonic sensors for collision avoidance and an RGB camera for object detection. Besides, the *Khepera IV* uses two microphones to localize sound sources such as voices. The mobile robot uses a modular approach and is extensible with extension-boards. The robot is mainly used in education and research.

Popular, especially as a demonstrator for the Robot Operation System (ROS), is the mobile robot *Turtlebot*, whose third generation, the *Turtlebot 3* [21] is currently available. The open-source robot has initially been developed by *Willow Garage*, who is also the core developer of ROS. The robot uses a modular approach and as standard a 360° Light Detection and Ranging (LiDAR) scanner, an ARM-based control unit, which controls, e.g., an Inertial Measurement Unit (IMU). A camera can be attached as an official extension. The control unit is a Raspberry Pi 3 Single-Board Computer (SBC). Noteworthy is the integration of ROS and ROS2, which makes the *Turtlebot* popular among new developers and beginners in the field of robotics. The *Turtlebot* is used in many use-cases, e.g., for Simultaneous Localization and Mapping (SLAM) applications to hardware extension to handle objects.

The *Autonomous Mini Robot (AMiRo)* is developed at Center for Cognitive Interaction Technology (CITEC) of Bielefeld University [22]. The mobile robots is implemented as a distributed and modular system with three main modules. Those modules, the *Di Wheel Drive*, the *Power Management*- and the *Light Ring* are each equipped with a Micro Controller Unit (MCU) which processes sensor- and actuator information under real-time conditions. The *Cognition*-module, which is an extension board for the *AMiRo*, hosts an ARM-CPU, which is more powerful than the MCUs of the basic modules. ROS [33] or Robotics Service Bus (RSB) [34] interfaces are provided for the *Cognition* extension. Another module for the modular approach of the *AMiRo* is the FPGA-based *Image Processing* module, which is typically used to proceed CV algorithms in a ultra-parallel approach. For communication within the distributed system the system provides e.g., a Controler Area Network (CAN) bus. The mobile robot is mainly used in education and research, e.g., in multi-robot applications using the color of the *Light Rings* to identify each *AMiRo*.

The *WolfBot* [23] is another mobile robot that is available as an open-source project. The robot uses an omnidirectional drive with three mecanum wheels. Sensors, e.g., infrared sensors, a camera, and a microphone, are handled by a BeagleBone [35] SBC, which is equipped with an ARM Cortex-A8 CPU. The robot can navigate via SLAM. Betthauser et al. emphasis the low energy consumption from 3.47 W to 6.27 W. Besides the hardware, the software of the *WolfBot* is also available as open-source. The mobile robot has been developed for research and education.

The robots *e-puck* [24] and *e-puck2* [36] are commercial mini robots which can be extended with various modules. The base of the *e-puck2* uses a STM32F4 MCU, an IMU, a simple camera and sensors for obstacle detection. The robot can be extended with more powerful computers, other cameras ans self-developed hardware. The mobile robot hosts several internal interfaces for communications. Many research projects can be found in the literature (see Table 1).

OmniMan is the name of another robot system, which is presented by Röhrig and Heß in [25]. The robot uses an omnidirectional drive with four mecanum wheels and a mechanical arm with a gripper to handle, e.g., organizer boxes. *OmniMan* is controlled by a PC. A laser scanner is used to avoid collisions to fulfill safety requirements and to map the environment. Besides, the real-time synchronization between the robot's movement with the PC is shown, and kinematic calculations are described in detail. The mobile robot is developed to study the cooperation of humans and robots.

Gartseev et al. introduce the *ArEduBot* in [26]. This mobile robot is based on the *iRobot Create* [37] platform, a hardware development kit by iRobot, one of the leading manufactures of robotic vacuum cleaners. In addition to the *iRobot Create* hardware, an Arduino board is installed. The focus of the publication is the development of a toolbox for Model-based development (MBD) using Mathworks MATLAB [30]. MATLAB's code generation produces real-time capable code, although the software is comparatively simple as the application only controls the mobile robot's driving and the avoidance of collisions. The sensor data can be analyzed and visualized after the application has stopped. The *ArEduBot* is used in education and research contexts.

The mobile robot *Savvy* [27] implements an omnidirectional drive via mecanum wheels. Due to its modular approach, the robot is prepared for future applications. New modules can easily be adapted into the robot's base. The architecture of the robot is distributed into a local real-time layer and a higher performance layer. The local real-time layer uses an STM32-based MCU and a mini-PC. The top layer runs ROS and provides functions for localization, SLAM, CV and a Human Machine Interface (HMI). This top layer is also used to coordinate a multi-robot scenario. According to Wu et al. *Savvy* is used to map unknown environments or follow pedestrians with a 3D-depth camera.

Meghana et al. introduce in [28] a mobile robot, which is mainly used to observe an outdoor area. For this purpose, the robots are equipped with a camera that can be turned around in 360°. The robot is especially useful in dark conditions, where it can detect movements with its infrared sensors within the camera. Those movements are used to send notifications; authorized persons can identify themselves with Radio-Frequency Identification (RFID) cards. An Arduino Uno board controls the movement of the robot and the camera. The camera data is transmitted via GSM mobile communications, where the number of objects is limited due to the low bandwidth of GSM.

Yaseen Ismael and Hedley from Newcastle University describes the development of an omnidirectional robot in [29]. The mobile robot is controlled by an Arduino board [38], which calculates the kinematic of the omnidirectional motors, the information from the ultrasonic sensors, and the path planning onboard and sends data for further analysis to a PC. Technical details about software implementations are not included in the publication. The robot is used for internal research projects.

3. Analysis of Types

The definition of the types of mobile robots is mostly done be the type of movement, the environment, the mobile robot is used in, or special characteristics of the mobile robot. In *Springer Handbook of Robotics* [7] five types of mobile robots are defined, which encompass these characteristics.

Mobile robots can be roughly categorized into three different modes of locomotion. A classical type of mobile robot, on which this work focuses, is the so-called Wheeled Mobile Robots (WMR). Siciliano and Khatib describes in [7] (p. 575 ff.) the group of WMR as widely spread mobile robots, which have many advantages, such as their simple structure, energy efficiency, high speed and low production costs.

The second group, the underwater robots or Unmanned Underwater Vehicle (UUV) are described in [7] (p. 595 ff.), which are mainly used to investigate environmental issues and to combat pollution of the seas. Underwater robots tend to be implemented as Remotly Operated Vehicle (ROV) and are therefore operated remotely. The data transfer to the robot is often done by wire. In addition to ROV, there are autonomously acting robots, so-called Autonomous Mobile Robot (AMR). These navigate autonomously and perform their task underwater autonomously. The group of AMR is also increasingly found in other areas of robotics, for example the group of WMR.

Flying robots, on the other hand, have to manage without any external cables at all. Here the autonomy has an even larger portion, than with underwater robots, since the carrying along of a cable for remote control is usually not wanted or technically not feasible. Flying robots belong to the group of Unmanned Arial Vehicle (UAV). Together with the infrastructure, system and human-machine interfaces, they are also called Unmanned Aircraft System (UAS). In [7] (p. 623 ff.) it is explained that especially the group of flying robots are complex systems, since they have six degrees of freedom, for example in comparison to WMR, and the processing of navigation itself is a complex task for a flying mobile robot. Furthermore, flying robots are strictly limited in terms of flight duration, payload and size. A major advantage is the ability to reach long distances, for example with fixed wing aircraft. Multicopters are inexpensive alternatives for lower altitudes and are suitable, for example, for monitoring critical infrastructures [39] or transporting cargo such as medical articles [40]. Multicopters are also more agile than fixed-wing aircraft and can, among other things, "stand still" at one point in the air due to their rotors.

Biomimetic robots can be described as another class of mobile robots [7] (p. 543 ff.). Biomimetic systems, or bionics, describes the transfer of models from nature to technical systems. In the case of mobile robots, bionics often refers to the mode of locomotion. For example, there are mobile robots in the form of worms or snakes that copy the way these animals move. Humanoid robots can also be called bionic robots, such as the NAO robot already presented, which are modeled after the human body. Humanoid robots are used especially in cooperation with humans, because according to [41] the acceptance of the robot by the human partner can be promoted by the familiar form.

The class of micro- and nano-robots refers to the compact design of the robots [7] (p. 671 ff). Typically, the size of robots in the micro/nano class is in the range of nanometers to millimeters. Mobile micro/nano robots are being researched for use in medicine, for example, where these robots are used in living organisms [42].

4. Analysis of Applications

For the category of applications of mobile robots the IEEE definition [43] is considered and checked by means of two already presented sources (see Table 2). It is shown that although the naming and the level of detail of the individual categories of application areas differ, this generally holds little potential for discussion. For example, ref. [5] (p. 1006 ff.) distinguishes between *transportation* and *self-driving cars*, while these two terms are summarized as *transportation* in [18]. Similarly, *telepresence* can be counted as part of the *Service* area. The area *Health care* is missing in [18], while *Education/ Teaching* is not listed in [5]. The application area *Military* is only listed in [43].

Table 2. Applications of mobile robots in the literature.

Reference	Application											
Source	**Industrial/ Agriculture**	**Transportation/ Logistics**	**Self-Driven Cars/ Autonomous Vehicles**	**Health Care**	**Disaster Response**	**Exploration**	**Service**	**Entertainment**	**Human Augmentation**	**Education/ Teaching**	**Military**	**Telepresence**
IEEE [43]	✓	✓	✓	✓	✓	✓	✓	✓	✓	✓	✓	✓
AI [5]	✓	✓	✓	✓	✓	✓	✓	✓	✓			
Review [18]		✓			✓	✓	✓			✓		

✓: Application listed.

5. Analysis of Capabilities

In the following section the capabilities of mobile robots are analyzed in the three steps. The capability of a mobile robot is the ability to fulfill that capability. For example, a mobile robot has the capability to navigate or is able to heal itself. First, search terms obtained from the three reference types are compared and evaluated. For this purpose capabilities (see Table 3) are noted, which are described or explained in the respective publications. The table shows which terms are used frequently and less frequently in the selected publications.

Table 3 shows that there must be a differentiated view on the individual terms, since the subjectively selected references do not allow for reliable statements about the relevance of the capabilities. The terms *Navigation* and *Autonomy* are undisputed, since they are addressed in almost all sources. *Self-healing*, on the other hand, is presented in the overviews and explained in a book, but the reference systems under consideration do not implement it.

Table 3. Capabilities of mobile robots in the literature

Reference		Capability										
Type	**Title** [source]	**Navigation**	**Autonomy**	**Optimization/ Learning**	**Multi-Robot Cooperation**	**Safety**	**Human-Robot Interaction**	**Security**	**Reliability**	**Energy Efficiency**	**Usability**	**Self-Healing**
Books see Section 2.1	Artificial Intelligence [5]	✓	✓	✓	✓		✓					
	Probabilistic Robotics [6]	✓	✓	✓								
	Springer Handbook of Robotics [7]	✓	✓	✓	✓	✓	✓	✓	✓	✓		✓
	Intro. to Auton. Mobile Robots [8]	✓	✓		✓		✓			✓		
	Robotics, Vision and Control [9]	✓	✓		✓		✓		✓		✓	
	Embedded Robotics [10]	✓	✓	✓	✓		✓			✓		
Surveys see Section 2.2	Modular Reconfigurable Robots [11]			✓					✓	✓		✓
	Challenges of Science Robotics [12]	✓	✓	✓	✓	✓	✓	✓	✓	✓		✓
	Multi-Robot Taxonomy [13]	✓	✓	✓	✓	✓		✓	✓	✓		
	Challenges for Mars Expl. [14]	✓	✓		✓		✓		✓	✓		✓
	Robot Teleop, Taxonomy [15]	✓	✓	✓	✓		✓		✓		✓	
	Review of Mobile Robots [16]	✓	✓	✓	✓	✓	✓			✓		
	ANR Requirements [17]	✓	✓	✓	✓		✓		✓	✓		
	Challenges of Mobile Robots [18]	✓	✓	✓						✓		
Reference systems see Section 2.3	Spot [19]	✓	◷	◷	◷	✓	✓				✓	
	Khepera IV [20]	✓	✓	✓	✓					✓		
	TurtleBot3 [21]	✓	✓		✓		✓					
	AMiRo [22]	✓	✓		✓						✓	
	WolfBot [23]	✓	✓		✓					✓		
	e-puck2 [24]	✓	✓		✓		✓				✓	
	OmniMan [25]	✓					✓			✓	✓	
	ArEduBot [26]	✓	✓								✓	
	Savvy [27]	✓	✓		✓		✓					
	Arduino Robot [28]	✓	✓				✓					
	Omnidirectional Mobile Robot [29]	✓	✓		◷		✓			✓		

✓: Capability addressed/ described. ◷: As future capability addressed/ described.

The next step in the analysis of the capabilities of mobile robots is therefore the systematic application of the already defined search terms with the digital library IEEE Xplore to get the total number of search hits to compare all defined search terms to each other. The syntax and search matrix of the search query in IEEE Xplore and the number of hits can be looked up in the Appendix (see Table A1). In analogy to Table 3, it can be stated that the classical capabilities of mobile robots, such as the *Navigation*, provide a particularly large number of search hits. The *Usability*, on the other hand, has little relevance in scientific publications.

To analyze the trend of the search terms in more detail, each search term is explained and analyzed individually in the following and last step of the analysis. The search covers the years from 2009 to 2019, so that a good impression of the development and relevance of the individual topics can be shown. Tables 3 and A1 are used to classify the individual capabilities and compare them with each other.

5.1. Navigation

Navigation is the search term with the most search hits (All search queries have been executed in IEEE Xplore in August 2020) witch about 34.400 hits (see Table A1) in connection with mobile robotics. The term navigation, i.e., a mobile robot's ability to maneuver independently in unknown terrain, includes localization, mapping, collision avoidance, and path planning. Figure 2 shows a high number of search hits over the years 2009 to 2016 and a strong increase since 2017, which is not due to the popularity of mobile robotics in scientific publications, since the increase is almost linear (see Figure 1). Rather, the increase shows the high importance of navigation in mobile robotics and the progress in this field, especially since 2016. Paden et al. gives an overview of the state of the art, and the methods and algorithms of navigation and cartography using the example of self-propelled cars [44]. The course of autonomous navigation, from route planning to steering and engine control, and the current limitations in accuracy are explained.

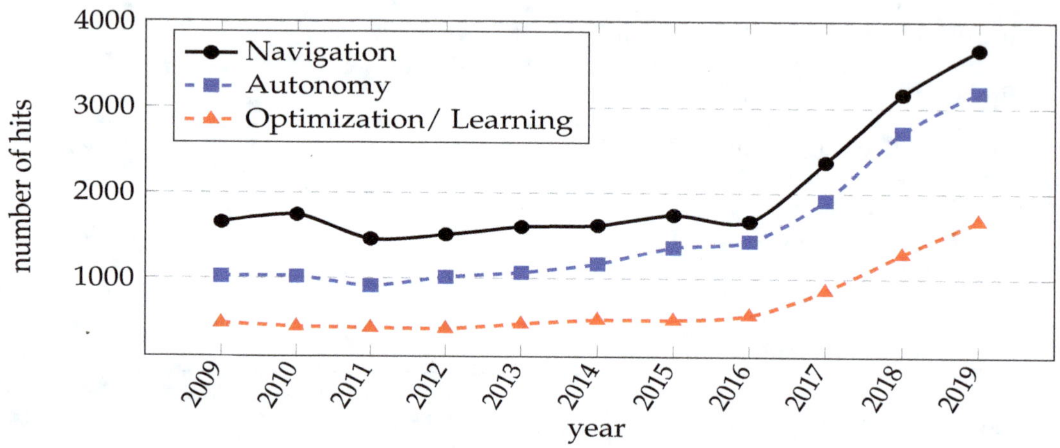

Figure 2. Development of search hits for *navigation*, *autonomy* and *optimization/ learning*.

5.2. Autonomy

Autonomy in mobile robotics refers to the ability to act autonomously, i.e., to make one's own decisions, for example, to perform a task. Similar to navigation, autonomous mobile robots have been on the scientific agenda for a long time. However, since 2016, the number of publications of the AMR has increased again (see Figure 2). The progress in the research of autonomy can be well illustrated with the development of the Mars Rovers [45]. Sojourner landed in 1997 and was the first mobile robot to operate autonomously on a foreign planet. With each successor (Spirit and Opportunity 2004 and Curiosity 2012), the degree of autonomy increased. This degree of autonomy can be measured, e.g., the SAE J3016 standard from the automotive industry is often used here [46]. SAE J3016 describes six categories from SAE Level 0 (no autonomy) to SAE Level 5 (full autonomy driving). There are also independent classifications of autonomy in robotics, which are compared, for example, in [47]. Furthermore, Beer et al. suggests a taxonomy from a gradation of 10 levels of autonomy.

5.3. Optimization/ Learning

Self-optimization and the capability of robots to learn new capabilities on their own is a research topic that appeared in scientific publications about 11,000 times (see Table A1). The development in the last ten years (see Figure 2), on the other hand, has been relatively constant. The almost linear increase from 2016 onward can be attributed to the increased number of scientific publications in mobile robotics. Although learning and optimization are summarized here, the optimization of existing functions and algorithms is more trivial than learning entirely new capabilities. The latter is considered a necessity to achieve real, complete, and lasting autonomy of technical systems. In [48], it is shown how mobile robots improve navigation in a known environment by learning routes. It is also interesting that although optimization and learning are addressed in the surveys and books, they are hardly implemented in the selected reference systems (see Table 3).

5.4. Multi-Robot Cooperation

The Multi-Robot Cooperation includes the cooperation of two robots and the interaction of whole swarms and the so-called ANR. Especially swarm robotics can be described as a trend of the last years, since search hits have more than tripled since 2016 (see Figure 3). In [49] heterogeneous swarms of robots are shown. A significant challenge in the development of multi-robot systems is the implementation of the capabilities of the distribution of the task(s) among each other and the Machine to Machine (M2M) interface, respectively the communication between the individual robots.

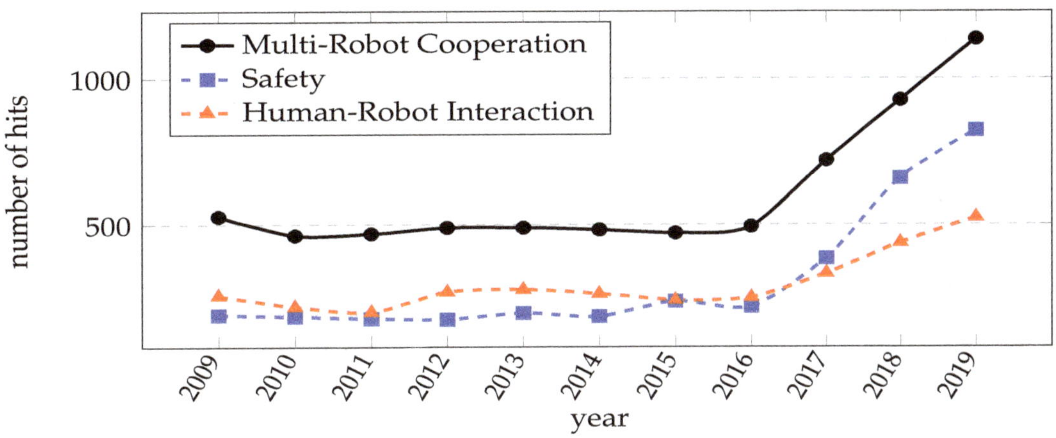

Figure 3. Development of search hits for *multi-robot cooperation*, *safety* and *human-robot interaction*.

5.5. Safety

Robots are to be classified as safety-critical systems because they are a potential danger for the users, the environment, and themselves, as Guiochet et al. explains in [50]. A safe mobile robot has the ability not to harm others, others, or itself. For this purpose, algorithms are implemented on the software side, and special material is used on the hardware side to protect the system and its environment. For the observance of the safety, it requires extensive tests [51]. Vasic and Billard show risk factors in the interaction of robots and humans [52]. The safety considered here can be described as a trend since the search hits are almost constant until 2016 and have quadrupled since then (see Figure 3).

5.6. Human-Robot Interaction

The search hits for Human-Robot Interaction are around 200 hits and increased from 2016 onward in line with the popularity of mobile robotics in scientific publications. Compared to the Multi-Robot Cooperation, the search hits are about half. However, the interaction between robots and humans is addressed in almost all selected references (see Table 3). The capability or the challenges of the interaction between humans and robots often lies in communication (HMI). The communication direction can also play a role here since the human-robot interaction is not necessarily bidirectional. In the literature, the terms Machine to Human (M2H) or Human to Machine (H2M) are also used. In [53], the necessity of Human-Robot Interaction and its implementation problems are described. Riek and Member explain in [54] methods for the coordination of movements of humans and mobile robots.

5.7. Security

While *safety* is generally concerned with harm the machine can potentially cause in its environment, *security* refers to the machine's resistance against attacks from outside—e.g., hacking attempts. The relevance in scientific publications has been increasing until 2017 (see Figure 4), although the search hits are a small part compared to navigation or autonomy (see Table A1). Also, the selected reference systems do not address this topic. In the surveys and the books, the topic is also not very present (see Table 3). Also, the number of search hits decreases again in 2018 and 2019. Nevertheless, ref. [55], for example, explains the importance of protecting against external attacks, especially for mobile robots. On the one hand, mobile robots usually collect data that should not be tapped by third parties in order to protect data privacy. On the other hand, mobile robots, in particular, are potentially dangerous, as they could cause physical damage. For example, an externally controlled UAV can cause enormous damage if it is deliberately flown into humans.

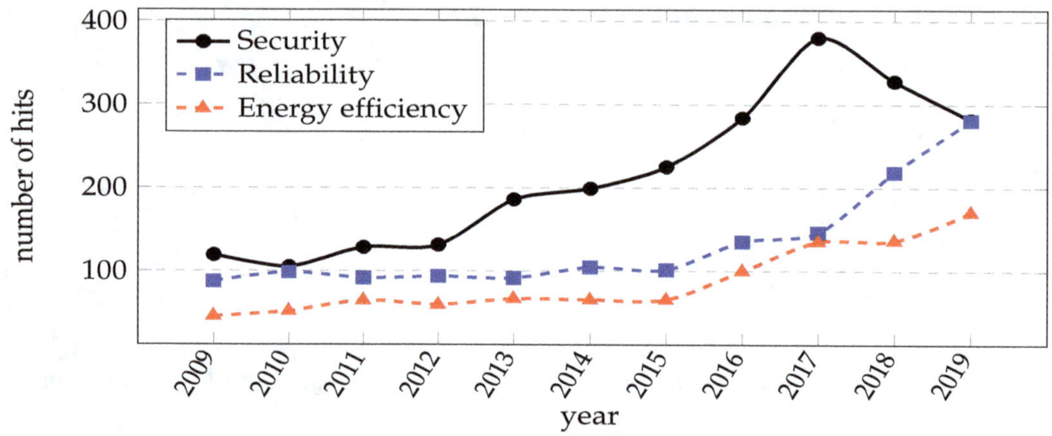

Figure 4. Development of search hits for *security*, *reliability* and *energy efficiency*.

5.8. Reliability

Reliability can be described as a trend in mobile robotics. The number of search hits increases from 2015 on, and in 2018 and 2019, there were massive leaps towards 300 search hits per year (see Figure 4). Similar to security, reliability is also meaningless for the selected reference systems (see Table 3). For example, Carlson and Murphy published an analysis of reliability in 2003, which clearly shows the need for improvements in this area [56]. The reliability of mobile robots has many positive and important aspects. For example, reliability contributes to mobile robots' acceptance and increases safety, which can be impaired by malfunctions.

5.9. Energy Efficiency

The ability to use energy effectively or to use Low Energy (LE) components is indispensable in many mobile robot applications. The number of search hits for this purpose is at a relatively constant low level (see Figure 4). The increase from 2015 onward is analogous to the increase in search hits for mobile robots. This topic is overrepresented in the selected references. Energy efficiency plays a comparatively large role, especially in the surveys (see Table 3). The increase in energy efficiency can be achieved in different areas. One possibility to use the available energy efficiency can be achieved by efficient use of the drive since it usually accounts for a large share of mobile robots' energy consumption. Künemund et al. show in [57] Methods and algorithms for efficient path planning.

5.10. Usability

Usability in mobile robots' use and development is almost irrelevant in scientific publications (see Figure 5). Although there have been isolated publications on this topic, the absolute number of publications is low, with a maximum of 50 search hits per year. In practice, usability plays a significant role, especially for such complex systems as robots. The need to present data in a simple form and make interaction with humans easy plays a significant role, especially in the interaction of mobile robots with elderly and sick people [58].

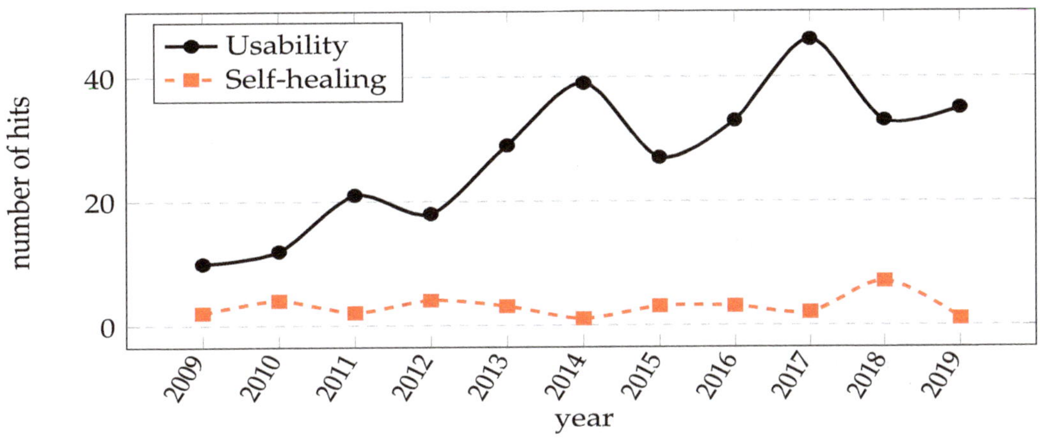

Figure 5. Development of search hits for *usability* and *self-healing*.

5.11. Self-Healing

Self-healing or independent repair of mobile robots is a capability that is needed to develop a fully autonomous robot. A mobile robot must be able to replace defective hardware or compensate for the defect in some other way in order to be able to act autonomously in the long term. The relevance in scientific publications disappears only slightly (see Figure 5). This leads to the conclusion that there are only very few solutions for this field of research. This research area seems to be an open challenge within mobile robotics, as this topic is addressed in [7]. Some of the few publications of Self-healing/Self-Repair show the use of material that can be restored after damage [59].

6. Analysis of Implementations

An analysis of the implementations follows. The implementations of a mobile robot in this context means the realization or technology used. The differentiation to the capabilities of a mobile robot is not always easy. In principle, the implementations should serve the capability or implement it. The *system monitoring* is needed to ensure the reliability of the mobile robot, although the *system monitoring* itself could also be a capability. The same is true for the *real-time*, which obviously can be called a capability. Here, however, the *real-time capability* is assigned as implementation for the fulfillment of safety.

The technology used in technical devices changes frequently, especially since mobile robotics is a very popular field of research and therefore new research results and insights are gained quickly. The analysis is performed analogous to the analysis of capabilities (see Section 5). First, the implementations from the three reference types are compared. Table 4 shows which implementations were explained or described in which reference.

When looking at Table 4, it becomes clear that here, too, a more differentiated analysis of the individual topics is required. While the *real-time capability* is omnipresent, research areas such as *monitoring* or *reconfigurability* are mentioned only in a few cases. In the following, each search term is secured in a literature search. First the absolute search hits are collected (see Appendix A, Table A1 and later each search term is described individually and analyzed in its temporal development. The syntax and the search matrix of the search query in IEEE Xplore can be looked up in the Appendix A (see Table A1).

Table 4. Implementations of mobile robots in the literature.

Reference		Implementation									
Type	**Title** [source]	**Real-Time capability**	**Machine Learning**	**Computer Vision**	**Cloud Computing**	**Monitoring**	**Modularity**	**Model-Based Development**	**Redundancy**	**Reconfigurability**	**Edge Computing**
Books see Section 2.1	Artificial Intelligence [5]	✓	✓	✓	✓	✓	✓			✓	
	Probabilistic Robotics [6]	✓	✓	✓							
	Springer Handbook of Robotics [7]	✓	✓	✓	✓	✓	✓	✓	✓	✓	
	Intro. to Auton. Mobile Robots [8]	✓		✓			✓				
	Robotics, Vision and Control [9]	✓		✓			✓	✓	✓		
	Embedded Robotics [10]	✓		✓			✓				✓
Surveys see Section 2.2	Modular Reconfigurable Robots [11]						✓			✓	
	Challenges of Science Robotics [12]	✓	✓	✓	✓	✓	✓		✓	✓	
	Multi-Robot Taxonomy [13]			✓		✓	✓	✓			
	Challenges for Mars Expl. [14]										
	Robot Teleop, Taxonomy [15]				✓	✓					
	Review of Mobile Robots [16]	✓	✓	✓			✓		✓	✓	
	ANR Requirements [17]	✓			✓	✓	✓			✓	
	Challenges of Mobile Robots [18]	✓	✓	✓				✓	✓		
Reference systems see Section 2.3	Spot [19]	✓		✓			✓				
	Khepera IV [20]			✓			✓				
	TurtleBot3 [21]			✓			✓				
	AMiRo [22]	✓	✓	✓			✓				
	WolfBot [23]	✓		✓							
	e-puck2 [24]	✓		✓		✓	✓				
	OmniMan [25]	✓									
	ArEduBot [26]	✓						✓			
	Savvy [27]	✓		✓			✓				
	Arduino Robot [28]	✓		✓							
	Omnidirectional Mobile Robot [29]										

✓: Implementation addressed/described.

6.1. *Real-Time Capability*

As already mentioned, real-time capability is a requirement that is absolutely necessary in mobile robotics. By adhering to a defined time condition, the safety and also the reliability of the system can be increased [60]. For example, the immediate execution of an action, such as stopping the motors within a predefined time after detection of an obstacle, can prevent accidents, defects or dangerous situations. Looking at the development of search hits since 2009 (see Figure 6) shows a growth since 2006 and a strong increase since 2016. This trend is accompanied by many search terms related to mobile robotics, as the latter itself has also gained popularity (see Figure 1).

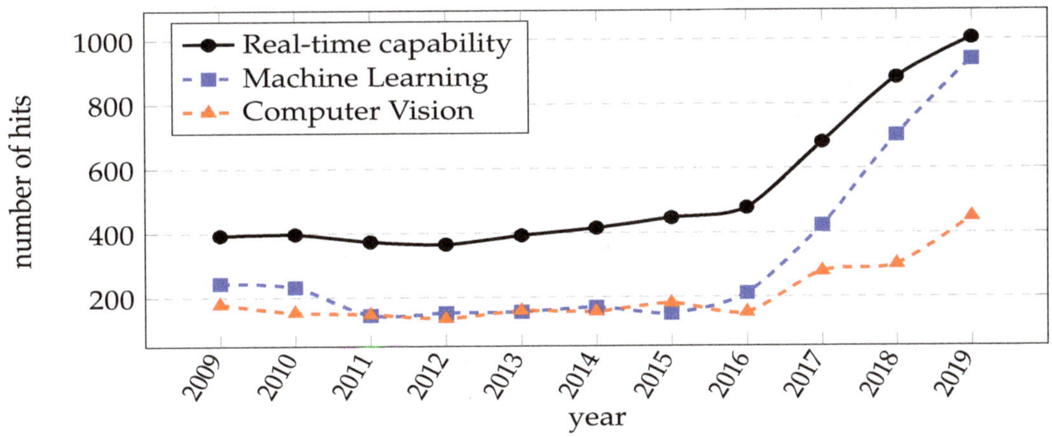

Figure 6. Development of search hits for *real-time capability*, *machine learning* and *computer vision*.

6.2. Machine Learning

The use or application of machine learning was addressed comparatively little in the selected sources (see Table 4). In the search hits, on the other hand, the topic is very present, with a slightly negative trend until 2015, but rising very strongly from 2015 onward (see Figure 6). Machine Learning can be used to evaluate large amounts of data, for example to recognize patterns in the data and to learn from them. The category and search matrix machine learning also includes *neural networks*, and *deep learning*. Neural networks are one of many ways to apply machine learning. Here, data is fed to a network of information, which is provided by various algorithms, for example for object recognition [61]. Deep Learning is a method which extends this information independently and thus extends its knowledge or the information base. Positive or negative weighting is calculated for each data input. The more data is processed, the better and more accurate the result will be, a deeper and broader network of information is formed. However, neural networks must first be trained by generating correct paths in the network. In robotics, this technology is used for example in navigation, but especially in object recognition [62].

6.3. Computer Vision

Object recognition also falls into the field of computer vision, but several technologies and methods can be used for this purpose. Accordingly, the course of search results is nearly congruent with the search results for neural networks (see Figure 6). For example, computer vision is used as an implementation to recognize objects whose position is needed for navigation. The computer vision has images as source, in mobile robotics usually as video stream(s), which are evaluated. Typical here is the checking for patterns, for which OpenCV [63] can be used. The source of the image material can be generated by various sensors, from classic 2D cameras to 360 degree LiDAR scanners. An overview of current hardware, methods and algorithms is given by Arnold et al. in [64].

6.4. Cloud Computing

With a few exceptions, research into cloud computing in combination with mobile robotics has only been underway since 2009. The trend since 2009 has been a sharp increase almost every year (see Figure 7). The search hits include not only the trend towards cloud computing, but also the approach of integrating servers into the technical implementation of mobile robots. Cloud computing or the use of servers can be used to add very large amounts of computing power and storage space to a technical system. This can be used to outsource algorithms and methods, such as neural networks, whose performance requirements can often not be provided locally on a mobile robot. In addition, cloud services in particular can be adapted if, for example, the performance requirements increase spontaneously. In [65] gives a short outline of the use of cloud computing in robotics. Ref. [66] explains in more detail the current state of the art and the open questions in the research area *cloud robotics*.

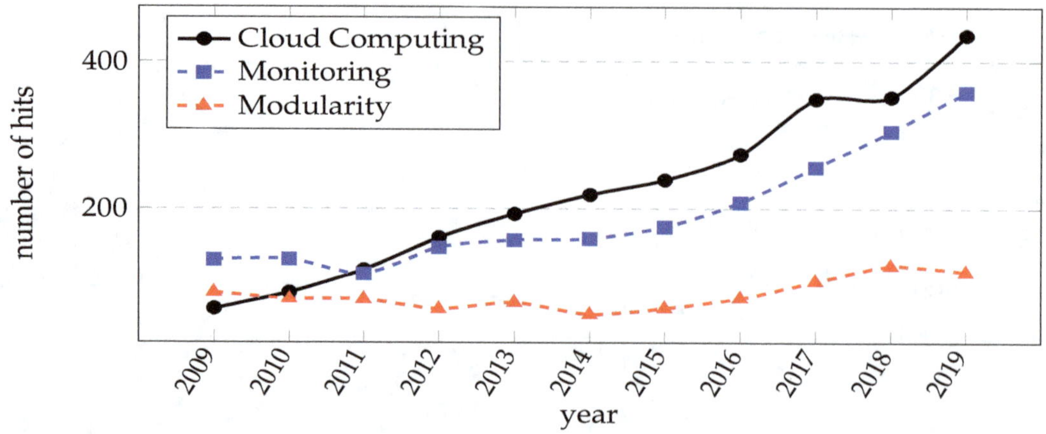

Figure 7. Development of search hits for *cloud computing*, *monitoring* and *modularity*.

6.5. Monitoring

The monitoring and verification of the system is a broad field of research, which is summarized here as *monitoring*. This includes the *validation* of the system, i.e., checking whether the system serves the defined purpose, as well as the *verification*, i.e., checking whether the system is implemented correctly, which is done for example by regular measurements. For example, it is possible to verify whether time schedules are adhered to in the system (*Time Verification*). The real-time compliance and the verification of other factors can be used to validate that the mobile robot meets the safety requirements. In addition to validation and verification, monitoring also includes the search term *diagnosis* of the system, as well as *self-awareness*. The search matrix also includes the research fields *monitoring* and *fault detection*, which also aim at monitoring and testing the system. The number of search hits for the research field of monitoring in mobile robotics is increasing almost linearly, and this increase will be even stronger from 2015 onward (see Figure 7). In [67] an example of monitoring the state of a mobile robot is shown, which shows users and developers with a Graphical User Interface (GUI) the current internal state. A model for the self-perception of robots is explained in [68], which detects internal errors and thus leads to a higher safety and reliability. The early detection of errors, especially in the drive of a WMR is shown in [69].

6.6. Modularity

Modular mobile robots are widely used, as seen in the selected reference systems (see Table 4). In the years 2009 to 2015, publications on modularity in mobile robotics are constant, from 2016 on there is a slight increase (see Figure 7), analogous to the increasing popularity of mobile robots in scientific publications. Modularity means that the technical system is built up in the form of a building set. The overall system is divided into individual smaller components. The subdivision is usually made according to service or function and in any case concerns the software, sometimes also the hardware. Modularity makes it much easier to adapt and expand the system than in a complete system without strictly defined interfaces between hardware and software components. In mobile robotics, a modular design is particularly in demand, since the location and the environment often change and new sensors or actuators may be required. The implementation of the modularity and expandability of the mobile robot AMiRo is presented by Herbrechtsmeier et al. in [70] and others., e.g., the structure of the ROS framework is also based on its modularity.

6.7. Model-Based Development

The development and maintenance of complex technical systems, such as mobile robots, is costly and a challenge for developers. The software is usually very extensive. In [71] (p. 46 ff.) it is explained that to cope with these requirements the MBD should be applied. By using these methods the developers are supported and the documentation is simplified. This leads for example to a higher usability for the developer. In [71] the author has already shown an implementation of methods for MBD, mainly by using MATLAB Simulink for the development of a mobile robot [72]. The MBD in scientific publications has a large increase in recent years (see Figure 8) and with about 1.400 search hits a comparatively relevant keyword for systematic literature search (see Table A1).

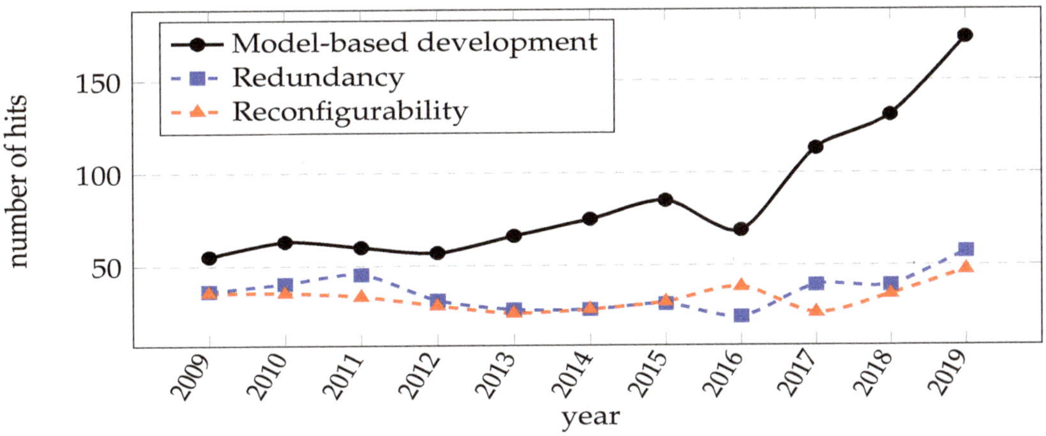

Figure 8. Development of search hits for *model-based development*, *redundancy* and *reconfigurability*.

6.8. Redundancy

Another implementation to increase the safety and reliability of a mobile robot is the use of redundancies. Redundancy describes the availability of information at more than one place in the software, or the availability of more than one part or component, the omission of which has no direct consequences. In the event of a failure, e.g., a defect, of a component or function, the redundant counterpart takes over. Redundancies can still be used to detect faulty sensors. In [73] it is shown how to use redundant sources to identify faulty data sets from the motor decoders. Redundancies are not addressed in the selected reference systems (see Table 4), but are mentioned in the surveys and books. The number of search hits is constantly below 50 search hits per year (see Figure 8).

6.9. Reconfigurability

The number of search hits for reconfigurability of mobile robots behaves analogously to the number of search hits for redundancy (see Figure 8). On the one hand, reconfigurability describes the hardware reconfigurability to adapt the configuration to the environmental conditions. According to [11] the development of these systems is a challenge for robots in general and describes mobile robots that can adapt their shape. The adaptation of the hardware can also be used to replace defective parts with existing hardware, which leads to self-healing of the robot [74]. However, the software of mobile robots can also be reconfigured, for example by changing the sequence of functions or by adjusting the timing conditions in the software. This software reconfiguration can also be used for the automated distribution of tasks in multi-robot cooperations [75].

6.10. Edge Computing

Edge and Fog Computing *(in the following only referred to as Edge Computing)* is a technology that will not achieve the first search hits in combination with mobile robotics until 2015 (see Figure 9). In 2016 and 2017, there are more isolated search hits. In 2018 and 2019, there is a clear trend. Edge computing is generally a trend that comes from the area of Internet of Things (IoT). It shifts the processing of applications or services from central nodes, such as the cloud, to the network's edges. This reduces the amount of data that needs to be sent to a server to analyze data and the cloud's dependence, as this analysis takes place on-site, and only the results are sent to the server. In [76] Edge Computing is used to reduce the SLAM algorithms between the local robots and the cloud in a so-called edge-fog layer.

This reduces the latency compared to processing the algorithms in the cloud and, thus, the application's reliability and security.

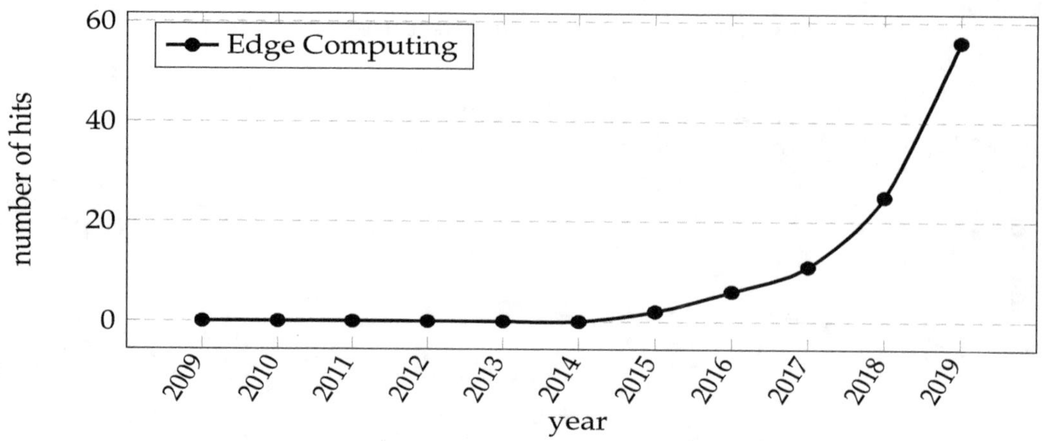

Figure 9. Development of search hits for *edge computing*.

7. Taxonomy

The taxonomy for mobile robots is divided into those four categories (types, applications, capabilities, and implementations) using the previous chapters to derive the taxonomy for mobile robots.

7.1. Types

The analysis of the types of mobile robots in Section 3 presents five types defined by *Springer Handbook of Robotics* [7] which are transferred into the taxonomy:

T1 Wheeled Mobile Robots (WMR)
T2 Unmanned Arial Vehicle (UAV)
T3 Unmanned Underwater Vehicle (UUV)
T4 Biomimetic
T5 Micro/ Nano

7.2. Applications

The analysis of the applications in Section 4 showed that mobile robots' application areas are not clearly defined. However, the differences lie largely in the level of detail of the listing. For the taxonomy of a mobile robot, the already presented applications from the IEEE are taken over from [43] into the taxonomy, since this provides the most complete listing of the shown ones:

A1 Industrial/ Agriculture
A2 Transportation/ Logistics
A3 Self-driven cars/ Autonomous vehicles
A4 Health care
A5 Disaster response
A6 Exploration
A7 Service
A8 Entertainment
A9 Human Augmentation
A10 Education/ Teaching
A11 Military
A12 Telepresence

7.3. Capabilities

The capabilities of mobile robots were explained in Section 5. From the analysis of these, capabilities can be derived, which today every robot should typically support. Not all of the shown capabilities are taken over from the selected sources, but only those whose relevance is justified by frequent search hits. Thus the capabilities *usability* and *self-healing* are not typical capabilities of mobile robots today. All other capabilities are transferred to the taxonomy:

C1 Navigation
C2 Autonomy
C3 Optimization/ Learning
C4 Multi-Robot Cooperation
C5 Safety
C6 Human-Robot Interaction
C7 Security
C8 Reliability
C9 Energy efficiency

7.4. Implementations

The implementations of mobile robots were examined in Section 6. This can be derived from current technologies and which implementations should be considered and supported in a mobile robot's design and development. Analogous to mobile robots' capabilities, only those implementations are included in the taxonomy, whose relevance is justified by the selected sources or the number of search hits. It was found that *redundancy*, *reconfigurability*, and *edge computing* are not found in a typical mobile robot today. All other implementations are taken over into the taxonomy:

I1 Real-time capability
I2 Machine Learning
I3 Computer Vision
I4 Cloud Computing
I5 Monitoring
I6 Modularity
I7 Model-based development

8. Conclusions

The complex field of mobile robotics was worked up in the form of a taxonomy. While the definition of the types and application areas of mobile robots can be determined with little effort, the elaboration of typical capabilities and implementations requires a comprehensive literature review. In particular, the trends of the last ten years are derived from search hits. Nine capabilities and seven implementations are defined, which the architecture of an archetypal mobile robot should support. This results in particular requirements for the robots' system and software architecture, such as integrating the cloud or compliance with real-time requirements. The realization of these capabilities and implementations are also a challenge for the developers of mobile robots. In the future, the list of capabilities and implementations is likely to be extended again. For example, integrating the current trend of using edge computing can be a challenge for future mobile robots.

Author Contributions: Conceptualization, U.J.; methodology, U.J. and C.W.; investigation, U.J.; resources, U.J.; data curation, U.J.; writing—original draft preparation, U.J.; writing—review and editing, D.H., M.S., A.S., C.R., P.S. and C.W.; visualization, U.J.; supervision, C.W. and P.S.; project administration, U.J. All authors have read and agreed to the published version of the manuscript.

Appendix A

Table A1. Search matrix and number of hits for capabilities and implementations of mobile robots.

Topic	Search Matrix	Hits
Navigation	("All Metadata":mobile AND robot AND (navigation OR mapping OR slam OR "collision avoidance" OR "path planning"))	little potential for
Autonomy	("All Metadata":mobile AND robot AND (autonomy OR autonomous))	24,626
Optimization/ Learning	("All Metadata":mobile AND robot AND (learning OR optimizing))	11,051
Multi-Robot Cooperation	("All Metadata":mobile AND robot AND (swarm OR "multi robot" OR "networked robots"))	9265
Safety	("All Metadata":mobile AND robot AND safety)	4617
Human-Robot Interaction	("All Metadata":mobile AND robot AND ("human machine" OR "machine human" OR "human robot" OR "robot human"))	4477
Security	("All Metadata":mobile AND robot AND security)	2907
Reliability	("All Metadata":mobile AND robot AND reliability)	2037
Energy efficiency	("All Metadata":mobile AND robot AND (energy AND (efficient OR efficiency)))	1237
Usability	("All Metadata":mobile AND robot AND usability)	379
Self-healing	("All Metadata":mobile AND robot AND ("self healing" OR "self repairing"))	45
Real-time capability	("All Metadata":mobile AND robot AND ("real time" OR realtime))	8874
Machine Learning	("All Metadata":mobile AND robot AND ("neural network" OR "neural networks" OR "machine learning" OR "deep learning"))	5768
Computer Vision	("All Metadata":mobile AND robot AND ("computer vision" OR cv OR "object recognition"))	4207
Cloud Computing	("All Metadata":mobile AND robot AND (cloud OR server))	2864
Monitoring	("All Metadata":mobile AND robot AND (verification OR validation OR diagnosis OR "self awareness" OR "system monitoring" OR "fault detection"))	2851
Modularity	("All Metadata":mobile AND robot AND modular)	1438
Model-based development	("All Metadata":mobile AND robot AND ("model based" OR "model driven"))	1346
Redundancy	("All Metadata":mobile AND robot AND redundancy)	739
Reconfigurability	("All Metadata":mobile AND robot AND (reconfiguration OR reconfigure))	581
Edge Computing	("All Metadata":mobile AND robot AND ("edge computing" OR "fog computing"))	130

All search queries have been executed in July and August 2020.

References

1. IEEE Xplore-Advanced Search. Available online: https://ieeexplore.ieee.org/search/advanced (accessed on 26 August 2020).
2. Randolph, J.J. A guide to writing the dissertation literature review. *Pract. Assess. Res. Eval.* **2009**, *14*, 13.
3. Brocke, J.V.; Simons, A.; Niehaves, B.; Niehaves, B.; Riemer, K.; Plattfaut, R.; Cleven, A. Reconstructing the giant: On the importance of rigour in documenting the literature search process. In Proceedings of the 17th European Conference on Information Systems, Verona, Italy, 8–10 June 2009.
4. Google Scholar-Advanced Search. Available online: https://scholar.google.com/?hl=en&as_sdt=0,5#d=gs_asd (accessed on 26 August 2020).
5. Russell, S.; Norvig, P. *Artificial Intelligence: A Modern Approach*, 3rd ed.; Pearson Education Limited: Essex, UK, 2016.
6. Thrun, S.; Burgard, W.; Fox, D. *Probabilistic Robotics*; The MIT Press: Cambridge, MA, USA, 2006.
7. Siciliano, B.; Khatib, O. *Springer Handbook of Robotics*, 2nd ed.; Springer Publishing Company, Incorporated: Berlin/Heidelberg, Germany, 2016.
8. Siegwart, R.; Nourbakhsh, I.R.; Scaramuzza, D. *Introduction to Autonomous Mobile Robots*, 2nd ed.; The MIT Press: Cambridge, MA, USA, 2011.
9. Corke, P. *Robotics, Vision and Control: Fundamental Algorithms in MATLAB*, 2nd ed.; Springer Publishing Company, Inc.: Berlin/Heidelberg, Germany, 2016.
10. Bräunl, T. *Embedded Robotics*, 3rd ed.; Springer: Berlin/Heidelberg, Germany, 2008. [CrossRef]
11. Yim, M.; Shen, W.M.; Salemi, B.; Rus, D.; Moll, M.; Lipson, H.; Klavins, E.; Chirikjian, G.S. Modular self-reconfigurable robot systems [Grand challenges of robotics]. *IEEE Robot. Autom. Mag.* **2007**, *14*, 43–52. [CrossRef]
12. Yang, G.Z.; Bellingham, J.; Dupont, P.E.; Fischer, P.; Floridi, L.; Full, R.; Jacobstein, N.; Kumar, V.; McNutt, M.; Merrifield, R.; et al. The grand challenges of science robotics. *Sci. Robot.* **2018**, *3*. [CrossRef] [PubMed]
13. Robin, C.; Lacroix, S. Multi-robot target detection and tracking: Taxonomy and survey. *Auton. Robot.* **2016**, *40*, 729–760. [CrossRef]
14. Huntsberger, T.; Rodriguez, G.; Schenker, P.S. Robotics challenges for robotic and human Mars exploration. In Proceedings of the 4th International Conference and Exposition on Robotics for Challenging Situations and Environments, Albuquerque, NM, USA, 27 February–2 March 2000; pp. 340–346.

15. Adamides, G.; Christou, G.; Katsanos, C.; Xenos, M.; Hadzilacos, T. Usability guidelines for the design of robot teleoperation: A taxonomy. *IEEE Trans. Hum. Mach. Syst.* **2015**, *45*, 256–262. [CrossRef]
16. Rubio, F.; Valero, F.; Llopis-Albert, C. A review of mobile robots: Concepts, methods, theoretical framework, and applications. *Int. J. Adv. Robot. Syst.* **2019**, *16*, 1–22. [CrossRef]
17. Chukwuemeka, C.; Habib, M. Development of autonomous networked robots (ANR) for surveillance: Conceptual design and requirements. In Proceedings of the 44th Annual Conference of the IEEE Industrial Electronics Society, Washington, DC, USA, 21–23 October 2018; pp. 3757–3763. [CrossRef]
18. Alatise, M.B.; Hancke, G.P. A Review on Challenges of Autonomous Mobile Robot and Sensor Fusion Methods. *IEEE Access* **2020**, *8*, 39830–39846. [CrossRef]
19. Dynamics, B. Spot Boston Dynamics. Available online: https://www.bostondynamics.com/spot (accessed on 10 March 2020).
20. Soares, J.M.; Navarro, I.; Martinoli, A. The Khepera IV Mobile Robot: Performance Evaluation, Sensory Data and Software Toolbox. In *Robot 2015: Second Iberian Robotics Conference*; Reis, L.P., Moreira, A.P., Lima, P.U., Montano, L., Muñoz-Martinez, V., Eds.; Springer International Publishing: Cham, Switzerland, 2016; pp. 767–781.
21. Open Source Robotics Foundation, Inc. Turtlebot. Available online: https://www.turtlebot.com/ (accessed on 13 March 2020).
22. Schöpping, T.; Korthals, T.; Hesse, M.; Rückert, U. AMiRo: A Mini Robot as Versatile Teaching Platform. *Adv. Intell. Syst. Comput.* **2019**, *829*, 177–188. [CrossRef]
23. Betthauser, J.; Benavides, D.; Schornick, J.; O'Hara, N.; Patel, J.; Cole, J.; Lobaton, E. WolfBot: A distributed mobile sensing platform for research and education. In Proceedings of the 2014 Zone 1 Conference of the American Society for Engineering Education, Bridgeport, CT, USA, 3–5 April 2014; pp. 1–8. [CrossRef]
24. Mondada, F.; Bonani, M.; Raemy, X.; Pugh, J.; Cianci, C.; Klaptocz, A.; Magnenat, S.; Zufferey, J.C.; Floreano, D.; Martinoli, A. The e-puck, a robot designed for education in engineering. In Proceedings of the 9th Conference on Autonomous Robot Systems and Competitions, Castelo Branco, Portugal, 7 May 2009; Volume 1, pp. 59–65.
25. Röhrig, C.; Heß, D. Mobile Manipulation for Human-Robot Collaboration in Intralogistics. In *IAENG Transactions on Engineering Sciences-Special Issue for the International Association of Engineers Conferences*; World Scientific: Singapore, 2020; pp. 1–20. [CrossRef]
26. Gartseev, I.B.; Lee, L.f.; Krovi, V.N. A Low-Cost Real-Time Mobile Robot Platform (ArEduBot) to support Project-Based Learning in Robotics & Mechatronics. In Proceedings of the 2nd International Conference on Robotics in Education (RiE 2011), Vienna, Austria, 15–16 September 2011; pp. 117–124.
27. Wu, J.; Lv, C.; Zhao, L.; Li, R.; Wang, G. Design and implementation of an omnidirectional mobile robot platform with unified I/O interfaces. In Proceedings of the 2017 IEEE International Conference on Mechatronics and Automation, ICMA 2017, Takamatsu, Japan, 6–9 August 2017; pp. 410–415. [CrossRef]
28. Meghana, S.; Nikhil, T.V.; Murali, R.; Sanjana, S.; Vidhya, R.; Mohammed, K.J. Design and implementation of surveillance robot for outdoor security. In Proceedings of the RTEICT 2017-2nd IEEE International Conference on Recent Trends in Electronics, Information and Communication Technology, Bangalore, India, 19–20 May 2017; pp. 1679–1682. [CrossRef]
29. Yaseen Ismael, O.; Hedley, J. Analysis, Design, and Implementation of an Omnidirectional Mobile Robot Platform. *Am. Sci. Res. J. Eng.* **2016**, *22*, 195–209.
30. MATLAB. Available online: https://www.mathworks.com/products/matlab.html (accessed on 6 October 2020).
31. Simulation and Model-Based Design. Available online: https://www.mathworks.com/products/simulink.html (accessed on 6 October 2020).
32. Spectrum, I. Boston Dynamics' Spot Robot Dog Goes on Sale. Available online: https://spectrum.ieee.org/automaton/robotics/industrial-robots/boston-dynamics-spot-robot-dog-goes-on-sale (accessed on 12 March 2020).
33. Quigley, M.; Conley, K.; Gerkey, B.; Faust, J.; Foote, T.; Leibs, J.; Wheeler, R.; Ng, A.Y. ROS: An open-source Robot Operating System. In Proceedings of the ICRA Workshop on Open Source Software, Kobe, Japan, 12–17 May 2009.
34. Wienke, J.; Wrede, S. A middleware for collaborative research in experimental robotics. In Proceedings of the 2011 IEEE/SICE International Symposium on System Integration, Kyoto, Japan, 20–22 December 2011; pp. 1183–1190. [CrossRef]

35. BeagleBoard Bone. Available online: https://beagleboard.org/bone (accessed on 13 October 2020).
36. GCtronic. E-Puck2. Available online: https://www.gctronic.com/doc/index.php/e-puck2 (accessed on 20 October 2020).
37. Create 2 Robot. Available online: https://edu.irobot.com/what-we-offer/create-robot (accessed on 13 October 2020).
38. Arduino. Arduino Home. Available online: https://www.arduino.cc/ (accessed on 19 March 2020).
39. Gerke, M.; Borgolte, U.; Masár, I.; Jelenciak, F.; Bahnik, P.; Al-Rashedi, N. Lighter-than-air UAVs for surveillance and environmental monitoring. In *Future Security Research Conference*; Springer: Berlin/Heidelberg, Germany, 2012; pp. 480–483.
40. Tom Plümmer. *Impfstoffe und Medikamente für abgelegene Orte*; Bundesministerium für Wirtschaft und Energie Broschüre: Berlin, Germany, 2019.
41. Goetz, J.; Kiesler, S.; Powers, A. Matching robot appearance and behavior to tasks to improve human-robot cooperation. In Proceedings of the 12th IEEE International Workshop on Robot and Human Interactive Communication, Millbrae, CA, USA, 31 October–2 November 2003; pp. 55–60.
42. Li, J.; de Ávila, B.E.F.; Gao, W.; Zhang, L.; Wang, J. Micro/nanorobots for biomedicine: Delivery, surgery, sensing, and detoxification. *Sci. Robot.* **2017**, 2. [CrossRef]
43. IEEE. Types of Robots: ROBOTS: Your Guide to the World of Robotics. Available online: https://robots.ieee.org/learn/types-of-robots/ (accessed on 10 March 2020).
44. Paden, B.; Čáp, M.; Yong, S.Z.; Yershov, D.; Frazzoli, E. A survey of motion planning and control techniques for self-driving urban vehicles. *IEEE Trans. Intell. Veh.* **2016**, *1*, 33–55. [CrossRef]
45. Bajracharya, M.; Maimone, M.W.; Helmick, D. Autonomy for Mars Rovers: Past, Present, and Future. *Computer* **2008**, *41*, 44–50. [CrossRef]
46. SAE International. Taxonomy and Definitions for Terms Related to Driving Automation Systems for On-Road Motor Vehicles. Available online: https://www.sae.org/standards/content/j3016_201806 (accessed on 11 August 2020).
47. Beer, J.M.; Fisk, A.D.; Rogers, W.A. Toward a Framework for Levels of Robot Autonomy in Human-Robot Interaction. *J. Hum. Robot Interact.* **2014**, *3*, 74–99. [CrossRef] [PubMed]
48. Levinson, J.; Askeland, J.; Becker, J.; Dolson, J.; Held, D.; Kammel, S.; Kolter, J.Z.; Langer, D.; Pink, O.; Pratt, V.; et al. Towards fully autonomous driving: Systems and algorithms. *IEEE Intell. Veh. Symp. Proc.* **2011**, 163–168. [CrossRef]
49. Dorigo, M.; Floreano, D.; Gambardella, L.M.; Mondada, F.; Nolfi, S.; Baaboura, T.; Birattari, M.; Bonani, M.; Brambilla, M.; Brutschy, A.; et al. Swarmanoid: A novel concept for the study of heterogeneous robotic swarms. *IEEE Robot. Autom. Mag.* **2013**, *20*, 60–71. [CrossRef]
50. Guiochet, J.; Machin, M.; Waeselynck, H. Safety-critical advanced robots: A survey. *Robot. Auton. Syst.* **2017**, *94*, 43–52. [CrossRef]
51. Broggi, A.; Buzzoni, M.; Debattisti, S.; Grisleri, P.; Laghi, M.C.; Medici, P.; Versari, P. Extensive tests of autonomous driving technologies. *IEEE Trans. Intell. Transp. Syst.* **2013**, *14*, 1403–1415. [CrossRef]
52. Vasic, M.; Billard, A. Safety issues in human-robot interactions. In Proceedings of the 2013 IEEE International Conference on Robotics and Automation, Karlsruhe, Germany, 6–10 May 2013; pp. 197–204.
53. Chen, J.Y.; Haas, E.C.; Barnes, M.J. Human performance issues and user interface design for teleoperated robots. *IEEE Trans. Syst. Man Cybern. Part C Appl. Rev.* **2007**, *37*, 1231–1245. [CrossRef]
54. Riek, L.D.; Member, S. Movement Coordination in Human—Robot Teams. *IEEE Trans. Robot.* **2016**, *32*, 909–919. [CrossRef]
55. Hochgeschwender, N.; Cornelius, G.; Voos, H. Arguing Security of Autonomous Robots. In Proceedings of the IEEE International Conference on Intelligent Robots and Systems, Macau, China, 3–8 November 2019; pp. 7791–7797. [CrossRef]
56. Carlson, J.; Murphy, R.R. Reliability analysis of mobile robots. In Proceedings of the 2003 IEEE International Conference on Robotics and Automation, Taipei, Taiwan, 14–19 September 2003; Volume 1, pp. 274–281. [CrossRef]

57. Künemund, F.; Hess, D.; Röhrig, C. Energy efficient kinodynamic motion planning for holonomic AGVs in industrial applications using state lattices. In Proceedings of the 47th International Symposium on Robotics ISR 2016, Munich, Germany, 21–22 June 2016; Volume 2016, pp. 459–466.
58. Pripfl, J.; Körtner, T.; Batko-Klein, D.; Hebesberger, D.; Weninger, M.; Gisinger, C.; Frennert, S.; Eftring, H.; Antona, M.; Adami, I.; et al. Results of a real world trial with a mobile social service robot for older adults. In Proceedings of the ACM/IEEE International Conference on Human-Robot Interaction, Christchurch, New Zealand, 7–10 March 2016; pp. 497–498. [CrossRef]
59. Terryn, S.; Brancart, J.; Lefeber, D.; Van Assche, G.; Vanderborght, B. Self-healing soft pneumatic robots. *Sci. Robot.* **2017**, *2*, 1–13. [CrossRef]
60. Khatib, O. Real-time obstacle avoidance for manipulators and mobile robots. In Proceedings of the 1985 IEEE International Conference on Robotics and Automation, St. Louis, MO, USA, 25–28 March 1985; Volume 2, pp. 500–505. [CrossRef]
61. Zou, A.M.; Hou, Z.G.; Fu, S.Y.; Tan, M. Neural networks for mobile robot navigation: A survey. In *Lecture Notes in Computer Science (Including Subseries Lecture Notes in Artificial Intelligence and Lecture Notes in Bioinformatics)*; Springer: Berlin/Heidelberg, Germany, 2006; Volume 3972 LNCS, pp. 1218–1226. [CrossRef]
62. Dubé, R.; Cramariuc, A.; Dugas, D.; Nieto, J.; Siegwart, R.; Cadena, C. SegMap: 3D Segment Mapping using Data-Driven Descriptors. *arXiv* **2018**, arXiv:1804.09557
63. Bradski, G.; Kaehler, A. *Learning OpenCV: Computer vision with the OpenCV library*; O'Reilly Media, Inc.: Sebastopol, CA, USA, 2008.
64. Arnold, E.; Al-Jarrah, O.Y.; Dianati, M.; Fallah, S.; Oxtoby, D.; Mouzakitis, A. A Survey on 3D Object Detection Methods for Autonomous Driving Applications. *IEEE Trans. Intell. Transp. Syst.* **2019**, *20*, 3782–3795. [CrossRef]
65. Guizzo, E. Robots with their heads in the clouds. *IEEE Spectr.* **2011**, *48*, 17–18. [CrossRef]
66. Wan, J.; Tang, S.; Yan, H.; Li, D.; Wang, S.; Vasilakos, A.V. Cloud robotics: Current status and open issues. *IEEE Access* **2016**, *4*, 2797–2807. [CrossRef]
67. Roennau, A.; Heppner, G.; Kerscher, T.; Dillmann, R. Fault diagnosis and system status monitoring for a six-legged walking robot. In Proceedings of the IEEE/ASME International Conference on Advanced Intelligent Mechatronics, AIM, Budapest, Hungary, 4–6 July 2011; pp. 874–879. [CrossRef]
68. Golombek, R.; Wrede, S.; Hanheide, M.; Heckmann, M. Learning a probabilistic self-awareness model for robotic systems. In Proceedings of the IEEE/RSJ 2010 International Conference on Intelligent Robots and Systems, IROS 2010, Taipei, Taiwan, 18–22 October 2010; pp. 2745–2750. [CrossRef]
69. Mellah, S.; Graton, G.; El Mostafa, E.; Ouladsine, M.; Planchais, A. Mobile robot additive fault diagnosis and accommodation. In Proceedings of the 2019 8th International Conference on Systems and Control (ICSC), Marrakesh, Morocco, 23–25 October 2019; pp. 241–246.
70. Herbrechtsmeier, S.; Korthals, T.; Schopping, T.; Rückert, U. AMiRo: A modular customizable open-source mini robot platform. In Proceedings of the 2016 20th International Conference on System Theory, Control and Computing (ICSTCC), Sinaia, Romania, 13–15 October 2016; pp. 687–692.
71. Kagermann, H.; Wahlster, W.; Helbig, J. *Umsetzungsempfehlungen für das Zukunftsprojekt Industrie 4.0*; Technical report; acatech Deutsche Akademie der Technikwissenschaften e.V.: München, Germany, 2013.
72. Lauschner, U.; Igel, B.; Krawczyk, L.; Wolff, C. Applying model-based principles on a distributed robotic system application. In Proceedings of the 2015 IEEE 8th International Conference on Intelligent Data Acquisition and Advanced Computing Systems: Technology and Applications, IDAACS 2015, Warsaw, Poland, 24–26 September 2015; Volume 2, pp. 893–897. [CrossRef]
73. Mendoza, J.P.; Simmons, R. Mobile Robot Fault Detection based on Redundant Information Statistics. In Proceedings of the 2012 International Conference on Intelligent Robots and Systems, Algarve, Portugal, 7–12 October 2012.
74. Meng, Y.; Zhang, Y.; Jin, Y. Autonomous self-reconfiguration of modular robots by evolving a hierarchical mechanochemical model. *IEEE Comput. Intell. Mag.* **2011**, *6*, 43–54. [CrossRef]
75. Tang, F.; Parker, L.E. ASyMTRe: Automated synthesis of multi-robot task solutions through software reconfiguration. In Proceedings of the 2005 IEEE International Conference on Robotics and Automation, Barcelona, Spain, 18–22 April 2005; pp. 1501–1508. [CrossRef]

76. Sarker, V.K.; Pena Queralta, J.; Gia, T.N.; Tenhunen, H.; Westerlund, T. Offloading SLAM for Indoor Mobile Robots with Edge-Fog-Cloud Computing. In Proceedings of the 2019 1st International Conference on Advances in Science, Engineering and Robotics Technology (ICASERT), Dhaka, Bangladesh, 3–5 May 2019; Volume 2019. [CrossRef]

2

3D Exploration and Navigation with Optimal-RRT Planners for Ground Robots in Indoor Incidents

Noé Pérez-Higueras *, Alberto Jardón , Ángel Rodríguez and Carlos Balaguer

Robotics Lab, Department of Systems Engineering and Automation, University Carlos III of Madrid, Avda. Universidad, 30, Leganés, 28911 Madrid, Spain; ajardon@ing.uc3m.es (A.J.); angrodri@ing.uc3m.es (A.R.); balaguer@ing.uc3m.es (C.B.)

* Correspondence: noe.perez@uc3m.es

Abstract: Navigation and exploration in 3D environments is still a challenging task for autonomous robots that move on the ground. Robots for Search and Rescue missions must deal with unstructured and very complex scenarios. This paper presents a path planning system for navigation and exploration of ground robots in such situations. We use (unordered) point clouds as the main sensory input without building any explicit representation of the environment from them. These 3D points are employed as space samples by an Optimal-RRTplanner (RRT*) to compute safe and efficient paths. The use of an objective function for path construction and the natural exploratory behaviour of the RRT* planner make it appropriate for the tasks. The approach is evaluated in different simulations showing the viability of autonomous navigation and exploration in complex 3D scenarios.

Keywords: mobile robotics; search and rescue robots; path planning; navigation; exploration; point clouds

1. Introduction

The development of autonomous robots capable of operating in dangerous situations for humans like inspection, surveillance or search and rescue, is a topic of intense investigation in the field of Robotics. Some examples of robotic systems devised for real rescue applications are a multi-flipper controlled platform for collapsed environments [1], the team of teleoperated robots employed in the earthquake in Italy in 2012 [2] or a review of terrestrial robotic systems for nuclear environments, like Fukushima, in Reference [3]. It is remarkable that most of the real robotic applications are teleoperated, which indicates the complexity of developing effective autonomous behaviour in such environments.

Autonomous navigation is a complex function for performing the referred tasks. Within such cluttered and diverse situations, robots must address several issues like localization, perception and representation of the environment, or planning optimal paths according to given criteria. Moreover, in the case of ground robots, an assessment of the terrain traversability is required [4].

For perception, 3D range sensors are usually employed. In outdoos tasks, like self-driving cars and 3D lidars (moving 2D lasers to cover a set of layers in different angles) are widely used since they can provide a big amount of accurate range data at almost the 360° around the vehicle at long distances. In other tasks that do not require long ranges, like indoors or in confined environments, cameras with depth perception are used. They are smaller and lighter devices than 3D lidars, and provide point clouds in a shorter field of view and ranges between 0.4 m and 10.0 m.

From data of 3D range sensors, different structures are derived for the representation of the environment (map) and different techniques can be employed to describe the terrain for robot navigation. Common approaches are triangular meshes [5], the analysis through basic shapes obtained by the Tensor Voting technique [6,7], or mainly voxel grids like octomaps [8–10]. Other methods capture spatial relationships between measurements of the environment naturally. They are called Hilbert maps [11].

However, the generation and/or maintenance of most of these structures can be computationally expensive and lead to a discretization of the space.

On the contrary, we propose working on the point clouds generated from the sensors since we perform a terrain analysis on-demand on robot-sized areas of the environment; similar to the approach of Krusi et al. [12]. Along with this idea, the use of RRT planners [13] seems to be an appropriate option, since we can directly employ the point cloud as a sampling space for the planner and perform the on-demand surface evaluation on it.

In addition to navigation, the autonomous exploration of an unknown area, as in search and rescue missions, is considered in this work. Generally, the exploration is performed by determining regions called "frontiers." The frontiers are regions on the boundary between known open space (no obstacles) and unexplored space. Therefore, by reaching the frontier regions we have more chance to gain information about the environment (map) [14].

Recently, several advances in exploration for aerial vehicles have been made, some examples are found in References [15–17]. However, ground robots are still necessary because of their larger autonomy and their higher payload capacity. In Reference [18], Octomaps are used for ground robot navigation applied to search and rescue missions. Another example of a navigation system for that task is Reference [19]. Moreover, other authors also make use of the exploratory benefits of the sampling-based methods, like RRT planners, for detecting frontiers on 2D grids [20–22] or 3D grids [23]. Our approach also takes advantage of the exploratory behavior of RRTs. Nevertheless, we do not need to discretize the space in a voxel structure, since we use the point cloud directly.

In Reference [24] a potential field approach for exploration is proposed. Good simulation results in 2D are presented. The approach is closely linked to a PoseSLAM algorithm since the uncertainty in the robot poses are taken into account to build an entropy information field for exploration and loop closure. In contrast, we propose a different 3D approach that detaches the autonomous exploration from the SLAM problem. Our approach is able to work with any SLAM algorithm able to provide an online point-cloud map. Another approach for 2D environments are presented in Reference [25], which is based on Growing Neural Gas (GNG) networks; or Reference [26], which proposes a safe and efficient navigation algorithm for exploration of planar scenarios.

The approach presented has been devised to assist rescuers in mine incident environments—the system can perform an autonomous exploration of the area, or the rescuer can indicate remotely a goal position to reach. Thus, the contribution of this paper is twofold—(1) a safe and effective path planning system for ground robot navigation in 3D environments, and (2) an autonomous exploration approach that makes use of the previous path planning system.

This paper is structured as follows. First, we describe the path planning approach based on point clouds and RRT* planners in Section 2.1. Section 2.2 presents the exploration approach based on the previous path planner. Then, Section 3 shows the results of a set of experiments in simulated environments. Finally, Section 4 summarizes the paper's contribution and outlines future work.

2. Methods and Algorithms

The methods and algorithms that have been developed are described here. First, the path planning system is presented. Second, the exploration approach based on the previous path planner is explained.

2.1. RRT—Based Path Planning in 3D Environments*

2.1.1. RRT* Planning on Point Clouds

RRT planners can cope with continuous state and action spaces and kinodynamic constraints They can be easily extended to spaces with higher dimensions [13]. In particular, we make use of Optimal RRT planners (RRT*) [27]; which, unlike regular RRTs, are asymptotically optimal.

They explore the configuration space to obtain optimal paths on cost spaces. Particularly, they sample the configuration space randomly and create a tree towards the goal based on a cost function evaluated for each node candidate of the tree.

In this work, we do not build and maintain any surface map, like polygonal meshes or octomaps to assess the traversability of the terrain. Instead, we make use of (unordered) point clouds obtained from 3D range sensors directly. Thus, we avoid the computational cost of maintaining such structures and the inherent space discretization that can lead to the loss of information.

We follow a similar approach to the work of Krüsi [12], in which nearest-neighbour search over the 3D space points is performed in particular robot-sized areas.

Figure 1 shows a general diagram of the proposed exploration and path planning system. We propose a scheme in which two different point clouds are provided to the RRT* planner—one as a sampling space, and a second one for traversability analysis. For exploration analysis, the global point cloud is downsampled and cropped to a local area a bit bigger than the one employed for traversability, as explained in Section 2.2. Moreover, Figure 2 presents a simple example of the two different point clouds in a tunnel environment employed by the path planning system.

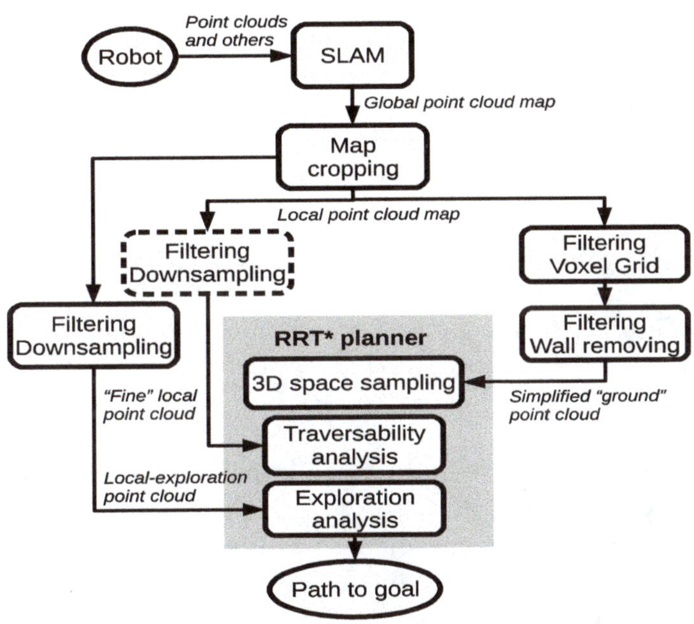

Figure 1. General diagram of the exploration and path planning system.

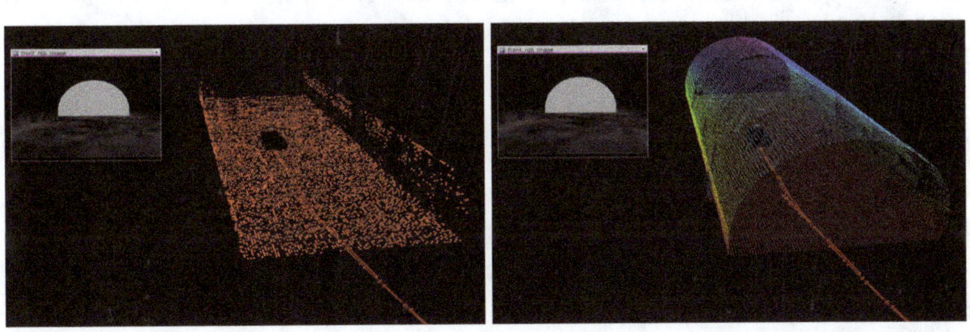

Figure 2. Pointclouds examples. Left: local point cloud used as sampling space for the RRT* planner in red color. Right: local point cloud employed by the RRT* for traversability analysis (multicolor).

Initially, the general point cloud map is cropped to a local robot-centred point cloud of size 10 m $\times$ 10 m. The reason behind is that we are employing short-range sensors (5–6 m) in indoor environments so that we do not need to evaluate regions farther than that distance. Furthermore, that sets a boundary of the size of the point cloud to be assessed and, therefore, the computation burden remains always controlled.

2.1.2. 3D Sampling Space for RRT* Planning

As sampling space for the planner, we employ the points of the local point cloud instead of sampling for the whole 3D space. However, unlike [12], we perform a filtering process over this point cloud. Firstly, a voxel grid structure is employed for downsampling the point cloud (in this paper we use voxels of 5 cm^3). Then, Principal Component Analysis (PCA) is used for computing the surface normals based on the covariance matrix of the N points contained in each voxel V_i.

With those data, the surface orientation is obtained and, therefore, we can filter out vertical surfaces like walls and the ceiling, by setting boundaries for the pitch and roll angles, δ and β respectively, of the surface:

$$|\delta| \leq \delta_{max} \in \mathbb{R}_{>0} \tag{1}$$

$$|\beta| \leq \beta_{max} \in \mathbb{R}_{>0}. \tag{2}$$

Thus, we prevent the RRT* planner from sampling from invalid areas like walls and ceilings, and consequently, to waste computing time in evaluating them.

It is worth noting that these filters and pre-evaluations are executed in parallel to the RRT planning, which takes the newest processed data available, and therefore, it is not a sequential process.

2.1.3. Terrain Analysis

The traversability analysis of the surface employed by the RRT* planner is based on the local point cloud, according to Figure 1. Also, a visual example can be seen in the right image of Figure 2. Depending on the complexity of the scenario, the local point cloud can be downsampled or not (dotted square of Figure 1). This has a clear effect on the RRT* computation time as will be discussed in Section 3.

For each sample taken by the planner, the points of the local point cloud that lies inside a sphere around the sample are analyzed. This sphere has the size required to circumscribe the robot shape, so the radius of the sphere, r, is a required parameter. It is noteworthy that we use in this work a sphere for the sake of generality of the algorithm but it could be changed to other shapes, like a rectangular box, that can fit better the robot dimensions or its touch surfaces.

Based on this analysis, a cost is assigned to the corresponding sample and a new node is added to the tree. Without loss of generality, we can assume that the RRT^* cost function for each point p can be expressed as a weighted linear combination of a set of J functions $f_j(p)$ and weights ω_j defining the task according to:

$$c(p) = \sum_{j=1}^{J} \omega_j f_j(p) = \omega^T f(p). \tag{3}$$

This function $f(p) = [f_1(p), f_2(p), \cdots, f_J(p)]^T$ is based on J measurable characteristics that describe the task to be performed and are called features [28]. The cost of a path is then the sum of the cost for all points in it.

The set of features most-employed for terrain analysis are the inclination (pitch δ and roll β), and also the roughness λ of the surface [12]. The roughness can be approximated by the smallest eigenvalue obtained, along with the pitch and roll, through the technique explained in Section 2.1.2.

We also propose the use of another set of secondary features that may help to evaluate the terrain:

- Number of points in the sphere, η. The more points we have in the area the more accurate and reliable the representation of the surface is. In our case, we consider that a value of one point per centimetre in a plane is good density, although this value can be changed according to the particular constraints. Therefore, an upper boundary of number of points has been calculated as $\eta_{max} = \pi * sphere\_radius^2 * 100$. In which the $sphere\_radius$ is presented in meters. This boundary is also used for normalization:

$$\eta = (1.0 - num\_points/\eta_{max}). \tag{4}$$

- Distance between the sample p_i and the mean of the set of points $\overline{p}$, named as d_m. If these two points are not close, some areas of the region could be poorly represented by very few points in several cases:

$$d_m = dist_{p_i - \overline{p}}/sphere\_radius. \tag{5}$$

- Standard deviation of the point set, σ. A high deviation could indicate that the points are dispersed along the patched region, and therefore, the probability of having voids without points could be smaller.

$$\sigma = (1.0 - stddev/\sigma_{max}). \tag{6}$$

Furthermore, we add another feature based on the distance from the sample to the goal d_g in order to reduce the path length. Finally, we have a feature set of seven normalized feature functions $f(p) = [\delta, \beta, \lambda, \eta, d_m, \sigma, d_g]$. In addition, the set of weights is also normalized such as $\sum_{j=1}^{J} \omega_j = 1$.

Therefore, for each RRT* space sample p_i, the terrain is evaluated following the next steps:

1. Obtaining the points around p_i through nearest-neighbours search from the local point cloud.
2. Calculating the pitch, roll and roughness values.
3. If δ, β and λ exceed the boundaries, the terrain is considered invalid, and the remaining steps are skipped.
4. If the terrain is valid, the remaining features (η, d_m, σ, d_g) are calculated.
5. The cost for the sample is computed according to (3), and a new node is added to the tree following the regular RRT* expansion and rewiring process.

2.2. 3D RRT*-Based Exploration

The proposed path planner is also employed for the task of autonomous exploration. Unlike other works [21,23], we take advantage of the RRT* path planner over the point cloud for finding frontier points instead of using a regular RRT over an occupancy grid or a voxel grid.

The RRT family of planners present a natural exploratory behavior of the space. That means the leaves of the tree are potential frontier points that must be analyzed. Our approach also presents the advantage that the paths to the leaves have been optimized regarding the terrain traversability criteria presented in Section 2.1.3. It is clear that in exploration mode, the feature of the distance to the goal is not used in the cost function for terrain assessment.

Once the planner has finished the expansion over the area, the leaves of the tree are evaluated and the most promising one is chosen as goal frontier. Since a detailed analysis of the terrain is not necessary for this evaluation, a local point cloud with lower resolution than the one employed for traversability analysis is used. That saves computation time. Furthermore, the size of this point cloud is a bit bigger than the traversability one since we need to take into account the points around the tree leaves. Finally, after this evaluation, the system returns the corresponding path to that goal.

Following the general diagram in Figure 1, a new diagram focused on the Exploration Analysis is presented in Figure 3. The different modules involved in this analysis are explained next.

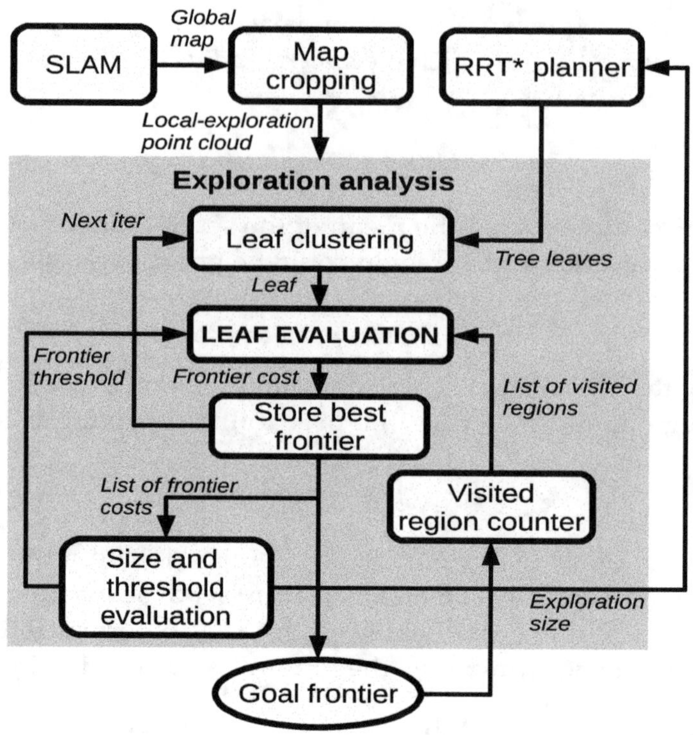

Figure 3. General diagram of the exploration system.

2.2.1. Leaf Clustering

The number of leaves of the tree can be large and some can be very close to each other in terms of distance. Consequently, we propose performing a clustering of the leaves, as in Reference [21], with the purpose of avoiding unnecessary computations and speed up the process. Nevertheless, instead of taking the mean point of the cluster as a frontier point, a different method is employed. For each cluster detected, the leaf with the minimum path cost is chosen as a potential frontier point and the rest of the leaves are discarded. This way, the robot can reach the area through the most-efficient-path found.

We have used a simple Euclidean clustering, which is defined by three parameters—the radius of search for candidates of clustering, and the minimum and the maximum number of points that can be contained in a cluster. In this work, we have employed a radius of 0.3 m, with a minimum of 2 points and a maximum of 4 points per cluster. With these values, our aim is not to achieve a big reduction in the number of potential frontier points but to simply join those that are very close.

2.2.2. Evaluation of Potential Frontiers

Following the diagram presented in Figure 3, after the leaf clustering, a process of evaluation for each selected leaf is performed. Figure 4 shows the different steps and validation checks carried out in that process which are listed next:

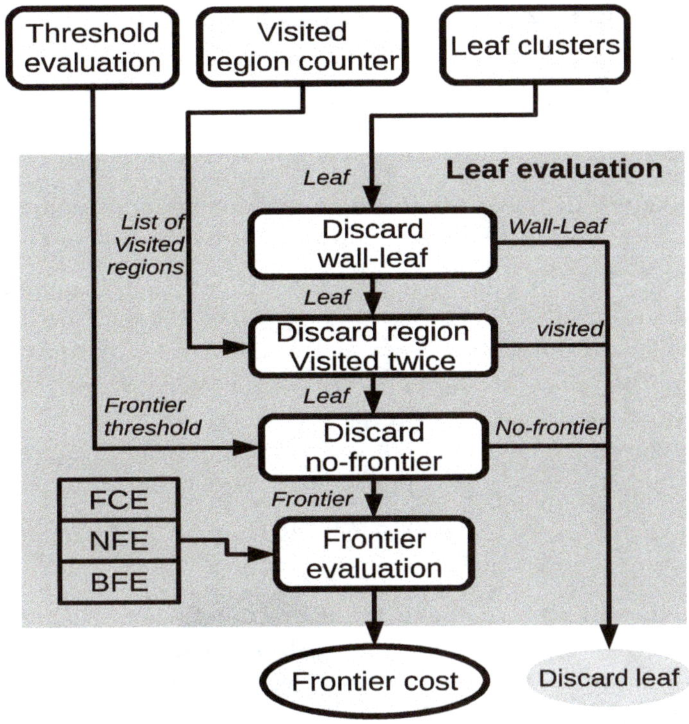

Figure 4. Diagram of the evaluation of potential frontiers.

1. Wall-leaf rejection.

   Firstly, the position of the leaf in the space is studied. The points inside a sphere of a radius equal to the inflated circumscribed sphere of the robot shape are assessed. The inflation radius has been chosen to be 0.3 m. The points are associated with a grid, and the normal vector to the surface formed by the points of each cell is computed. This way, we can count the number of cells of a vertical surface. If this number exceeds a pre-defined threshold, the leaf is considered that is very close to a wall and it is discarded as a possible frontier. An example is shown in Figure 5.

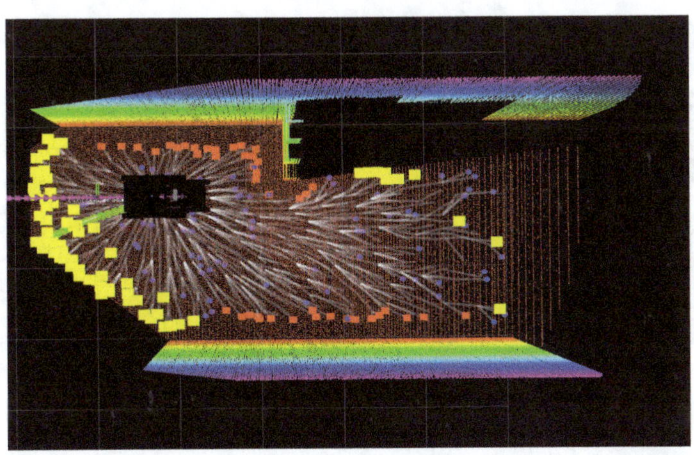

Figure 5. Example of wall-leaf rejection and frontier rejection. The tree leaves in red colour have been discarded as possible frontiers by the wall-leaf detection module. The leaves in yellow color have been accepted as frontiers by the no-frontier rejection module.

2. Visited region rejection.

   Secondly, the position of the leaf is compared with the position of all the previous visited frontiers. If it is very close to a region that has been visited twice, the leaf is also discarded. To do that, the list of visited regions (with the number of visits) is provided by the counter of visited regions, which is explained in Section 2.2.3. This module plays an important role when there are frontier areas in which it is not possible to gain more information. In those cases, it is able to lead the robot to abandon the areas already explored.

3. No-frontier rejection.

   The third step checks whether the leaf can be considered as a frontier or not. To determine that, the points in a sphere of radius 1.5 m around the leaf (this value can be configured) are counted. Moreover, the standard deviation of the points is also computed. Therefore, the lesser number of points are detected in the area where the more promising the frontier point for exploration is. The number of points (*points*) and the standard deviation (*stddev*) are normalized by choosing upper bounds (*max_points, max_stddev*) employed for the normalization. Finally, a normalized frontier cost F_{cost} for the leaf p is computed as:

$$F_{cost}(p) = 0.7 * \frac{points(p)}{max\_points} + 0.3 * \frac{stddev(p)}{max\_stddev}. \tag{7}$$

   A threshold value Δ_f is set, so that if the frontier cost F_{cost} is higher than the threshold, then the leaf is not considered as a frontier. An example is shown in Figure 5 with a threshold value of $\Delta_f = 0.4$. The threshold evaluation determines the value of this threshold dynamically as explained in Section 2.2.4.

4. Frontier evaluators.

   Once we have determined the set of tree leaves that are potential frontiers, the selection of the most promising frontier is performed by the frontier evaluators.

   We propose a novel evaluator based on a cost function, C_{exp}, that we have called **Cost Function Exploration (CFE)**. This cost is obtained as the weighted sum of the cost of the path to the frontier $C(\zeta_{p_f})$, the cost of the frontier $F_{cost}(p_f)$, and a "return" cost $R(p_f)$, which penalizes the frontiers that are close to areas that the robot has already visited:

$$C_{exp}(p_f) = \omega_c\ C(\zeta_{p_f}) + \omega_F\ F_{cost}(p_f) + \omega_R\ R(p_f). \tag{8}$$

   Unlike other works based on regular RRTs [21,23], our navigation cost $C(\zeta_{p_f})$ includes the terrain evaluation and not only the length of the path to the frontier.

   Finally, the aim of the return cost R is to prevent the robot from re-visiting areas where the robot already was close by. This is particularly useful in our setup, which involves tunnel scenarios and the evaluation on a local point cloud instead of the global map. In these cases, the robot is biased to explore forward the tunnel instead of going back.

   The return cost is calculated by computing the distance between the frontier point and the closest point of the trajectory followed by the robot so far. This cost decreases linearly from 1 to 0 in a pre-defined distance range from 0 up to 2 meters:

$$R(p_f) = max\_dist - d(p_f, t_r)/max\_dist, \tag{9}$$

   where $d(p_f, t_r)$ is the Euclidean distance between the frontier point p_f and the closest point of the robot trajectory t_r. If this distance is greater than the maximum distance threshold ($max\_dist = 2\ m$), the cost returned is zero.

   We also compare this evaluator with two methods widely-used in the literature, as in Reference [22], which have been adapted to work in 3D over point clouds:

- **Nearest Frontiers Exploration (NFE).** This is an adaptation of the well-known Nearest Frontier approach [14] to 3D point clouds. It is based on proximity criteria by selecting the frontier with the smallest Euclidean distance to the robot ignoring the existence of obstacles.
- **Biggest frontier Exploration (BFE).** It is based on size criteria. The frontier with less information (F_{cost}) is selected as the goal.

2.2.3. Counter of Visited Regions

The counter of visited regions is in charge of storing all the exploration goals selected so far. Moreover, it counts the number of visits that each goal region has received. A visit is counted when the distance between the new frontier goal and any of the previous goals is lower than a distance threshold, d_{reg}. In this work, we have employed a distance of 1.5 m. Then, in the leaf evaluation, if the leaf is close to a region visited twice, the leaf is discarded as a possible frontier.

2.2.4. Evaluation of exploration size and frontier threshold

This module analyzes the current exploration situation with the aim of dynamically adapting, firstly, the size of the exploration area. The objective is to save computation time by trying to keep a "small" area for planning and secondly, to change the value of the frontier threshold, which determines the leaves that are considered frontiers according to the frontier costs computed at each iteration of the exploration.

To do that, the module uses a list of the frontier costs computed in the last exploration (*cost_array*). These costs are compared with a cost boundary κ. The value of the cost boundary is computed regarding the current frontier threshold value ($\kappa = \Delta_f - 0.05$). Furthermore, the minimum size and the maximum size of the planning area for the RRT* planner is provided.

The process is described in more detail in Algorithm 1. If the percentage of frontier costs that are equals or higher than the cost boundary is above a pre-defined limit ϵ, the current size of exploration area employed by the RRT* planner is increased as well as the time employed for planning (from Line 2 to Line 8). If that is not fulfilled, and the current planning size is higher than the minimum size, then the size and the time are decreased (Line 9). Contrary, if the array of costs is empty, which means that we did not find any frontier, the planning size and time is also increased only if the maximum planning size was not reached (Line 12). Finally, if the maximum planning size was reached, the frontier threshold Δ_f is increased with the aim of detecting "smaller" frontiers (Line 14).

Algorithm 1 Size-and-threshold evaluator algorithm.

```
Input: cost_array, κ, Δ_f, ε, min_size, max_size
1  if cost_array NOT empty then
2      count ← 0 foreach c ∈ cost_array do
3          if c ≥ κ then
4              count ← count + 1
5          end
6      end
7      θ ← count/cost_array.size if θ > ε AND current_size < max_size then
8          RRTSize ← RRTSize + size_inc  RRTTime ← RRTTime + time_inc
9      else if current_size > min_size then
10         RRTSize ← RRTSize − size_inc  RRTTime ← RRTTime − time_inc
11     end
12 else if current_size < max_size then
13     RRTSize ← RRTSize + size_inc  RRTTime ← RRTTime + time_inc
14 else
15     Δ_f ← Δ_f + thres_inc  κ ← Δ_f − 0.05
16 end
```

3. Results

3.1. Implementation and Simulations

The proposed path planning and exploration system has been implemented in C++ under ROS (http://www.ros.org/) [29]. Moreover, the Point Cloud Library (PCL) (http://pointclouds.org/) has been employed for point cloud processing. The developed software is publicly available on Github under BSD license (https://github.com/noeperez/indires_navigation).

It is not part of this work the evaluation or development of SLAM algorithms. Therefore, a publicly available SLAM algorithm that provides an online map based on point clouds has been employed in this evaluation. Specifically, an ICP-based SLAM system which provides a real-time tracker and mapper in 2D and 3D has been used [30,31].

For evaluation of the approach, three different environments have been designed and simulated in Gazebo Robot Simulator (http://gazebosim.org/) along with the tracked robot employed in this work. Figure 6 shows a capture of these scenarios. The first environment (left image) is a tunnel with different obstacles in it. The tunnel has a width of around 4 m. The second environment (center image) is a squared area of 17 m × 17 m that presents a bumpy terrain with different ramps and obstacles. The third one (right image), is a set of connected corridors with ramps and fallen rocks inside. The dimensions of the bounding box that limits the space is around 35 m × 14.5 m.

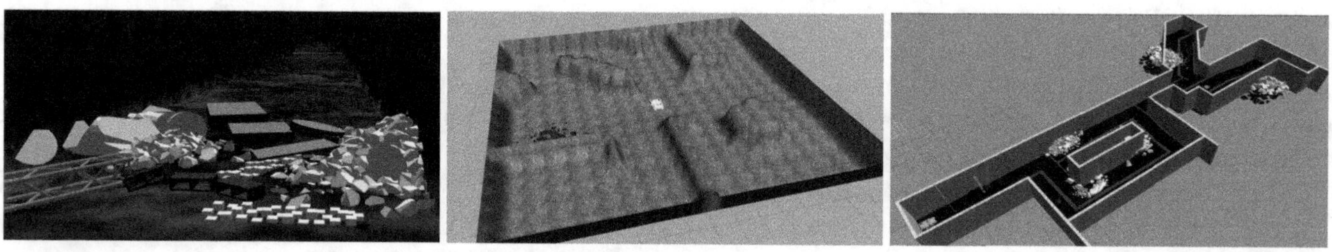

Figure 6. Scenarios for evaluation. Left: tunnel environment with obstacles for navigation (navigation map). Center and right: environments for evaluation of the exploration and navigation (exploration maps 1 and 2 respectively).

In the real robot we used only an RGB-D camera placed on top of the robot so that we used the same sensor and configuration to perform the evaluation in these experiments. The employed field of view was 80° H × 49.5° V with a distance range of 0.4 m–5.0 m. By using a wider field of view, like 3D lidars, that can cover the 360° around the robot approximately, probably better results can be obtained. However, in indoor and confined environments, like a collapsed tunnel, a large distance range is not required and a lot of unnecessary data would delay the computation of the algorithms. We prove that the performance of the system is still very good under these field-of-view constraints.

All the experiments were executed in a computer with an i7-6700HQ processor and 16GB of RAM.

3.2. Navigation Results

The tunnel with obstacles presented in the left image of Figure 6 was employed here to evaluate some aspects of the navigation system as the size and resolution of the point clouds, and the cost function employed for path building.

The robot was placed at the beginning of the tunnel and had to try to reach a goal 5 m away by crossing the obstacles in between. In this scenario, the RRT* planner is allowed to plan a path for 5 s.

3.2.1. Size of the Point Cloud for Sampling Space

In the work of Krusi [12], the whole point cloud is used as a sampling space for a RRT planner. On the contrary, we use a filtered local point cloud with a fixed size and without walls and ceiling what presents a remarkable impact on the performance of the algorithm.

From 10 path-planning repetitions in the tunnel scenario, we are able to reduce the number of points of the original point cloud in a 93% on average. Besides filtering most of the points of the walls and the ceiling, the point cloud is downsampled to a resolution of 0.05 cm. Thus, we prevent the algorithm from wasting computation time on sampling and evaluating most of the invalid areas. This poses a clear advantage over other approaches, like that in Reference [12], that use the whole point cloud.

Note that the number of points available for sampling must be rich enough to describe the area so that space can be appropriately explored by the tree expansion. Therefore, the resolution of the downsampled space must be properly chosen and very low resolutions must be avoided.

3.2.2. Analysis of the Resolution of the Point Cloud for Traversability Analysis

The resolution of the point cloud employed for terrain analysis also has a significant impact on the algorithm performance. The denser the point cloud is the more accurate the analysis of the surface is. However, the computing burden is also higher since there are more points to be processed.

We have performed a comparison of the size of the RRT* tree (number of tree nodes) and the costs of the paths according to different resolutions of point cloud for traversability. The planning has been repeated 10 times in the tunnel scenario (Figure 7).

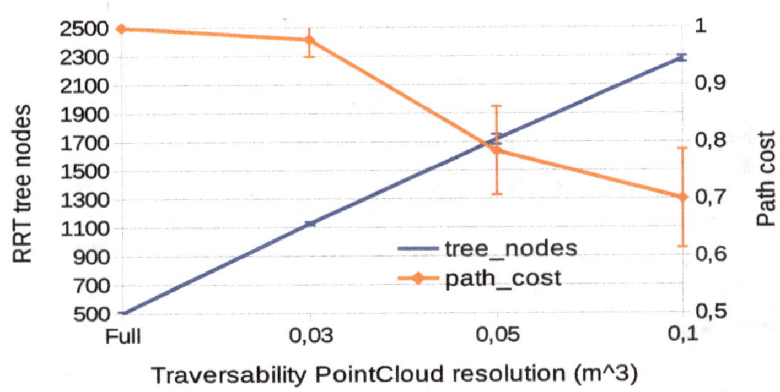

Figure 7. Comparison of pointcloud resolution for traversability versus size of RRT* tree and path costs in 5 *secs* of planning in the tunnel scenario.

On the one hand, with the highest-density point cloud, the algorithm is not able to expand a tree that reaches the goal in the given time (5 s). That is represented with the highest cost possible (1). For a point cloud of 0.03 m^3, the algorithm is unable to find a path in some of the 10 cases. In case of using a resolution of 0.1 m^3, the costs of the paths are small. Though, the paths obtained overcome almost all the obstacles since they are not properly represented with such low resolution. According to these experiments in the tunnel environment, a resolution of 0.05 m^3 seems to be appropriate, although higher resolutions can improve the results if more processing power or time can be given to the planner.

On the other hand, as expected, the lesser number of points to be processed (low resolution) the more speedy the algorithm is (bigger RRT tree). However, the representation of the environment is less accurate if we use fewer points to describe it.

3.2.3. Evaluation of the Cost Function for Path Building

To evaluate the path planning capabilities, we compare a RRT* cost function based on the regular features for terrain analysis (inclination and roughness) employed in some approaches [5,12], with the proposed augmented cost function described in Section 2.1.3. The weights of the two cost functions have been hand-tuned to the best behavior found. Specifically, we have employed in this work the following weight sets:

- Basic cost function: $\omega_\delta = 0.4$, $\omega_\beta = 0.4$, $\omega_\lambda = 0.2$.

- Augmented cost function: $\omega_\delta = 0.1$, $\omega_\beta = 0.1$, $\omega_\lambda = 0.1$, $\omega_\eta = 0.175$, $\omega_{d_m} = 0.175$, $\omega_\sigma = 0.175$, $\omega_{d_g} = 0.175$.

We also evaluate the tree expansion according to a different set of threshold values for the pitch, roll and roughness that determine whether a sample is considered valid or not. Three sets for evaluation are considered without regard to the robot characteristics, as presented in Table 1.

Table 1. Set of boundaries for RRT* expansion.

	Strict	Medium	Relaxed
pitch	0.7	0.87	1.3
roll	0.7	0.87	1.3
roughness	0.2	0.8	3.0

Figure 8 shows a visual comparison of the paths obtained in the tunnel scenario with different combinations of the boundaries and the cost functions. As can be observed, the proposed cost function improves the paths of the basic cost function. Regarding the tree expansion, the strict boundaries only allow the expansion on the flat floor while the relaxed boundaries allow overcoming almost all the obstacles. Therefore, proper values for the boundaries must be chosen according to the robot capabilities of overcoming obstacles and climbing slopes.

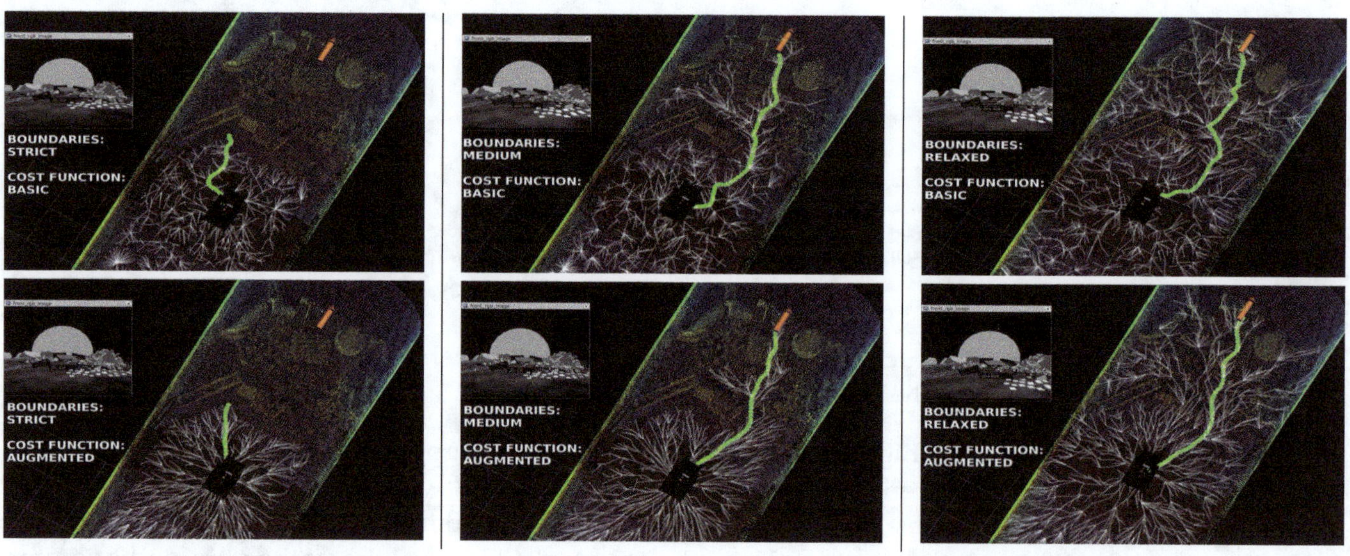

Figure 8. Visual comparison of different tree-expansion boundaries and cost functions.

3.3. Exploration Results

The exploration process employed in this evaluation is the following. While planning and evaluating, the robot is still on the spot waiting for obtaining a new exploration goal. Therefore, the processing of the next exploration goal is not launched until the robot has reached its previous goal, and consequently, the robot waits there until the new goal has been computed.

The maximum linear velocity that the vehicle can reach is 0.3 m/s, and the maximum angular velocity is 0.6 rad/s. The values of the exploration parameters that have been used can be consulted in Table 2. These parameters have been hand-tuned to the best performance achieved.

Table 2. Parameters employed in the exploration evaluation.

$\omega_C = 0.305\ [0,1]$	$\omega_F = 0.39\ [0,1]$	$\omega_R = 0.305\ [0,1]$
$min\_size = 5.0\ m$	$max\_size = 13.0\ m$	$size\_inc = 2.0\ m$
$\Delta_f = 0.4\ [0,1]$	$thresh\_inc = 0.1\ [0,1]$	$time\_inc = 0.8\ seg$
$\kappa = 0.35\ [0,1]$	$\epsilon = 0.85\ [0,1]$	$d_{reg} = 1.5\ m.$

3.3.1. Frontier Evaluators Comparison

For comparison we have executed the proposed approach (CFE) and the state-of-the-art methods (NFE and BFE), 3 times for each of the two exploration environments (exploration maps 1 and 2 of Figure 6). We compare the percentage of the surface area that has been explored by each algorithm regarding a ground-truth map that has been built by teleoperating the robot carefully. The duration of the manual exploration has lasted around 20 min for map1, and 25 min for map 2 approximately.

The results are presented in Figure 9 for map 1 and map 2 respectively. The percentage of area explored and their standard deviations along the time elapsed are shown. As can be seen, in both environments, the system is able to lead the vehicle to overcome all the obstacles and cover more than the 90% of the area in a reasonable time in all the cases. Furthermore, the three frontier evaluators reach similar performance in terms of area covered.

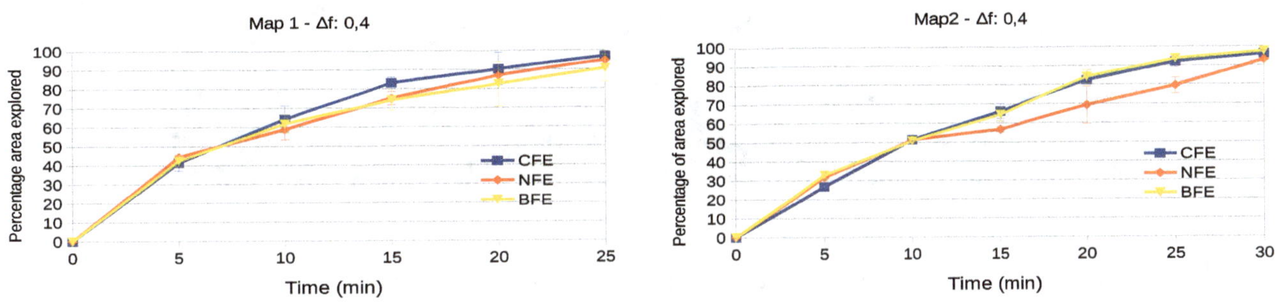

Figure 9. Percentage of ground truth area covered by the proposed approaches. Left: map 1. Right: map 2.

However, we can observe some differences regarding the distance traveled by the robot, as stated in Table 3. As expected, the BFE method leads the robot to travel more distance since it always tries to reach the "bigger" frontier detected no matter the distance. On the contrary, the NFE tries to reach the closest frontier at any moment, so, the traveled distance is a bit shorter. Similar distance is reached by the CFE approach, which tries to present a balanced behavior between the other two approaches, according to the chosen weights of its cost function.

Table 3. Comparison of distance traveled in the two environments.

Dist. (m)	CFE	NFE	BFE
Map 1	74.08 ± 5.84	74.59 ± 11.80	79.10 ± 12.84
Map 2	87.63 ± 3.01	86.30 ± 9.95	99.98 ± 5.01

The parameters chosen have an important impact on the performance of the approaches. The values of the frontier threshold Δ_f, or the minimum and maximum planning sizes are also very relevant as we analyze next.

3.3.2. Frontier Threshold Evaluation

One of the most important parameters is the frontier threshold Δ_f. Figure 10 shows an example of the frontiers detected regarding four different values for Δ_f.

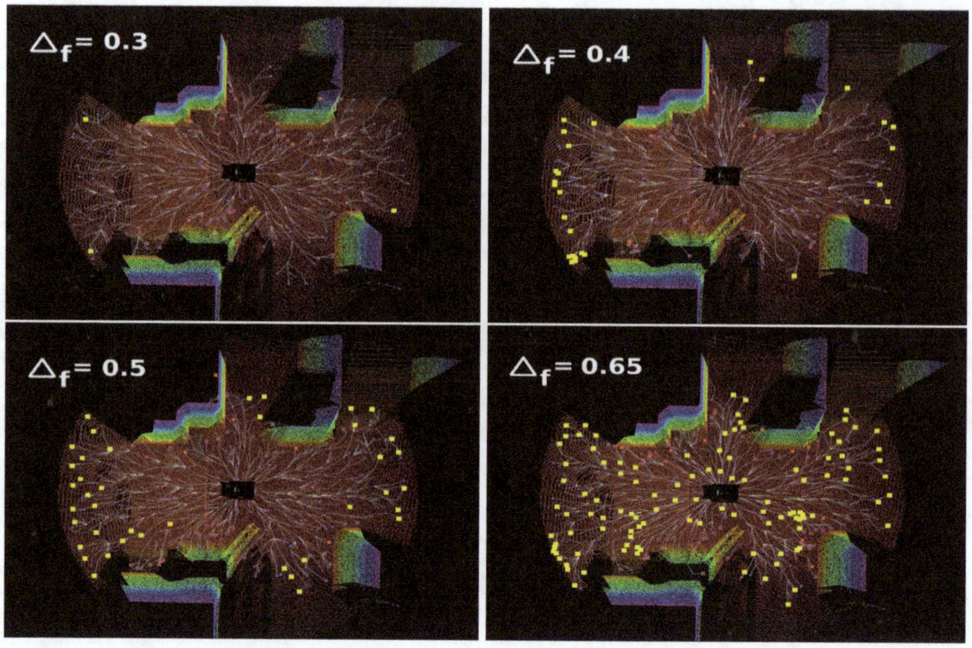

Figure 10. Frontiers detected (yellow squares) according to different values of frontier threshold Δ_f.

As can be observed, it determines how exhaustive the exploration is. If it is very low, only the "biggest" frontiers are detected, and some areas can remain poorly explored. Moreover, if the planning size does not cover the whole area to be explored (most of the cases), the robot can fall into a blocked situation since new frontiers can not be detected. On the contrary, if its value is high, a lot of regions are considered as frontiers and the exploration system will perform a thorough exploration. The dynamical change of this value is also linked to the current planning size, which is also adapted according to the situation. For these reasons, the module of adaptive planning size and frontier threshold is required.

An example of the functioning of this module is presented in Figure 11. The planning radius around the robot and the frontier threshold are shown for each exploration iteration performed in the exploration maps 1 and 2 respectively. For the shake of clarity, we show only the results of the CFE approach with an initial frontier threshold of $\Delta_f = 0.4$ since the performance of the module does not depend on the frontier evaluator employed. As can be seen, the planning radius is adapted dynamically with the aim of finding frontier points. When the maximum radius is reached (13 m) and no frontiers are found, the frontier threshold is increased too.

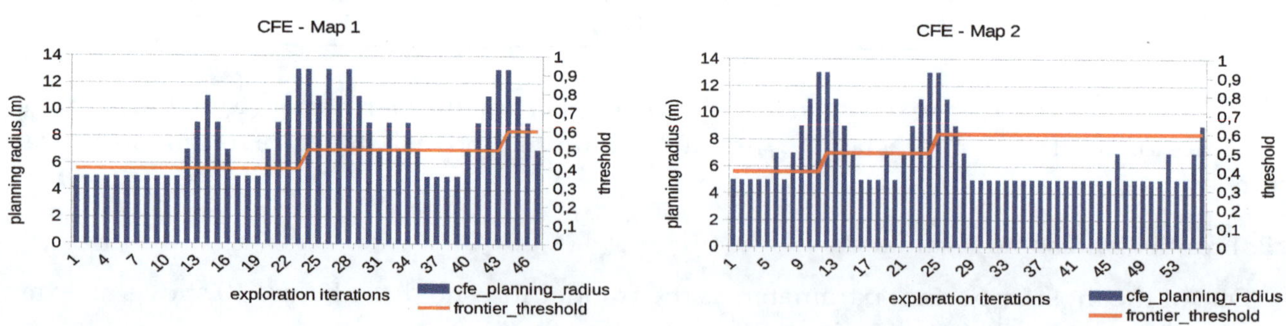

Figure 11. Adaptation of the planning radius and frontier threshold during the exploration of the maps 1 and 2 using the CFE approach.

Table 4 shows the results in the percentage of area explored and distance travelled in 25 min of exploration for the three proposed frontier evaluators according to two initial values of Δ_f in the exploration map 1.

Table 4. Comparison of exploration in map 1 with different frontier threshold values.

		CFE	NFE	BFE
$\Delta_f = 0.4$	$Exp.(\%)$	97.00 ± 2.51	94.76 ± 4.47	90.41 ± 7.14
	$Dist.(m)$	74.08 ± 5.84	74.59 ± 11.80	79.10 ± 12.84
$\Delta_f = 0.6$	$Exp.(\%)$	90.01 ± 6.47	75.36 ± 3.55	92.63 ± 1.23
	$Dist.(m)$	72.07 ± 2.89	41.09 ± 2.33	$95.42 \pm 7.90.$

On the one hand, in case of using an initial threshold of 0.4, the adaptive-frontier-threshold module increased the threshold during the exploration to reach a final value 0.6 for all the approaches. Moreover, the three proposed approaches present good and very similar metrics.

On the other hand, if the initial threshold is 0.6, many frontiers are detected during the whole exploration and no need of increasing it is required. In this case, the NFE approach worsens its performance. In the same time period, it is able to cover only the 75% of the area since it has a lot of close frontiers to visit. On the contrary, with more frontiers detected, the BFE seems to be affected only in the travelled distance, which has been increased. Finally, the CFE presents a similar performance in both cases. It always tries to choose a "good" frontier according to the cost function from the set of frontiers detected.

Overall, it seems that starting the exploration with a low threshold value and letting the threshold evaluator module to increase when necessary is an appropriate technique. However, this must be fixed according to the complexity of the environment and how exhaustive the exploration is required. Regarding the frontier evaluators, the CFE approach seems to present a slightly better performance than the other two, since it is able to cover the complete area without travelling big distances.

3.3.3. Visited Regions Evaluation

The module of rejection of visited regions can lead the robot to abandon the areas already explored twice and, this way, to avoid block situations in which the robot tries to repeat the exploration of regions where it is not possible to gain more information. This can occur more frequently when the frontier threshold Δ_f is high and a lot regions without a high level of information are considered as frontiers. Hence, instead of staying in the same area trying to gain more information, the robot is pushed to move to other regions.

For each exploration iteration, Figure 12 shows the number of regions that were already visited twice and therefore, the close leaves were discarded as possible frontiers. Each region is an sphere with radius $d_{reg} = 1.5$ m. Again, only the FCE approach is shown since the performance of the visited-region rejection does not depend on the frontier evaluator employed.

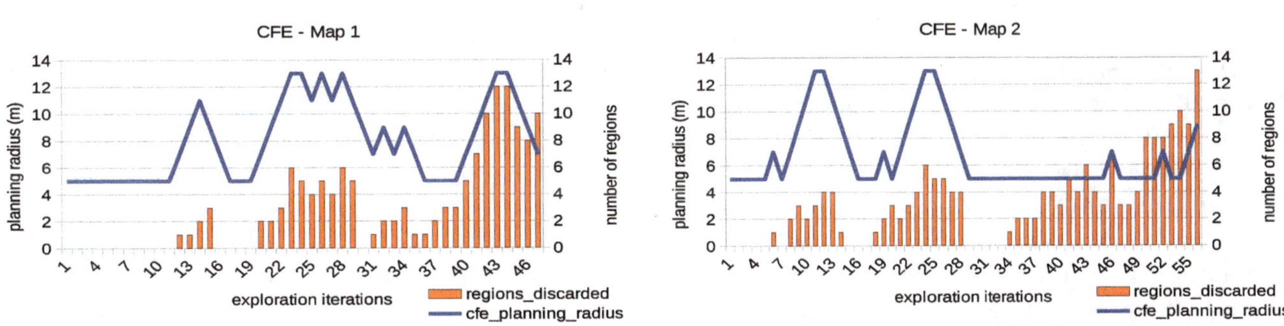

Figure 12. Number of regions discarded for each exploration iteration of the maps 1 and 2 using the CFE approach.

As expected, the number of regions discarded is higher when the planning radius is bigger, and also when the area of the given map is almost completely explored.

4. Conclusions

This paper presented a navigation and exploration system for ground robots in complex indoor 3D environments like a mine. The approach has a twofold function since it is employed for autonomous path planning and also for exploration.

Some improvements and novelties are included in the path planning regarding the current approaches of the state of the art. The use of different point clouds with bounded sizes and resolutions directly in the different stages of the RRT* planning present advantages in efficiency computation time.

To employ the expanded RRT* tree for exploration, which includes feasible paths to the leaves in the 3D environment, seems to be an efficient approach. Unlike regular methods, the proposed approach can dynamically adapt the size of the area employed for planning, which saves computation resources. Moreover, the modules for frontier threshold evaluation and the counter of visited regions can adapt the system to the situations found and to lead the robot to perform a complete exploration in complex 3D environments. Furthermore, the proposed frontier evaluator presents good results as well as the adaptation of the state-of-the-art approaches.

The experiments in simulation in diverse 3D environments present good performance of the different modules and deliver promising results for the application of the system in real robots.

Future work will consider experimentation with real robots and environments and the study of learning approaches that can help to fit the set of different parameters of the modules and the weights of the cost functions.

Author Contributions: conceptualization, methodology, software, validation, formal analysis, investigation, writing—original draft preparation, N.P.; visualization, supervision, A.J.; writing—review and editing, A.R.; project administration, funding acquisition, C.B. All authors have read and agreed to the published version of the manuscript.

References

1. Menna, M.; Gianni, M.; Ferri, F.; Pirri, F. Real-time autonomous 3D navigation for tracked vehicles in rescue environments. In Proceedings of the 2014 IEEE/RSJ International Conference on Intelligent Robots and Systems, Chicago, IL, USA, 14–18 September 2014, pp. 696–702.
2. Kruijff, G.J.M.; Tretyakov, V.; Linder, T.; Pirri, F.; Gianni, M.; Papadakis, P.; Pizzoli, M.; Sinha, A.; Emanuele, P.; Corrao, S.; Priori, F.; Febrini, S.; Angeletti, S. Rescue Robots at Earthquake-Hit Mirandola, Italy: A Field Report. In Proceeding of the IEEE International Symposium on Safety, Security, and Rescue Robotics (SSRR), College Station, TX, USA, 5–8 November 2012, pp. 1–8. doi:10.1109/SSRR.2012.6523866. [CrossRef]
3. Tsitsimpelis, I.; Taylor, C.J.; Lennox, B.; Joyce, M.J. A review of ground-based robotic systems for the characterization of nuclear environments. *Prog. Nucl. Energy* **2019**, *111*, 109–124. doi:10.1016/j.pnucene.2018.10.023. [CrossRef]
4. Papadakis, P. Terrain traversability analysis methods for unmanned ground vehicles: A survey. *Eng. Appl. Artif. Intell.* **2013**, *26*, 1373–1385. doi:10.1016/j.engappai.2013.01.006. [CrossRef]
5. Garrido, S.; Malfaz, M.; Blanco, D. Application of the fast marching method for outdoor motion planning in robotics. *Rob. Auton. Syst.* **2013**, *61*, 106–114. doi:10.1016/j.robot.2012.10.012. [CrossRef]
6. Liu, M.; Siegwart, R. Navigation on point-cloud-A Riemannian metric approach. In Proceedings of the 2014 IEEE International Conference on Robotics and Automation (ICRA), Hong Kong, China, 31 May–7 June 2014, pp. 4088–4093. doi:10.1109/ICRA.2014.6907453. [CrossRef]
7. Liu, M. Robotic online path planning on point cloud. *IEEE Trans. Cybern.* **2016**, *46*, 1217–1228. doi:10.1109/TCYB.2015.2430526. [CrossRef] [PubMed]

8. Hornung, A.; Wurm, K.M.; Bennewitz, M.; Stachniss, C.; Burgard, W. OctoMap: An efficient probabilistic 3D mapping framework based on octrees. *Auton. Robot.* **2013**, *34*, 189–206. doi:10.1007/s10514-012-9321-0. [CrossRef]
9. Song, S.; Jo, S. Online inspection path planning for autonomous 3D modeling using a micro-aerial vehicle. In Proceedings of the 2017 IEEE International Conference on Robotics and Automation (ICRA), Singapore, 29 May–3 June 2017, pp. 6217–6224. doi:10.1109/ICRA.2017.7989737. [CrossRef]
10. Oleynikova, H.; Taylor, Z.; Fehr, M.; Nieto, J.; Siegwart, R. Voxblox: Building 3D Signed Distance Fields for Planning. arXiv **2016**, arXiv-1611.
11. Ramos, F.; Ott, L. Hilbert maps: Scalable continuous occupancy mapping with stochastic gradient descent. *Int. J. Rob. Res.* **2016**, *35*, 1717–1730, [CrossRef]
12. Philipp Krüsi, P.; Furgale, P.; Bosse, M.; Siegwart, R. Driving on Point Clouds: Motion Planning, Trajectory Optimization, and Terrain Assessment in Generic Nonplanar Environments. *J. Field Rob.* **2016**, *34*, 940–984. doi:10.1002/rob.21700. [CrossRef]
13. Lavalle, S.M. Rapidly-Exploring Random Trees: A New Tool for Path Planning. Technical Report; CiteSeerX, 1998. doi:10.1.1.35.1853. [CrossRef]
14. Yamauchi, B. A Frontier-Based Approach for Autonomous Exploration. In Proceedings of the IEEE International Symposium on Computational Intelligence, Robotics and Automation, Monterey, CA, USA, 10–11 July 1997, pp. 146–151.
15. Witting, C.; Fehr, M.; Bähnemann, R.; Oleynikova, H.; Siegwart, R. History-Aware Autonomous Exploration in Confined Environments Using MAVs. In Proceedings of the 2018 IEEE/RSJ International Conference on Intelligent Robots and Systems, IROS 2018, Madrid, Spain, 1–5 October 2018, 2018, pp. 1–9. doi:10.1109/IROS.2018.8594502. [CrossRef]
16. Song, S.; Jo, S. Surface-Based Exploration for Autonomous 3D Modeling. In Proceedings of the 2018 IEEE International Conference on Robotics and Automation (ICRA), Brisbane, QLD, Australia, 21–25 May 2018, pp. 1–8. doi:10.1109/ICRA.2018.8460862. [CrossRef]
17. Dang, T.; Papachristos, C.; Alexis, K. Visual Saliency-Aware Receding Horizon Autonomous Exploration with Application to Aerial Robotics. In Proceedings of the 2018 IEEE International Conference on Robotics and Automation (ICRA), Brisbane, QLD, Australia, 21–25 May 2018, pp. 2526–2533. doi:10.1109/ICRA.2018.8460992. [CrossRef]
18. Dornhege, C.; Kleiner, A. A Frontier-Void-Based Approach for Autonomous Exploration in 3D. *Adv. Rob.* **2013**, *27*, 459–468. [CrossRef]
19. Colas, F.; Mahesh, S.; Liu, M.; Siegwart, R. 3D Path Planning and Execution for Search and Rescue Ground Robots. In Proceedings of the IEEE/RSJ International Conference on Intelligent Robots and Systems (IROS), Tokyo, Japan, 3–7 November 2013, pp. 722–727.
20. Kantaros, Y.; Schlotfeldt, B.; Atanasov, N.; Pappas, G.J. Asymptotically Optimal Planning for Non-Myopic Multi-Robot Information Gathering. In Proceedings of Robotics: Science and Systems (RSS), Freiburg, Germany, 22–26 June 2019. doi:10.15607/RSS.2019.XV.062. [CrossRef]
21. Umari, H.; Mukhopadhyay, S. Autonomous robotic exploration based on multiple rapidly-exploring randomized trees. In Proceedings of the 2017 IEEE/RSJ International Conference on Intelligent Robots and Systems (IROS), Vancouver, BC, Canada, 24–28 September 2017, pp. 1396–1402. doi:10.1109/IROS.2017.8202319. [CrossRef]
22. Pimentel, J.M.; Alvim, M.S.; Campos, M.F.M.; Macharet, D.G. Information-Driven Rapidly-Exploring Random Tree for Efficient Environment Exploration. *J. Intell. Rob. Syst.* **2018**, *91*, 313–331. [CrossRef]
23. Bircher, A.; Kamel, M.; Alexis, K.; Oleynikova, H.; Siegwart, R. Receding horizon path planning for 3D exploration and surface inspection. *Auton. Robot.* **2018**, *42*, 291–306. doi:10.1007/s10514-016-9610-0. [CrossRef]
24. Vallvé, J.; Andrade-Cetto, J. Potential Information Fields for Mobile Robot Exploration. *Robot. Auton. Syst.* **2015**, *69*, 68–79. doi:10.1016/j.robot.2014.08.009. [CrossRef]
25. Wang, D.; Duan, Y.; Wang, J. Environment exploration and map building of mobile robot in unknown environment. *Int. J. Simul. Proc. Model.* **2015**, *10*, 241–252. doi:10.1504/IJSPM.2015.071381. [CrossRef]
26. Savkin, A.; Li, H. A safe area search and map building algorithm for a wheeled mobile robot in complex unknown cluttered environments. *Robotica* **2017**, *36*, 1–23. doi:10.1017/S0263574717000194. [CrossRef]

27. Karaman, S.; Frazzoli, E. Sampling-based algorithms for optimal motion planning. *Int. J. Rob. Res.* **2011**, *30*, 846–894. doi:10.1177/0278364911406761. [CrossRef]
28. Pérez-Higueras, N.; Caballero, F.; Merino, L. Teaching Robot Navigation Behaviors to Optimal RRT Planners. *Int. J. Soc. Robot.* **2018**, *10*, 235–249. doi:10.1007/s12369-017-0448-1. [CrossRef]
29. Koubaa, A. (Ed.) Robot Operating System (ROS): The Complete Reference (Volume 3); In*Studies in Computational Intelligence*; Springer: Berlin/Heidelberg, Germany, 2018; Volume 778. doi:10.1007/978-3-319-91590-6. [CrossRef]
30. Pomerleau, F.; Colas, F.; Siegwart, R.; Magnenat, S. Comparing ICP variants on real-world data sets. *Auton. Robot.* **2013**, *34*, 133–148. [CrossRef]
31. Pomerleau, F.; Magnenat, S.; Colas, F.; Liu, M.; Siegwart, R. Tracking a Depth Camera: Parameter Exploration for Fast ICP. In Proceedings of the 2011 IEEE/RSJ International Conference on Intelligent Robots and Systems, San Francisco, CA, USA, 25–30 September 2011, pp. 3824–3829.

3

Biologically-Inspired Learning and Adaptation of Self-Evolving Control for Networked Mobile Robots

Sendren Sheng-Dong Xu [1], Hsu-Chih Huang [2,*], Tai-Chun Chiu [1] and Shao-Kang Lin [2]

[1] Graduate Institute of Automation and Control, National Taiwan University of Science and Technology, Taipei City 10607, Taiwan; sdxu@mail.ntust.edu.tw (S.S.-D.X.); ifuzzy2015@gmail.com (T.-C.C.)

[2] Department of Electrical Engineering, National Ilan University, Yilan City 26047, Taiwan; cacs2015@gmail.com

* Correspondence: hchuang@niu.edu.tw

Abstract: This paper presents a biologically-inspired learning and adaptation method for self-evolving control of networked mobile robots. A Kalman filter (KF) algorithm is employed to develop a self-learning RBFNN (Radial Basis Function Neural Network), called the KF-RBFNN. The structure of the KF-RBFNN is optimally initialized by means of a modified genetic algorithm (GA) in which a Lévy flight strategy is applied. By using the derived mathematical kinematic model of the mobile robots, the proposed GA-KF-RBFNN is utilized to design a self-evolving motion control law. The control parameters of the mobile robots are self-learned and adapted via the proposed GA-KF-RBFNN. This approach is extended to address the formation control problem of networked mobile robots by using a broadcast leader-follower control strategy. The proposed pragmatic approach circumvents the communication delay problem found in traditional networked mobile robot systems where consensus graph theory and directed topology are applied. The simulation results and numerical analysis are provided to demonstrate the merits and effectiveness of the developed GA-KF-RBFNN to achieve self-evolving formation control of networked mobile robots.

Keywords: biologically-inspired; self-learning; formation control; mobile robots

---

1. Introduction

Networked mobile robots that are capable of self-learning have received growing attention in the mobile robotics research community [1–3]. This emerging technology has surpassed the conventional robotic system by taking advantage of robot collaboration, system robustness, scalability, and greater flexibility. This modern robotic system has been commonly applied in manufacturing, military applications, surveillance, etc. to perform complex tasks [4–6]. Some self-learning strategies have been proposed to develop motion controllers for networked mobile robots [6,7]. Among them, an RBFNN incorporating the gradient descent method is regarded as a powerful tool for designing the self-learning controllers of networked mobile robots [7,8].

The RBFNN introduced by Broomhead and Lowe is a three-layer feedforward artificial neural network in which radial basis functions are used as activation functions [9,10]. This methodology takes advantage of fast learning capability and universal approximation. To date, it is a useful neural network architecture for addressing many engineering problems [11,12]. However, traditional RBFNNs adopt a gradient descent approach for training the neural network that is not capable of noise reduction [10–12]. In other words, these studies did not consider the uncertainty and noise induced in the process and measurement phases. This paper presents a Kalman filter based RBFNN and its application to self-learning control of networked mobile robots.

The Kalman filter is a state estimation technique introduced by R.E. Kalman [13]. It is a classic state estimation technique used widely in engineering applications, including spacecraft navigation, motion planning in robotics, signal processing and wireless sensor networks [14–16] because of its ability to extract useful information from noisy data. It is an optimal estimator for evaluating the internal state of a dynamic system under certain process patterns and/or measurement disturbances in the physical environment [13,17,18]. The objective of the Kalman filter is to minimize the mean squared error between the actual and estimated data. Although the proposed KF-RBFNN is useful for designing learning control schemes with noise reduction, the initial network parameters influences the system performance, namely that the selection of centers, widths and output weights for the Gaussian functions is an important consideration.

Parameter-tuning of an RBFNN is challenging when using this neural network to solve multidimensional optimization problems. Over the years, several methods have been developed for addressing this RBFNN optimization problem [19–21]. However, these traditional RBFNN methods may cause the output to converge to local optimum when the dimensionality of the problem increases [22]. Since RBFNN optimization can be formulated as a search problem, biologically, algorithms are new paths for optimizing RBFNNs. This is a successful hybridization of RBFNNs and evolutionary algorithms. Although there are some metaheuristic algorithms used to develop evolutionary RBFNNs [23–26], there has been no attempt to present an evolutionary KF-RBFNN using a GA to achieve learning control of networked mobile robotic systems.

The GA is one of the most popular evolutionary algorithms for solving optimization problems [27–29]. Although GAs have been widely applied to various optimization problems, these biologically inspired algorithms suffer from premature convergence. In other words, these traditional computing paradigms may converge to local optimum. This paper contributes to the development of a modified GA to improve the search diversity by including the Lévy flight approach. An adaptive determination of crossover and mutation probabilities in the GA is proposed via the Lévy flights. This random walk is very efficient in exploring the search space of the optimization problem. The proposed modified GA metaheuristics is then applied to the design of an optimal GA-KF-RBFNN for self-tuning motion control of networked mobile robots.

Of the increasing demands on networked mobile robot systems, formation using a leader-follower control strategy is one of the most important and is becoming increasing crucial [30–32]. It is a coordinated control in which the leader robot follows a desired trajectory while the follower robots maintain a specified geometrical pattern [32]. Although some studies have addressed this control problem by considering graph theory and consensus control approaches [33–35], these networked mobile robot systems suffer from communication delay problem. This paper presents a pragmatic self-learning optimal GA-KF-RBFNN formation control method for networked mobile robot systems that avoids the communication delay problem.

This paper is structured as follows: a biologically-inspired Kalman filter based RBFNN control technique, called GA-FA-RBFNN control is introduced in Section 2. Section 3 employs the proposed GA-FA-RBFNN to develop a networked mobile robot system to achieve self-evolving formation control. In Section 4, several simulation results are reported to demonstrate the effectiveness of the proposed methods. Finally, Section 5 concludes this paper.

2. Biologically-Inspired Kalman Filter Based RBFNN Control

2.1. Kalman Filter Algorithm

This section aims to describe the Kalman filter algorithm by which measurements are taken, and the state is estimated at discrete time points. The Kalman filter deals with the general problems encountered in estimating the state of a discrete-time controlled process, which is governed by the following state-space Equation (1) at time index k:

$$\begin{aligned} x(k) &= Ax(k-1) + BU(k-1) + w(k) \\ y(k) &= Cx(k) \\ z(k) &= Hy(k) + v(k) \end{aligned} \tag{1}$$

where A, B, and C are matrices in the state-space Equation (2). $w(k)$ is the process noise and $v(k)$ is the measurement noise. $z(k)$ is the measured signal and H is the sensor matrix. The probability of the process noise $w(k)$ is $p(w)$ and the probability of measurement noise $v(k)$ is $p(v)$. The process noise covariance of $p(w)$ is Q and the measured noise covariance of $p(v)$ is R. In Kalman filtering, $p(w)$ and $p(v)$ are independent white noises with normal probability distributions, expressed by:

$$\begin{aligned} p(w) &\sim N(0, Q) \\ p(v) &\sim N(0, R) \end{aligned} \tag{2}$$

Figure 1 presents the structure of Kalman filter algorithm in which the estimated state $\hat{x}(k-1)$ and the error covariance $P(k-1)$ are included [13]. As shown in Figure 1, The Kalman filter algorithm consists of two phases: time update phase (predictor) and measurement update phase (corrector). The following summarizes the two important phases in classical Kalman filter algorithm.

1. Time update phase:
   a. At time step $k-1$, calculate $\hat{x}(k-1)$ and $P(k-1)$.
   b. Update the estimation of state vector $\hat{x}^-(k)$ and the estimation of error covariance matrix $\hat{P}^-(k)$.
2. Measurement update phase:
   a. Update the optimal gain $K(k)$ of Kalman filter.
   b. Update the estimation of state vector $\hat{x}(k)$ using $z(k)$, $\hat{x}^-(k)$ and $K(k)$.
   c. Update the estimation of error covariance matrix $p(k)$ by utilizing $K(k)$ and $P^-(k)$ for next iteration in the Kalman filter algorithm process.

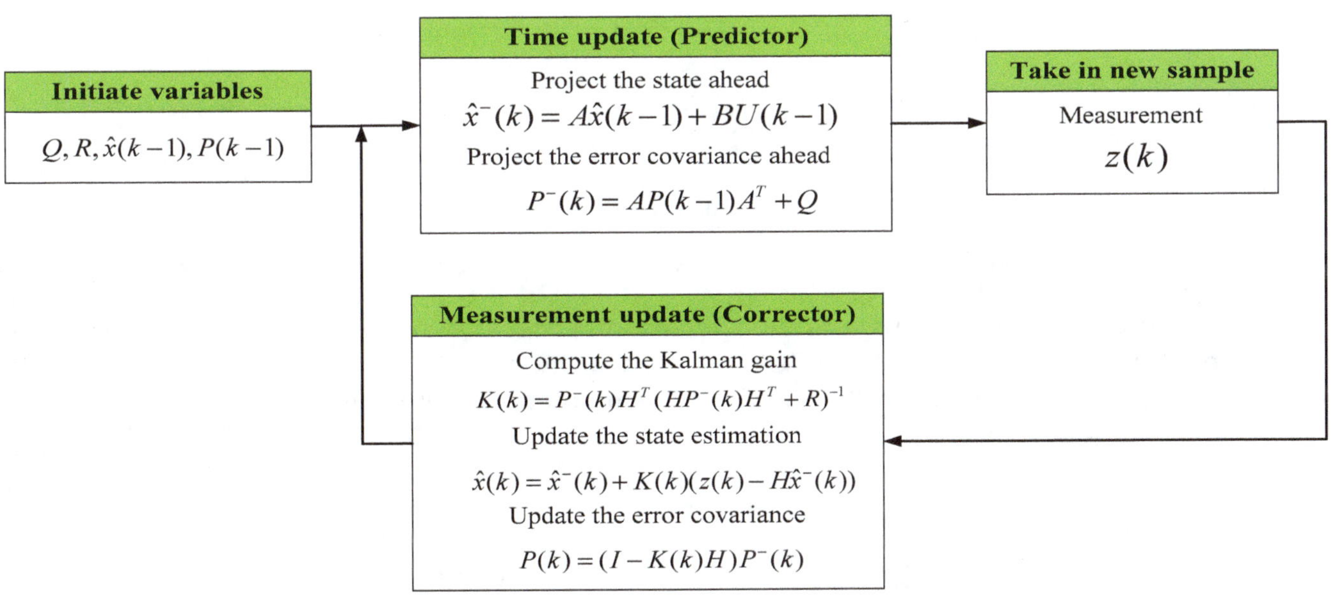

Figure 1. Structure of the classical Kalman filter.

2.2. *Classical RBFNN*

Figure 2 presents the structure of a classical RBFNN. This feed forward multilayer neural network has three layers, comprising the input layer, hidden layer and output layer. As shown in Figure 2, the inputs of the hidden layer in the RBFNN structure are the linear combinations of the weights and the input vector $[x_1\ x_2,\ x_3, \ldots \ldots, x_n]^T$. These vectors are then mapped by means of a radial basis functions in each node. Finally, the output layer of RBFNN generates a vector y_p for m outputs by linear combination of the outputs of the hidden nodes. This kind of artificial neural network has been regarded as a powerful tool that can approximate any continuous function with satisfactory accuracy [11]. In Figure 2, the output of the RBFNN is expressed by:

$$y_m = \sum_{j=1}^{m} w_j h_j \tag{3}$$

where h_j is the radial basis vector. This vector is described by using the following Gaussian function:

$$h_j = \exp(\frac{-\|X - C_j\|^2}{2b_j^2}), j = 1, 2, \ldots, m \tag{4}$$

where $\|\bullet\|$ is the Euclidean norm operation, $C_j = [c_{j1}, c_{j2}, \ldots, c_{jm}]^T$ is the center vector of the jth node, $B = [b_1, b_2, \ldots, b_m]^T$ is the basis width vector. $W = [w_1, w_2, \ldots, w_m]^T$ is the weight vector in the RBFNN. Typically, this neural network is initialized with a randomly determined of RBFNN parameters, including C_j, B, and W.

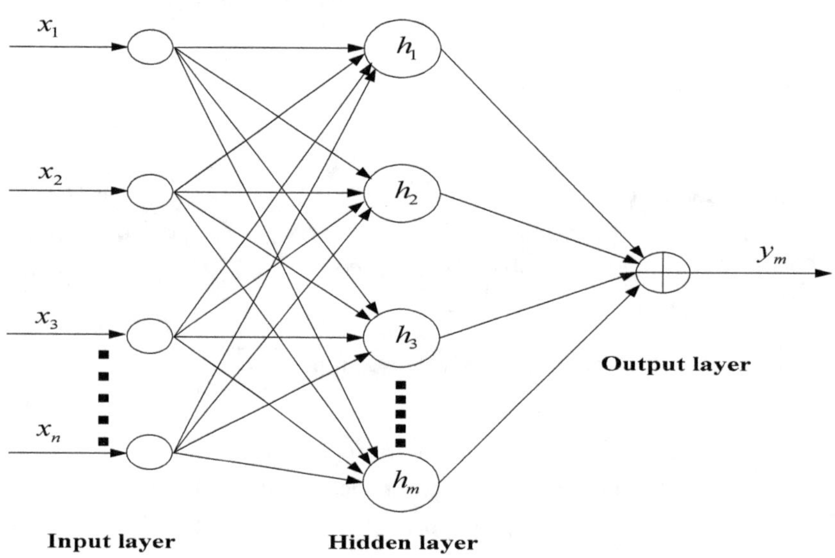

Figure 2. Structure of the classical RBFNN (Radial Basis Function Neural Network).

Gradient descent is an effective method for training RBFNN networks compared with other conventional training approaches [9,10]. In gradient descent training with one neuron in the output layer, the weights are updated at each time step by using the following rules:

$$w_j(k+1) = w_j(k) + \eta e(k) h_j(k), \tag{5}$$

$$C_{ji}(k+1) = C_{ji}(k) + \eta e(k) w_j h_j \frac{x_i(k) - C_{ji}(k)}{b_j^2(k)}, \tag{6}$$

$$b_j(k+1) = b_j(k) + \eta e(k) w_j h_j \frac{\|X(k) - C_j(k)\|^2}{b_j^3(k)}, \tag{7}$$

where $e(k)$ represents the error at the kth time step and η denotes the learning rate. Since the initial parameters for an RBFNN using the gradient descent method are determined either by a trial-and-error approach or randomly set, convergence to a local optimum solution is inevitable [10]. To improve the learning performance of the RBFNN, this paper has developed a Kalman filter to train the RBFNN network structure based on a gradient descent approach. The proposed KF-RBFNN is applied to the self-learning control of networked mobile robots.

2.3. *Kalman Filter Based RBFNN Control*

Figure 3 depicts the block diagram of the RBFNN-based control. In Figure 3, the error between the real output $y(k)$ and the estimated output of the neural network $y_m(k)$ are considered to develop a self-learning RBFNN. The cost function or performance index is defined by the squared estimation error:

$$J(k) = \frac{1}{2}[y(k) - y_m(k)]^2. \tag{8}$$

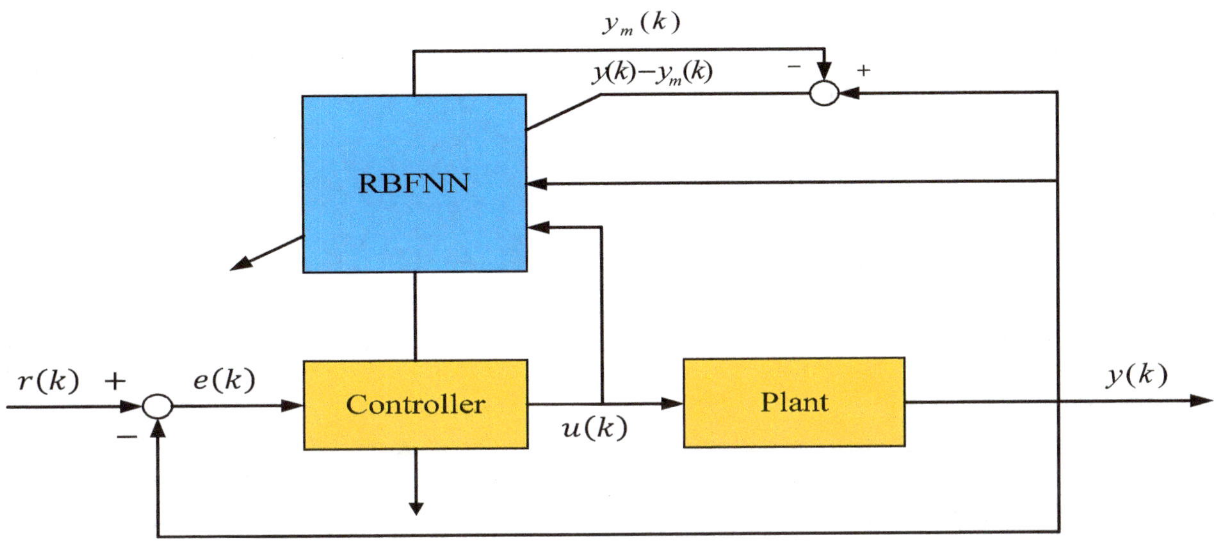

Figure 3. Block diagram of the RBFNN control scheme.

Considering the gradient descent method in Equations (5)–(7), the structure parameters: w_j, C_{ji}, and b_j are updated online at every sampling point, expressed by:

$$\begin{array}{l}
\Delta w_j = [y(k) - y_m(k)]h_j \\
w_j(k) = w_j(k-1) + \eta\Delta w_j + \varsigma[w_j(k-1) - w_j(k-2)] \\
\Delta b_j = [y(k) - y_m(k)]w_j h_j \frac{\|X - C_j\|^2}{b_j^3} \\
b_j(k) = b_j(k-1) + \eta\Delta b_j + \varsigma[b_j(k-1) - b_j(k-2)] \\
\Delta c_{ji} = [y(k) - y_m(k)]w_j \frac{x_j - c_{ji}}{b_j^2} \\
c_{ji}(k) = c_{ji}(k-1) + \eta\Delta c_{ji} + \varsigma[c_{ji}(k-1) - c_{ji}(k-2)]
\end{array}, \tag{9}$$

where ς denotes the momentum factor and η is the learning rate of the neural network.

Based on Figure 3, the proposed KF-RBFNN control scheme depicted in Figure 4 considers the process noise $w(k)$ and measurement noise $v(k)$. In the proposed intelligent KF-RBFNN control scheme, the KF-RBFNN serves as an on-line learning and adapting mechanism in the intelligent controller. As shown in Figure 4, the effects of the process uncertainty $w(k)$ and the measurement noise $v(k)$ in the control scheme can be reduced via the implementation of the Kalman filter. To retrain the uncertainty and noise, the measured output $z(k)$ is employed to derive an estimation $\hat{y}(k)$ of the output $y(k)$ in the proposed KF-RBFNN. Both the control signal $u(k)$ and output $y_m(k)$ of the RBFNN are fed into the neural network for on-line learning. Moreover, the estimated output value $y_m(k)$ of the RBFNN is then utilized to update the control parameters to achieve self-learning control using the Kalman filter.

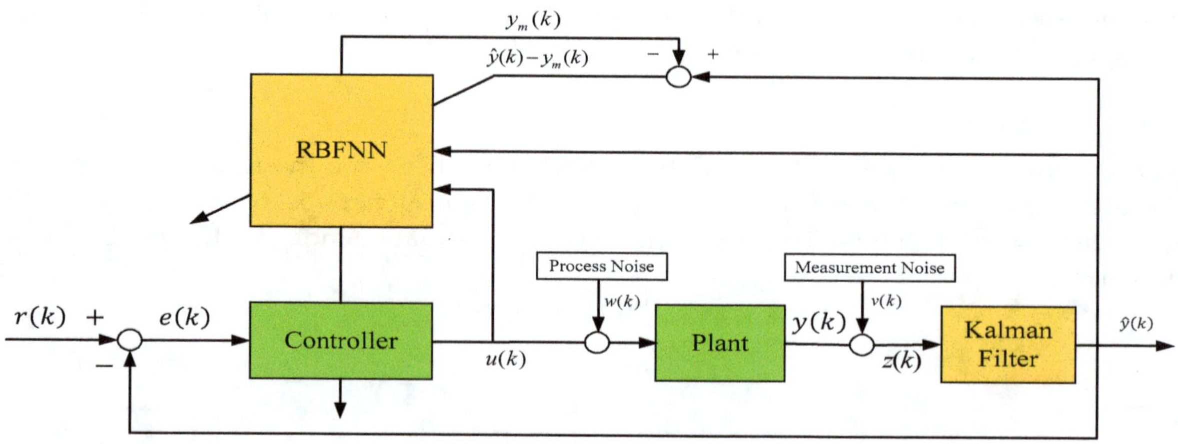

Figure 4. Block diagram of the KF-RBFNN control scheme.

2.4. Evolutionary KF-RBFNN Control

2.4.1. Modified GA with Lévy Flight

The GA developed by John Holland is a search algorithm that is inspired by Charles Darwin's theory of natural evolution. This evolutionary algorithm reflects the process of natural selection where the fittest individuals are selected for reproduction, thereby producing offspring of the next generation. This stochastic optimization technique starts with a set of solutions (chromosomes), called a population. This paradigm employs probabilistic rules to evolve a population from one generation to the next via the genetic operators: reproduction, crossover, and mutation. This paradigm is widely used to solve multidimensional optimization problems. When applying a GA to deal with optimization problems, an initial population of feasible solutions is generated. Each feasible solution is encoded as a chromosome string. These chromosomes are evaluated using a predefined fitness function or objective function based on the optimization problems. Figure 5 presents the flowchart of a GA. The initial population is randomly generated and the fitness function is defined before the GA evolutionary process begins. This study employs tournament selection, single-point crossover and single-point mutation strategies to develop a modified GA paradigm.

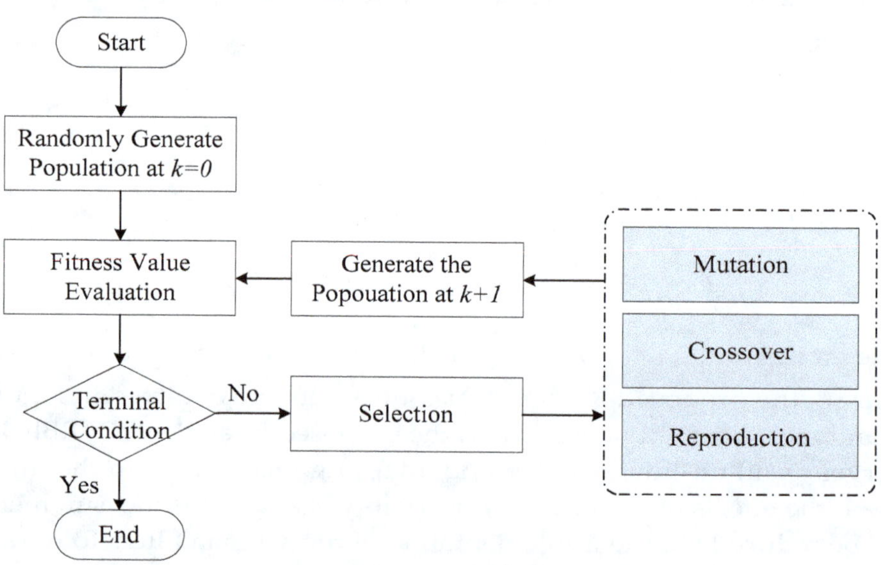

Figure 5. Flowchart of evolutionary GA.

Crossover and mutation are important operations for generating new individuals in the GAs. The performance of a GA is sensitive to the control parameter setting. The probabilities of crossover and mutation are significant parameters that influence the convergence performance of a GA. The use of unsuitable probability for crossover and mutation can result in poor convergence performance. More precisely, choosing suitable parameter values is a problem dependent task and requires previous experiences. Most studies adopt fixed crossover and mutation probabilities; this paper employs the Lévy flight which is a specialized random walk to increase the search diversity, expressed by:

$$Levy(\beta) = \left| \frac{\Gamma(\beta+1) \times \sin\left(\frac{\pi\beta}{2}\right)}{\Gamma\left(\frac{1+\beta}{2}\right) \times \beta \times 2^{\left(\frac{\beta-1}{2}\right)}} \right|^{\frac{1}{\beta}}, \tag{10}$$

where Γ denotes the gamma function and β is a constant $(1 < \beta \leq 3)$. The proposed modified GA is then applied to optimally set the initial parameters of the KF-RBFNN.

2.4.2. GA-Based KF-RBFNN

In the proposed KF-RBFNN, the accuracy and performance are influenced by the selections of the radial functions that are defined by a center vector B, width vector C_j, and weight vector W. In other words, proper tuning of these parameters is an important part of the optimal KF-RBFNN designs. This study employs the modified GA to develop a parameter tuning process that optimizes the KF-RBFNN. When applying the GA to address this issue, a chromosome is defined by the RBFNN parameter sequence $Chromosome = \{C_j, B, W\}$ and the optimal RBFNN structure can be evolved via the GA process with Lévy flight. The following fitness function (root mean square error, RMSE) for the N_s sample is used to evaluate the GA chromosomes.

$$Fitness = \sqrt{\frac{1}{N_s}\sum_{k=1}^{N_s}\left(y_p{}^*(k) - y_p(k)\right)^2} \tag{11}$$

where $y_p(k)$ is the output and $y_p{}^*(k)$ is the predicted output at the kth sampling time. The following describes the GA process for KF-RBFNN optimization.

Step 1: Initialize the GA computing with Lévy flight.

Step 2: Each GA chromosome in the population contains genes to represent the KF-RBFNN parameters, meaning that $Chromosome = \{C_j, B, W\}$.

Step 3: Construct the KF-RBFNN using $Chromosome = \{C_j, B, W\}$ and evaluate the performance using the fitness function (11).

Step 4: Perform GA crossover and mutation with the probabilities set by Lévy flight.

Step 5: Update the GA population.

Step 6: Check the termination criterion. Go to Step 3 or output the optimized GA individual $Chromosome^* = \left\{C_j^*, B^*, W^*\right\}$ for the proposed GA-KF-RBFNN.

3. Application to Self-Evolving Control of Networked Mobile Robots

3.1. Modeling and Lyapunov-Based Control

Figure 6 depicts the geometrical structure of a mobile robot with four Swedish wheels for the proposed networked mobile robot system. Compared to the conventional differential-drive (non-holonomic) mobile robots, this kind of mobile robot with omnidirectional capability has superior mobility. The kinematic model of the four-wheeled Swedish mobile robot is expressed by:

$$\begin{bmatrix} v_1(t) \\ v_2(t) \\ v_3(t) \\ v_4(t) \end{bmatrix} = \begin{bmatrix} r\omega_1(t) \\ r\omega_2(t) \\ r\omega_3(t) \\ r\omega_4(t) \end{bmatrix} = T(\theta(t)) \begin{bmatrix} \dot{x}(t) \\ \dot{y}(t) \\ \dot{\theta}(t) \end{bmatrix}, \tag{12}$$

where:

$$T(\theta(t)) = \begin{bmatrix} -\sin(\delta+\theta) & \cos(\delta+\theta) & L \\ -\cos(\delta+\theta) & -\sin(\delta+\theta) & L \\ \sin(\delta+\theta) & -\cos(\delta+\theta) & L \\ \cos(\delta+\theta) & \sin(\delta+\theta) & L \end{bmatrix},$$

δ is $\pi/4$; r represents the radius of the Swedish wheel; L denotes the distance from the Swedish wheel's center to the geometric center of the mobile robot; $v_i(t)$ and $\omega_i(t), i = 1,2,3,4$ respectively denote the linear and angular velocities of each omnidirectional wheel. $[x(t)\ y(t)\ \theta(t)]^T$ is the pose vector that includes the position and orientation of the mobile robot measured at time t.

In mobile robotic research, robots with over three degrees-of-freedom (DOFs) are classified as redundant robots because they provide redundancy. Note that $T(\theta(t))$ in Equation (12) is singular for any θ in this redundant mobile robot system. This study adopts the pseudo inverse matrix approach to address the redundant control problem of mobile robots. Considering the left pseudo-inverse matrix $T^{\#}(\theta(t))$ of $T(\theta(t))$ by using $T^{\#}(\theta(t))P(\theta(t)) = I$, the matrix $T^{\#}(\theta(t))$ is expressed by:

$$T^{\#}(\theta(t)) = \begin{bmatrix} \frac{-\sin(\delta+\theta)}{2} & \frac{-\cos(\delta+\theta)}{2} & \frac{\sin(\delta+\theta)}{2} & \frac{\cos(\delta+\theta)}{2} \\ \frac{\cos(\delta+\theta)}{2} & \frac{-\sin(\delta+\theta)}{2} & \frac{-\cos(\delta+\theta)}{2} & \frac{\sin(\delta+\theta)}{2} \\ \frac{1}{4L} & \frac{1}{4L} & \frac{1}{4L} & \frac{1}{4L} \end{bmatrix}. \tag{13}$$

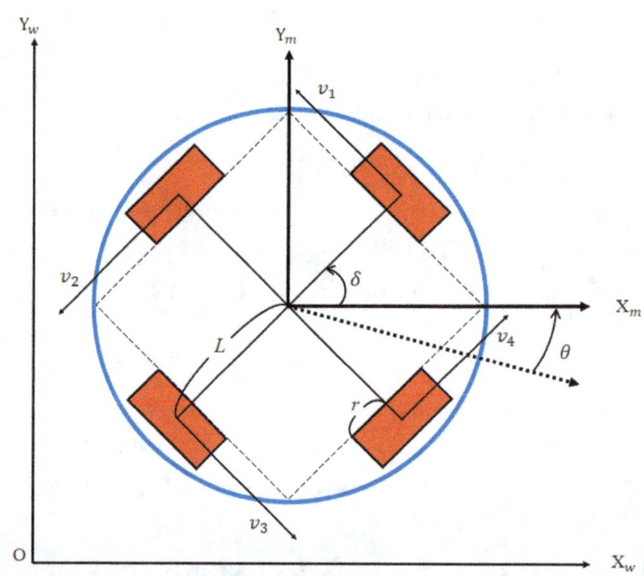

Figure 6. Geometry of the omnidirectional mobile robot with four Swedish wheels.

Combining Equations (12) and (13), the kinematics of the four-wheeled omnidirectional mobile robot is derived as follows:

$$\begin{bmatrix} \dot{x}(t) \\ \dot{y}(t) \\ \dot{\theta}(t) \end{bmatrix} = T^{\#}(\theta(t)) \begin{bmatrix} v_1(t) \\ v_2(t) \\ v_3(t) \\ v_4(t) \end{bmatrix}. \tag{14}$$

After the kinematics analysis of the Swedish mobile robots, the next step is to design a motion control law and prove its stability using Lyapunov theory. The current pose of the omnidirectional Swedish mobile robot at time t is defined as $S = \begin{bmatrix} x(t) & y(t) & \theta(t) \end{bmatrix}^T$, and the desired reference trajectory of the Swedish mobile robot is expressed as $S_r = \begin{bmatrix} x_r(t) & y_r(t) & \theta_r(t) \end{bmatrix}^T$. With the two pre-defined vectors, the tracking error of the mobile robot is given by:

$$S_e = \begin{bmatrix} x_e(t) \\ y_e(t) \\ \theta_e(t) \end{bmatrix} = \begin{bmatrix} x(t) \\ y(t) \\ \theta(t) \end{bmatrix} - \begin{bmatrix} x_r(t) \\ y_r(t) \\ \theta_r(t) \end{bmatrix} = S - S_r, \tag{15}$$

which gives:

$$\dot{S}_e = \begin{bmatrix} \dot{x}_e(t) \\ \dot{y}_e(t) \\ \dot{\theta}_e(t) \end{bmatrix} = \begin{bmatrix} \dot{x}(t) \\ \dot{y}(t) \\ \dot{\theta}(t) \end{bmatrix} - \begin{bmatrix} \dot{x}_r(t) \\ \dot{y}_r(t) \\ \dot{\theta}_r(t) \end{bmatrix} = T^{\#}(\theta(t)) \begin{bmatrix} r\omega_1(t) \\ r\omega_2(t) \\ r\omega_3(t) \\ r\omega_4(t) \end{bmatrix} - \begin{bmatrix} \dot{x}_r(t) \\ \dot{y}_r(t) \\ \dot{\theta}_r(t) \end{bmatrix}. \tag{16}$$

The goal of control law design is to derive the angular velocity vector $\begin{bmatrix} \omega_1(t) & \omega_2(t) & \omega_3(t) & \omega_4(t) \end{bmatrix}^T$ for tracking the desired differentiable trajectory $\begin{bmatrix} x_r(t) & y_r(t) & \theta_r(t) \end{bmatrix}^T$ with asymptotical stability. Based on the PID (Proportional-Integral-Derivative) control strategy, the following redundant control law is proposed:

$$\begin{bmatrix} v_1(t) \\ v_2(t) \\ v_3(t) \\ v_4(t) \end{bmatrix} = T(\theta(t)) \left(-K_P \begin{bmatrix} x_e(t) \\ y_e(t) \\ \theta_e(t) \end{bmatrix} - K_I \begin{bmatrix} \int_0^t x_e(\tau)d\tau \\ \int_0^t y_e(\tau)d\tau \\ \int_0^t \theta_e(\tau)d\tau \end{bmatrix} - K_D \begin{bmatrix} \dot{x}_e(t) \\ \dot{y}_e(t) \\ \dot{\theta}_e(t) \end{bmatrix} + \begin{bmatrix} \dot{x}_r(t) \\ \dot{y}_r(t) \\ \dot{\theta}_r(t) \end{bmatrix} \right), \tag{17}$$

where K_P, K_I and K_D are the control matrices. They are diagonal and positive, thus $K_P = diag[k_{xp}\ k_{yp}\ k_{\theta p}]$, $K_I = diag[k_{xi}\ k_{yi}\ k_{\theta i}]$, and $K_D = diag[k_{xd}\ k_{yd}\ k_{\theta d}]$. By substituting Equations (17) into (16), the closed-loop error system is obtained:

$$\dot{S}_e = \begin{bmatrix} \dot{x}_e(t) \\ \dot{y}_e(t) \\ \dot{\theta}_e(t) \end{bmatrix} = \left(-K_P \begin{bmatrix} x_e(t) \\ y_e(t) \\ \theta_e(t) \end{bmatrix} - K_I \begin{bmatrix} \int_0^t x_e(\tau)d\tau \\ \int_0^t y_e(\tau)d\tau \\ \int_0^t \theta_e(\tau)d\tau \end{bmatrix} - K_D \begin{bmatrix} \dot{x}_e(t) \\ \dot{y}_e(t) \\ \dot{\theta}_e(t) \end{bmatrix} \right). \tag{18}$$

To prove the asymptotical stability of the closed-loop error system in (18) via Lyapunov theory, the following Lyapunov function is selected:

$$V(t) = \tfrac{1}{2}\begin{bmatrix} x_e(t) & y_e(t) & \theta_e(t) \end{bmatrix} \begin{bmatrix} x_e(t) \\ y_e(t) \\ \theta_e(t) \end{bmatrix} + \tfrac{1}{2}\begin{bmatrix} \int_0^t x_e(\tau)d\tau & \int_0^t y_e(\tau)d\tau & \int_0^t \theta_e(\tau)d\tau \end{bmatrix} K_I \begin{bmatrix} \int_0^t x_e(\tau)d\tau \\ \int_0^t y_e(\tau)d\tau \\ \int_0^t \theta_e(\tau)d\tau \end{bmatrix} + \tfrac{1}{2}\begin{bmatrix} x_e(t) & y_e(t) & \theta_e(t) \end{bmatrix} K_D \begin{bmatrix} x_e(t) \\ y_e(t) \\ \theta_e(t) \end{bmatrix} > 0$$

one obtains:

$$\dot{V}(t) = -\begin{bmatrix} x_e(t) & y_e(t) & \theta_e(t) \end{bmatrix} K_P \begin{bmatrix} x_e(t) \\ y_e(t) \\ \theta_e(t) \end{bmatrix} < 0.$$

Since $\dot{V}$ is negative definite, the asymptotical stability is therefore proven. The proposed motion control law can steer the mobile robot to achieve $S \to S_r$ as $t \to \infty$.

3.2. GA-KF-RBFNN Self-Learning Control

Figure 7 depicts the GA-KF-RBFNN self-learning control scheme, in which the noise from the process and measurement phases are included. As shown in Figure 7, the proposed GA-KF-RBFNN is employed to online adjust the parameters $K_P = diag[k_{xp}\ k_{yp}\ k_{\theta p}]$, $K_I = diag[k_{xi}\ k_{yi}\ k_{\theta i}]$, and $K_D = diag[k_{xd}\ k_{yd}\ k_{\theta d}]$ of the mobile robot. It is worthy to mention that the control matrices in Equation (17) are online adjusted at every sampling point to achieve tracking control. This GA-KF-RBFNN evolutionary online tuning method with noise reduction outperforms the traditional off-line and hand-tuning approaches.

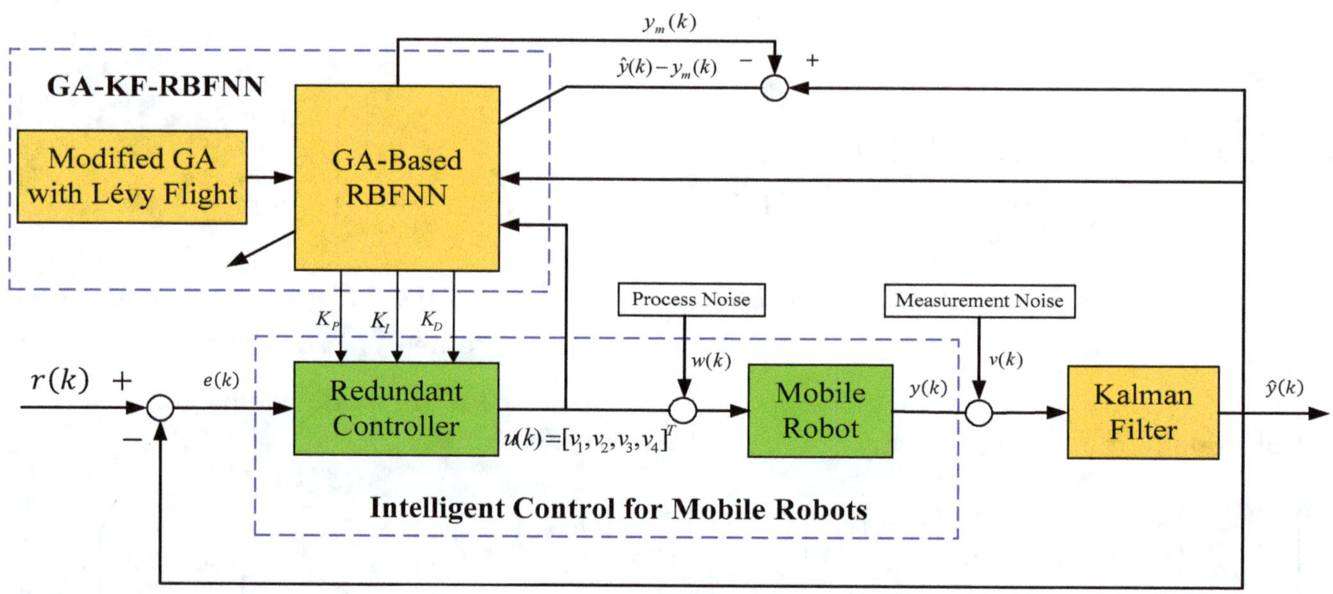

Figure 7. Block diagram of the GA-KF-RBFNN redundant control scheme for mobile robot.

In Figure 7, the control law $u(k) = [v_1, v_2, v_3, v_4]^T$ and the output $y_m(k)$ of the RBFNN are fed into the network for on-line learning. Moreover, the estimated output value $y_m(k)$ of the RBFNN is then employed to update the control matrices $K_P = diag[k_{xp}\ k_{yp}\ k_{\theta p}]$, $K_I = diag[k_{xi}\ k_{yi}\ k_{\theta i}]$, and $K_D = diag[k_{xd}\ k_{yd}\ k_{\theta d}]$ of the four-wheeled omnidirectional mobile robot to achieve the auto-tuning control with a Kalman filter.

3.3. Leader-Follower Formation Control of Networked Mobile Robots

The leader-follower model is the main trend of networked mobile robotics, where a leader robot and several follower robots are included in a multi-robot system [32]. For networked mobile robots, formation control is an important topic that the leader robot tracks the desired trajectory while the follower robots maintain the formation shape. Figure 8 depicts a leader-follower networked mobile robotic system to achieve triangular formation control with three robots.

In this paper, all mobile robots are independently controlled by using Equation (17) to accomplish leader-follower formation control, and the control parameters are self-tuned via the GA-KF-RBFNN paradigm. Compared to traditional consensus multiple robot systems with directed graph topology, the proposed broadcast leader-follower networked mobile robot system circumvents the delay problem. The position and orientation of the robots are broadcasted via the network. To maintain the desired formation shape, the geometrical relationship of the leader robot and follower robots in Figure 7 is calculated. Since the data flow is broadcasted online to every robot, the communication delay issues that occur in the consensus multiple robot system are therefore avoided. The proposed GA-KF-RBFNN self-evolving formation control for networked mobile robots not only reduces the system noises, but also avoids the communication delay.

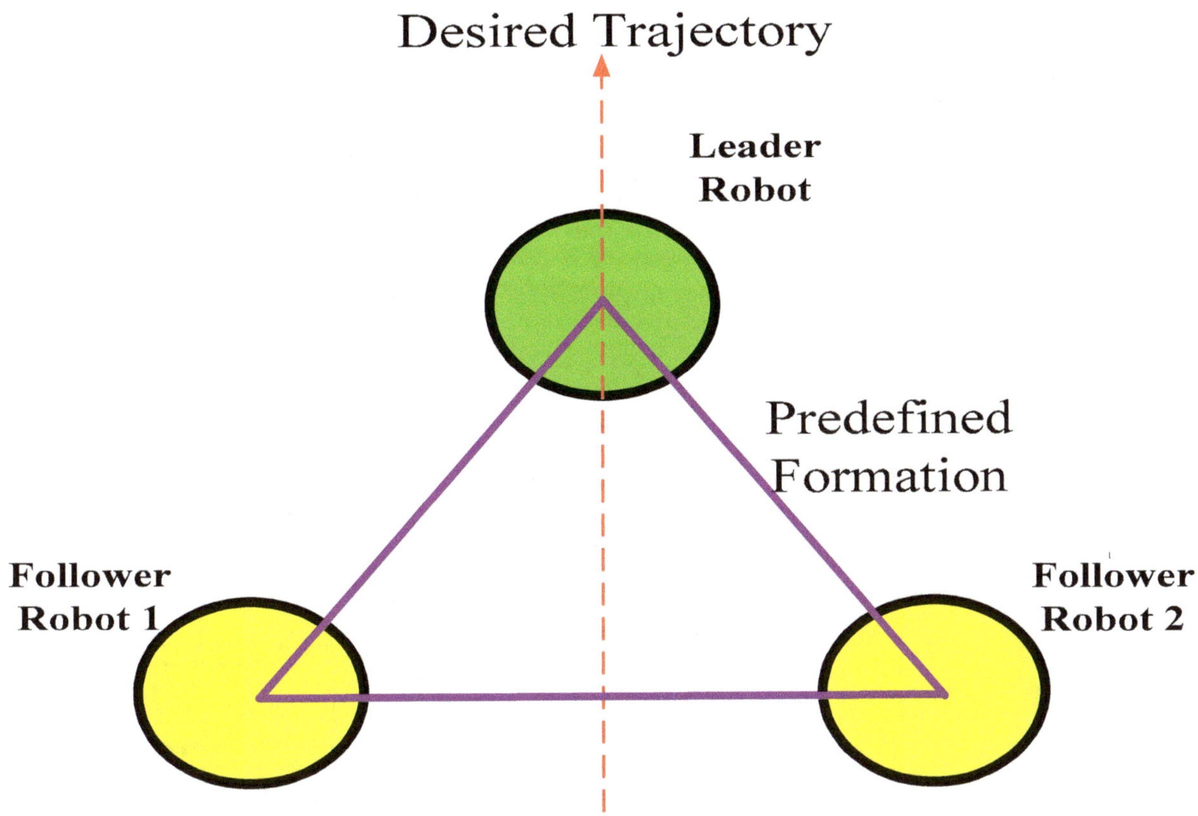

Figure 8. Leader-follower formation control with three mobile robots.

4. Simulations, Comparative Analysis, and Discussion

This section aims to conduct several simulations to examine the effectiveness of the proposed methods. The proposed networked mobile robot system consists of three four-wheeled Swedish omnidirectional mobile robots, including one leader and two follower mobile robots in a broadcast communication environment. The desired formation shape is triangular as shown in Figure 8. The system parameters in the proposed networked mobile robot system are $L = 0.25$ m and $r = 5.08$ cm.

The first simulation is conducted to demonstrate the performance of the circular formation control using the proposed GA-KF-RBFNN control approach. The number of iterations for the modified GA is 150, and the probabilities of crossover and mutation are determined by the Lévy flight. The circular trajectory for the leader robot is expressed as $\begin{bmatrix} x_r(t) & y_r(t) & \theta_r(t) \end{bmatrix}^T = \begin{bmatrix} 1.75\cos(\omega_i t)\text{ m} & 1.75\sin(\omega_i t)\text{ m} & \pi/4\text{ rad} \end{bmatrix}^T$, $\omega_i = 0.35$ rad/ sec. Figure 9 depicts the simulation result of the circular formation control. The three omnidirectional mobile robots are initially placed at different poses in the workspace. The desired trajectory for leader robot is a circular trajectory and the two follower robots aim to maintain a triangular formation. The tracking error of the leader mobile robot is presented in Figure 10. As shown in Figure 10, the leader mobile robot successfully tracks the desired circular trajectory in 4 s. Figure 11 depicts the formation error of follower robot #1 and Figure 12 depicts the formation error of follower robot #2. These simulation results demonstrate that the proposed GA-KF-RBFNN optimization is capable of accomplishing the self-learning formation control of networked mobile robotic systems.

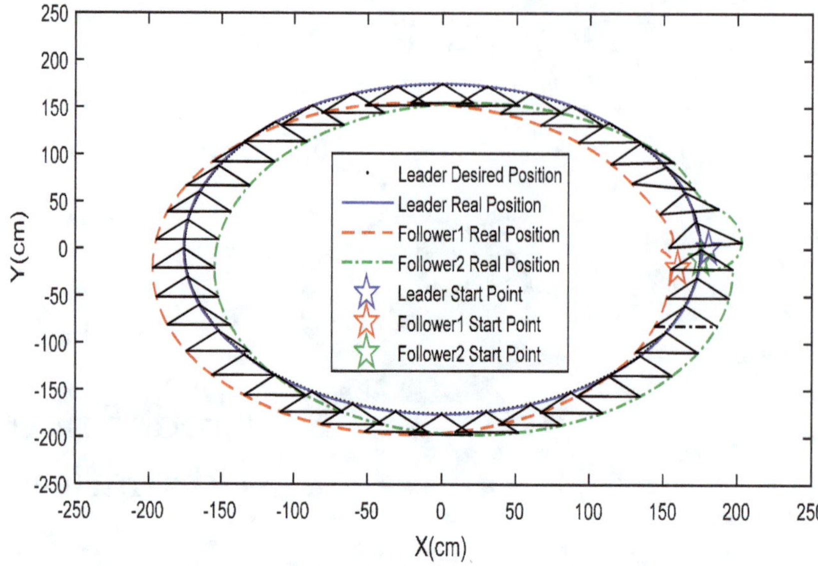

Figure 9. Simulation result of circular formation control.

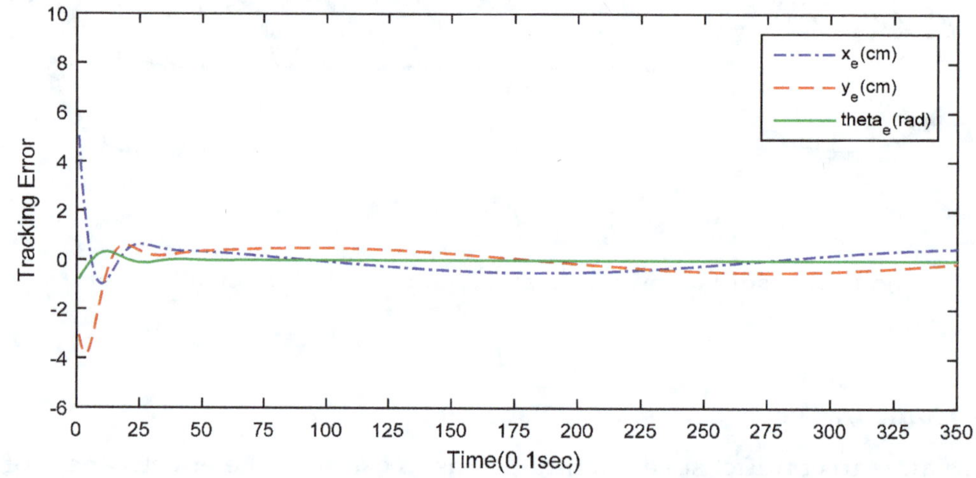

Figure 10. Tracking error of the leader mobile robot for circular formation control.

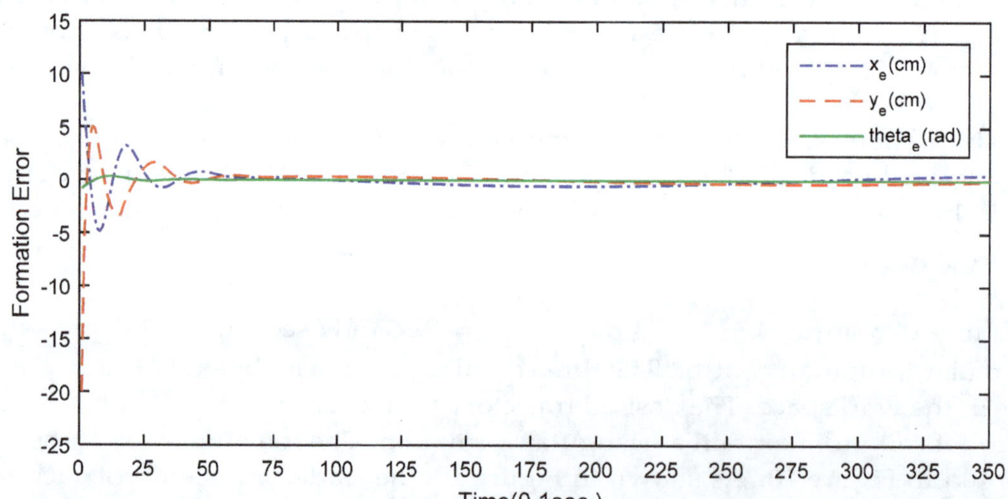

Figure 11. Formation error of follower robot #1 for circular formation control.

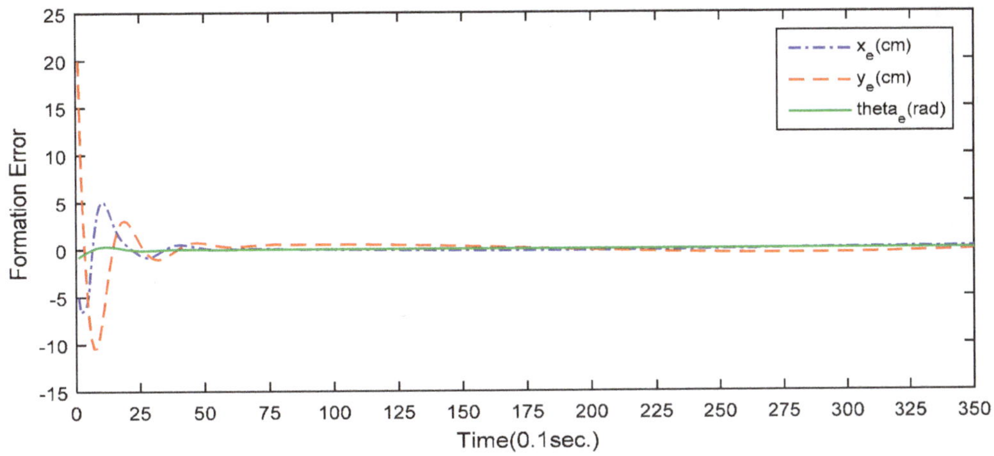

Figure 12. Formation error of follower mobile robot #2 for circular formation control.

The second simulation is provided to show the effectiveness of the proposed GA-KF-RBFNN formation control for daisy curve tracking. The desired trajectory of the leader robot is a special daisy curve with three petals, expressed by $\begin{bmatrix} x_r(t) & y_r(t) & \theta_r(t) \end{bmatrix}^T = \begin{bmatrix} \frac{1}{2} + b(a + r' \cos(p\omega_i t)) \cos(\omega_i t) \text{ m} & \frac{1}{2} + b(a + r' \cos(p\omega_i t)) \sin(\omega_i t) \text{ m} & \pi/4 \text{ rad} \end{bmatrix}^T$, $b = 0.2$ m, $a = 0.6$ m, $r' = 0.42$ m, $p = 3$, and $\omega_i = 0.35$ rad/sec. To illustrate the noise reduction capability of the proposed GA-KF-RBFNN self-learning controller, a Gaussian noise is added into the process and measurement phases. Figure 13 presents the simulation result of daisy curve formation control using the proposed GA-KF-RBFNN and Figure 14 depicts the tracking error of the leader robot. Moreover, Figures 15 and 16 present the formation error of follower robot #1 and follower robot #2, respectively, for daisy curve formation control. Both the tracking performance and formation behavior are guaranteed. These simulation results clearly indicate that the proposed metaheuristic GA-KF-RBFNN self-evolving control scheme with noise reduction achieves the formation control task for networked mobile robots. This approach outperforms the traditional consensus control methods where the uncertainty and self-adaptation are not considered.

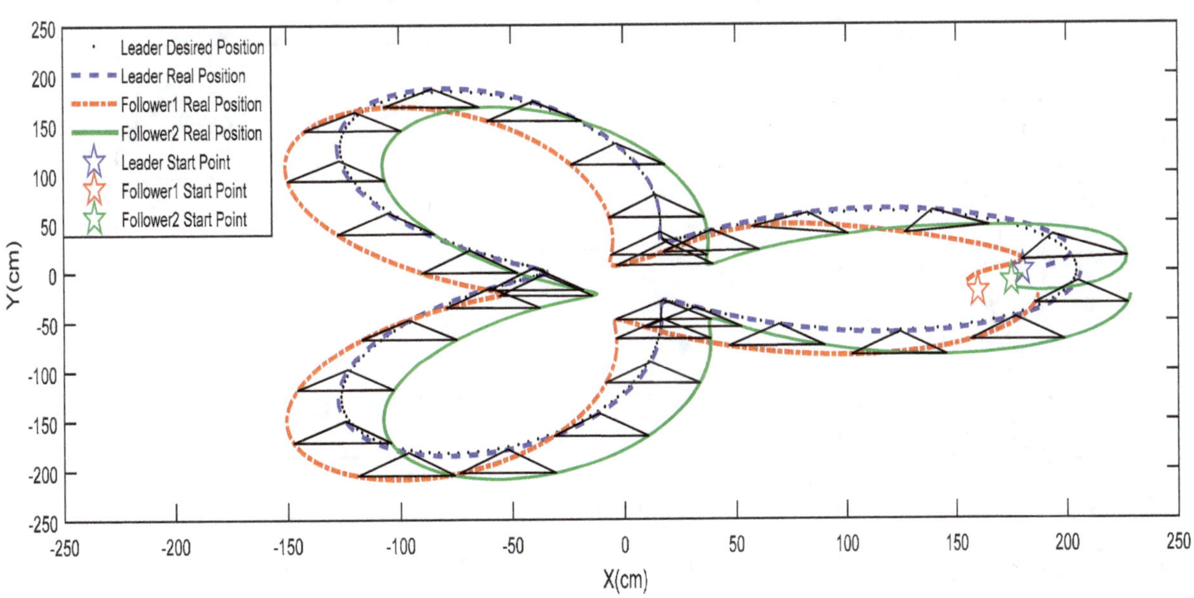

Figure 13. Simulation result for the daisy curve formation control.

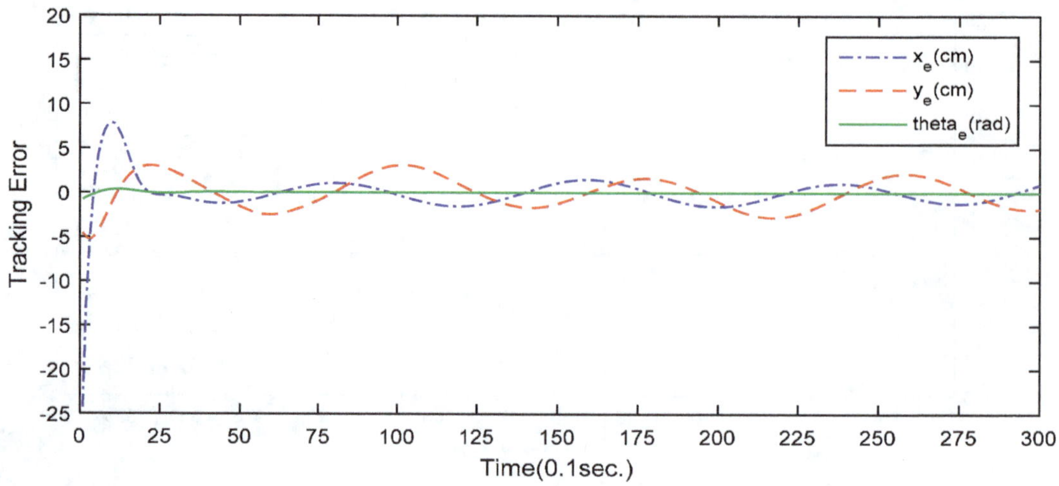

Figure 14. Tracking error of the leader mobile robot for daisy curve formation control.

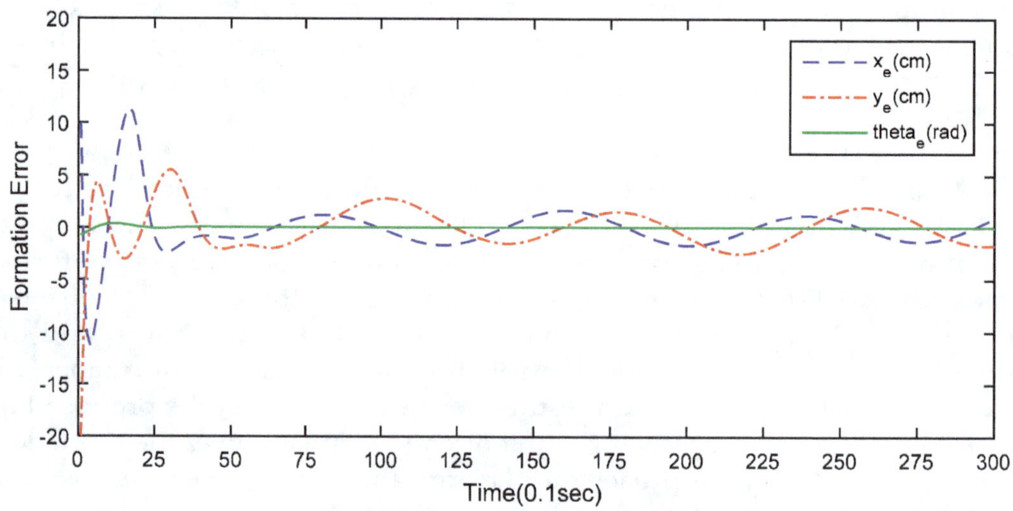

Figure 15. Formation error of follower robot #1 for daisy curve formation control.

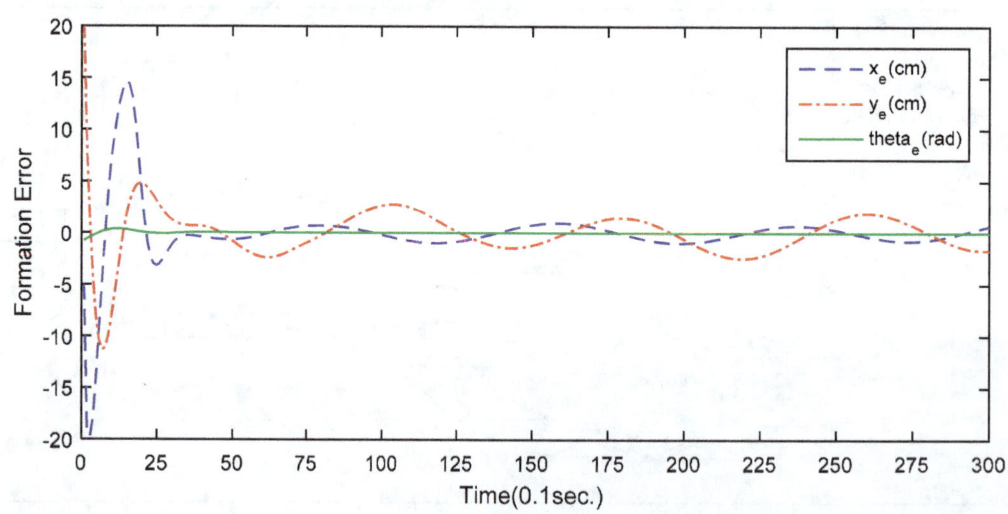

Figure 16. Formation error of follower robot #2 for daisy curve formation control.

In order to provide a comparative analysis to illustrate the merits of the proposed methods over other existing approaches, Table 1 lists the comparison of the proposed GA-KF-RBFNN formation control and traditional controllers. As shown in Table 1, the proposed biologically-inspired

GA-KF-RBFNN learning and adaptation for self-evolving of networked mobile robots is superior to the traditional formation control methods. With the modified GA metaheuristics and Kalman filter, both the noise and delay problem are avoided. The proposed GA-KF-RBFNN formation control strategy is applicable to all kinds of mobile robots, including different-drive [36] and three-wheeled mobile robotic systems.

Table 1. A comparative analysis of the formation controllers for networked robots.

	Evolutionary Strategy	Noise Reduction	Distributed Formation	Leader-Follower Approach	Consensus Delayed Formation	Omnidirectional Capability
[1,2]	No	No	Yes	Yes	Yes	No
[3–6]	No	No	Yes	Yes	Yes	No
[7–9]	No	No	Yes	Yes	Yes	No
[10–13]	No	No	Yes	Yes	Yes	No
[31–34]	No	No	Yes	Yes	Yes	No
This Study	Yes	Yes	Yes	Yes	No	Yes

5. Conclusions

This paper has presented a biologically-inspired GA-KF-RBFNN learning and adaptation method for the self-evolving control of networked mobile robots. The Kalman filter algorithm is employed to develop a self-learning RBFNN by considering uncertainty and noises. Moreover, the structure of the proposed KF-RBFNN is optimally initialized by means of the modified GA in which a Lévy flight strategy is applied. With the derived kinematic model of a four-wheeled omnidirectional mobile robot and broadcast leader-follower model, the GA-KF-RBFNN is utilized to design a self-evolving motion control law for a networked mobile robotic system. This approach overcomes the problem of communication delay found in conventional consensus networked robotic systems. Simulation results illustrate the merits of the proposed intelligent networked mobile robot system which uses a GA-KF-RBFNN to achieve self-learning formation control and consider the uncertainty and noise.

Author Contributions: Conceptualization, methodology, analysis, writing—review and editing, S.S.-D.X. and H.-C.H.; Experiments, T.-C.C. and S.-K.L.

References

1. Liu, Y.; Gao, J.; Shi, X.; Jiang, C. Decentralization of virtual linkage in formation control of multi-agents via consensus strategies. *Appl. Sci.* **2018**, *8*, 2020. [CrossRef]
2. Keviczky, T.; Borrelli, F.; Fregene, K.; Godbole, D.; Balas, G. Decentralized receding horizon control and coordination of autonomous vehicle formations. *IEEE Trans. Control Syst. Technol.* **2008**, *16*, 19–33. [CrossRef]
3. Qian, D.; Xi, Y. Leader-follower formation maneuvers for multi-robot systems via derivative and integral terminal sliding mode. *Appl. Sci.* **2018**, *8*, 1045. [CrossRef]
4. Qu, Z.; Wang, J.; Hull, R.A.; Martin, J. Cooperative control design and stability analysis for multi-agent systems with communication delays. In Proceedings of the 2006 IEEE International Conference of Robotics and Automation, Orlando, FL, USA, 15–19 May 2006; pp. 970–975.
5. Kantaros, Y.; Zavlanos, M.M. Global planning for multi-robot communication networks in complex environments. *IEEE Trans. Robot.* **2016**, *32*, 1045–1061. [CrossRef]
6. Pan, L.; Lu, Q.; Yin, K.; Zhang, B. Signal source localization of multiple robots using an event-triggered communication scheme. *Appl. Sci.* **2018**, *8*, 977. [CrossRef]
7. Ferrari-Trecate, G.; Galbusera, L.; Marciandi, M.P.E.; Scattolini, R. Model predictive control schemes for consensus in multi-agent systems with single and double integrator dynamics. *IEEE Trans. Autom. Control* **2009**, *54*, 2560–2572. [CrossRef]

8. Dong, F.; Lei, X.; Chou, W. The adaptive radial basis function neural network for small rotary-wing unmanned aircraft. *IEEE Trans. Ind. Electron.* **2014**, *61*, 4808–4815.
9. Yang, C.; Li, Z.; Cui, R.; Xu, B. Neural network-based motion control of an underactuated wheeled inverted pendulum model. *IEEE Trans. Neural Netw. Learn. Syst.* **2014**, *25*, 2004–2016. [CrossRef] [PubMed]
10. Khosravi, H. A novel structure for radial basis function networks-WRBF. *Neural Process. Lett.* **2012**, *35*, 177–186. [CrossRef]
11. Dash, C.S.K.; Dash, A.P.; Cho, S.B.; Wang, G.N. DE+RBFNs based classification: A special attention to removal of inconsistency and irrelevant features. *Eng. Appl. Artif. Intell.* **2013**, *26*, 2315–2326. [CrossRef]
12. Chang, G.W.; Chen, C.I.; Teng, Y.F. Radial-basis-function-based neural network for harmonic detection. *IEEE Trans. Ind. Electron.* **2010**, *57*, 2171–2179. [CrossRef]
13. Tong, C.C.; Ooi, E.T.; Liu, J.C. Design a RBF neural network auto-tuning controller for magnetic levitation system with Kalman filter. In Proceedings of the 2015 IEEE/SICE International Symposium on System Integration (SII), Nagoya, Japan, 11–13 December 2015; pp. 528–533.
14. Hamid, K.R.; Talukder, A.; Islam, A.K. Implementation of fuzzy aided Kalman filter for tracking a moving object in two-dimensional space. *Int. J. Fuzzy Log. Intell. Syst.* **2018**, *18*, 85–96. [CrossRef]
15. Ko, N.Y.; Jeong, S.; Bae, Y. Sine rotation vector method for attitude estimation of an underwater robot. *Sensors* **2013**, *16*, 1213. [CrossRef]
16. Lu, X.; Wang, L.; Wang, H.; Wang, X. Kalman filtering for delayed singular systems with multiplicative noise. *IEEE/CAA J. Autom. Sin.* **2016**, *3*, 51–58.
17. Wang, S.; Feng, J.; Tse, C. Analysis of the characteristic of the Kalman gain for 1-D chaotic maps in cubature Kalman filter. *IEEE Signal Process. Lett.* **2013**, *20*, 229–232. [CrossRef]
18. Foley, C.; Quinn, A. Fully probabilistic design for knowledge transfer in a pair of Kalman filters. *IEEE Signal Process. Lett.* **2018**, *25*, 487–490. [CrossRef]
19. Xie, S.; Xie, Y.; Huang, T.; Gui, W.; Yang, C. Generalized predictive control for industrial processes based on neuron adaptive splitting and merging RBF neural network. *IEEE Trans. Ind. Electron.* **2019**, *66*, 1192–1202. [CrossRef]
20. Gutiérrez, P.A.; Hervas-Martinez, C.; Martínez-Estudillo, F.J. Logistic regression by means of evolutionary radial basis function neural networks. *IEEE Trans. Neural Netw.* **2011**, *22*, 246–263. [CrossRef] [PubMed]
21. Huang, F.; Zhang, W.; Chen, Z.; Tang, J.; Song, W.; Zhu, S. RBFNN-based adaptive sliding mode control design for nonlinear bilateral teleoperation system under time-varying delays. *IEEE Access* **2019**, *7*, 11905–11912. [CrossRef]
22. Zao, Y.; Zheng, Z. A robust adaptive RBFNN augmenting backstepping control approach for a model-scaled helicopter. *IEEE Trans. Control Syst. Technol.* **2015**, *23*, 2344–2352.
23. Tian, J.; Li, M.; Chen, F.; Feng, N. Learning subspace-based RBFNN using coevolutionary algorithm for complex classification tasks. *IEEE Trans. Neural Netw. Learn. Syst.* **2016**, *27*, 47–61. [CrossRef] [PubMed]
24. Chang, G.W.; Shih, M.F.; Chen, Y.Y.; Liang, Y.J. A hybrid wavelet transform and neural-network-based approach for modelling dynamic voltage-current characteristics of electric arc furnace. *IEEE Trans. Power Deliv.* **2014**, *29*, 815–824. [CrossRef]
25. Zhao, Y.; Ye, H.; Kang, Z.S.; Shi, S.S.; Zhou, L. The recognition of train wheel tread damages based on PSO-RBFNN algorithm. In Proceedings of the 2012 8th International Conference on Natural Computation, Chongqing, China, 29–31 May 2012; pp. 1093–1095.
26. Wei, Z.Q.; Hai, X.Z.; Jian, W. Prediction of electricity consumption based on genetic algorithm-RBF neural network. In Proceedings of the 2010 2nd International Conference on Advanced Computer Control, Shenyang, China, 27–29 March 2010; Volume 5, pp. 339–342.
27. Wei, H.; Tang, X.S. A genetic-algorithm-based explicit description of object contour and its ability to facilitate recognition. *IEEE Trans. Cybern.* **2015**, *45*, 2558–2571. [CrossRef] [PubMed]
28. Huang, H.C. FPGA-based hybrid GA-PSO algorithm and its application to global path planning for mobile robots. *Przeglad Elektrotechniczny* **2012**, *7*, 281–284.
29. Ding, S.F.; Xu, L.; Su, C.Y.; Jin, F.X. An optimizing method of RBF neural network based on genetic algorithm. *Neural Comput. Appl.* **2012**, *21*, 333–336. [CrossRef]
30. Chou, C.J.; Chen, L.F. Combining neural networks and genetic algorithms for optimising the parameter design of the inter-metal dielectric process. *Int. J. Prod. Res.* **2012**, *50*, 1905–1916. [CrossRef]

31. Marshall, J.A.; Broucke, M.E.; Francis, B.A. Formations of vehicles in cyclic pursuit. *IEEE Trans. Autom. Control* **2004**, *49*, 1963–1974. [CrossRef]
32. Oh, K.K.; Park, M.C.; Ahn, H.S. A survey of multi-agent formation control. *Automatica* **2015**, *53*, 424–440. [CrossRef]
33. Alonso-Mora, J.; Baker, S.; Siegwart, R. Multi-robot navigation in formation via sequential convex programming. In Proceedings of the IEEE/RSJ International Conference on Intelligent Robots and Systems, Hamburg, Germany, 28 September–2 October 2015.
34. Lee, H.C.; Roh, B.S.; Lee, B.H. Multi-hypothesis map merging with sinogram-based PSO for multi-robot systems. *Electron. Lett.* **2016**, *52*, 1213–1214. [CrossRef]
35. Tsai, C.C.; Chen, Y.S.; Tai, F.C. Intelligent adaptive distributed consensus formation control for uncertain networked heterogeneous Swedish-wheeled omnidirectional multi-robots. In Proceedings of the Annual Conference of the Society of Instrument and Control Engineers of Japan (SICE), Tsukuba, Japan, 20–23 September 2016; pp. 154–159.
36. Tiep, D.K.; Lee, K.; Im, D.Y.; Kwak, B.; Ryoo, Y.J. Design of fuzzy-PID controller for path tracking of mobile robot with differential drive. *Int. J. Fuzzy Log. Intell. Syst.* **2018**, *18*, 220–228. [CrossRef]

4

A Trajectory Planning Method for Polishing Optical Elements based on a Non-Uniform Rational B-Spline Curve

Dong Zhao and Hao Guo *

Key Laboratory of Mechanism Theory and Equipment Design of The State Ministry of Education, Tianjin University, Tianjin 300072, China; dongzhao@tju.edu.cn

* Correspondence: guohaohalo@163.com.

Abstract: Optical polishing can accurately correct the surface error through controlling the dwell time of the polishing tool on the element surface. Thus, the precision of the trajectory and the dwell time (the runtime of the trajectory) are important factors affecting the polishing quality. This study introduces a systematic interpolation method for optical polishing using a non-uniform rational B-spline (NURBS). A numerical method for solving all the control points of NURBS was proposed with the help of a successive over relaxation (SOR) iterative theory, to overcome the problem of large computation. Then, an optimisation algorithm was applied to smooth the NURBS by taking the shear jerk as the evaluation index. Finally, a trajectory interpolation scheme was investigated for guaranteeing the precision of the trajectory runtime. The experiments on a prototype showed that, compared to the linear interpolation method, there was an order of magnitude improvement in interpolation, and runtime, errors. Correspondingly, the convergence rate of the surface error of elements improved from 37.59% to 44.44%.

Keywords: hybrid robot; curve fitting; fair optimisation; trajectory interpolation

---

1. Introduction

With the rapid development of astronomy, space exploration, and advanced optical instruments, optical elements are being increasingly widely used. The application demands for high-quality and high-efficiency elements present distinct higher requirements for the process technology used in such elements [1]. Computer-controlled optical surfacing (CCOS) has been successfully applied in industrial production, as it can precisely correct the surface error by converting the dwell time of the polishing tool into the feed-rate along the polishing trajectory. Therefore, a trajectory planning method is the key factor affecting high-quality, high-efficiency polishing.

Despite the extent of research on path planning in some fields [2–4], investigation of optical polishing has been rather limited. Although the problem of discontinuous surfaces between the adjacent mm-sized short line segments has attracted wide concern when using parametric curve theory [5–7], frequent acceleration and deceleration result in poor realisation precision of dwell time, i.e., the runtime of trajectory (hereafter referred to as runtime), which further influences the polishing quality of elements and the convergence rate of surfaces [8]. The main focus of this paper is to investigate an interpolation scheme to overcome the above problem, which has two main progressive aspects: fitting and interpolation of parametric curves.

The fitting of parametric curves is a process of converting discrete short line segments into parametric curves. At present, many scholars have carried out research based on the dominant points using a non-uniform rational B-spline (NURBS). Park [9,10] proposed a method for determining the dominant points according to the discrete curvature. Zhou [11] and Xu [12] improved this method by taking concave-convex turning points and extreme points on the curvature curve as dominant points. To improve the fitting precision, Zhao [13] proposed curve fitting taking squared distance minimisation (SDM) as the evaluation index. Although the dominant points based method is easy with regard to calculation and interpolation, it may result in the loss of runtime at non-dominant points. Thus, only the global fitting method is suitable to the optical polishing. To do so, Yang [14] proposed an optimisation algorithm by establishing the evaluation function for deviation of the fitted distance. Li [15] and Lin [16] classified the trajectory into the different forms of NURBS and then employed a piecewise fitting method for real-time implementation. Based on Gaussian elimination and the continuous short block (CSB) look-ahead algorithm, Tsai [17] and Wang [18] realised the on-line transformation from short line segments to NURBS: however, for the global fitting method, the simplified strategy that setting all the weight factors as 1 eliminates the regulating effect on the fairness of curves and easily causes curvature saltation on the trajectory. Therefore, generating a fair trajectory based on NURBS is the first problem facing the polishing trajectory planning technique.

The interpolation of parametric curves is a process that discretises the NURBS into numerical control (NC) commands. Speed planning, as the critical step in the interpolation process, has been an area of research for numerous scholars: this can be classified into the time-optimal approach and the non-time-optimal approach. The time-optimal approach deals with the planning problem by taking the minimisation of the motion time as the objective to promote manufacturing efficiency [19,20]. For example, Timar [21] proposed a speed planning scheme for NC interpolation by the use of the optimal control theory. On this basis, Sencer [22] and Lu [23] considered the driving capacity constraint and trajectory precision constraint in the interpolation, respectively. The non-time-optimal approach usually deals with the planning problem by taking the minimisation of speed fluctuations as the objective [24]. Various methods, such as the feed-rate evolutionary algorithm [25], the equidistance quaternion method [26], and the improved Adams-Malton algorithm [27] are employed to decrease speed fluctuations and guarantee steady, continuous-trajectory operation. Although the effectiveness of the method has been validated experimentally, is cannot be applied directly to optical polishing because the effect of acceleration and deceleration on the realisation precision of runtime is not considered.

Driven by the practical needs to improve the quality of optical polishing, this paper presents a systematic trajectory planning method that particularly enhances the realisation precision of runtime. Following this introduction, Section 2 calculates all the control points of NURBS through numerical solution to overcome the problem of calculation efficiency. In Section 3, an optimisation algorithm of fairing NURBS is established by taking the shear jerk of trajectory as an evaluation index. Section 4 then proposes an interpolation scheme with which to minimise the realisation error of runtime by planning the feed-rate of trajectory according to the given runtime between adjacent NC codes. Section 5 reports experiments on a prototype machine which shows that the proposed trajectory planning method is more accurate than the linear interpolation method. Conclusions are drawn in Section 6.

2. Trajectory Fitting Based on the NURBS Curve

In this section, according to the given NC codes, all the control points of NURBS are solved, which provides the necessary mathematical model for the interpolation of parametric curves. The basic settings are as follows:

(1) Considering the stability, ease of use and calculation efficiency, cubic NURBS is employed as the fitting tool. (2) The uniform parametric method is used for trajectory fitting, because the polishing elements have a large radius of curvature and the chord lengths between NC codes are distributed uniformly. (3) All weight factors are set to 1. In this case, the NURBS can be treated as a cubic B-spline to simplify the calculation process.

According to the basic theory of the NURBS, a segment of NURBS $\boldsymbol{C}(u)(0 \leq u \leq 1)$ can be determined based on the four adjacent control points $\boldsymbol{d}_k(k = 0,1,2,3)$. As shown in Figure 1, $\boldsymbol{C}(0)$

and $C(1)$ separately refer to the start point and the end point of the NURBS segment. Based on the aforementioned setting, the NURBS segment can be directly written as:

$$C(u) = \frac{1}{6}\begin{bmatrix} u^3 & u^2 & u & 1 \end{bmatrix}\begin{bmatrix} -1 & 3 & -3 & 1 \\ 3 & -6 & 3 & 0 \\ -3 & 0 & 3 & 0 \\ 1 & 4 & 1 & 0 \end{bmatrix}\begin{bmatrix} d_0 \\ d_1 \\ d_2 \\ d_3 \end{bmatrix} \tag{1}$$

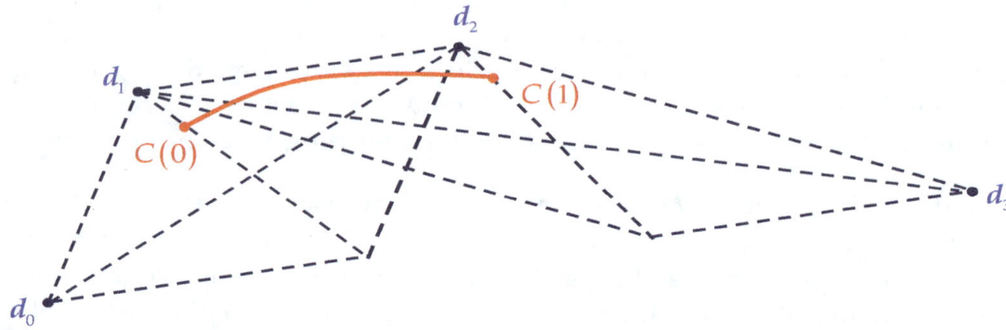

Figure 1. Schematic diagram of the parametric curve.

Thus, point C_i can be expressed as:

$$C_i = \frac{1}{6}(d_{i-1} + 4d_i + d_{i+1}) \tag{2}$$

where C_i denotes the ith NC code in the $n+1$ lines of NC codes and also is taken as the start point of the ith NURBS segment. d_{i-1}, d_i, d_{i+1} respectively denote the three control points corresponding to C_i. To calculate C_1 and C_{n+1}, it is defined that on the condition that $i < 1$, then $d_i = d_1$ and on the condition that $i > n+1$, then $d_i = d_{n+1}$. Equations (3) and (4) can thus be obtained:

$$C_1 = \frac{1}{6}(d_1 + 4d_1 + d_2) = \frac{5}{6}d_1 + \frac{1}{6}d_2 \tag{3}$$

$$C_{n+1} = \frac{1}{6}(d_n + 4d_{n+1} + d_{n+1}) = \frac{1}{6}d_n + \frac{5}{6}d_{n+1} \tag{4}$$

Rewriting Equations (2)–(4) in matrix notation yields:

$$Ad = C$$
$$A = \begin{bmatrix} 5 & 1 & 0 & 0 & 0 & \cdots & 0 \\ 1 & 4 & 1 & 0 & 0 & \cdots & 0 \\ 0 & 1 & 4 & 1 & 0 & \cdots & 0 \\ \vdots & \vdots & \vdots & \vdots & \vdots & \cdots & \vdots \\ 0 & 0 & 0 & 0 & 1 & 4 & 1 \\ 0 & 0 & 0 & 0 & 0 & 1 & 5 \end{bmatrix}, \; d = \begin{bmatrix} d_1 \\ d_2 \\ d_3 \\ \vdots \\ d_n \\ d_{n+1} \end{bmatrix}, C = \begin{bmatrix} 6C_1 \\ 6C_2 \\ 6C_3 \\ \vdots \\ 6C_n \\ 6C_{n+1} \end{bmatrix} \tag{5}$$

If the traditional solution methods such as Gauss elimination method and LU decomposition are directly applied to the Equation (5), it may lead to some undesirable phenomena (such as excessive calculation time) due to the large quantity of NC codes for polishing. Hence, the SOR iterative algorithm [28] is used to find stable numerical solutions of Equation (5) as described below.

The coefficient matrix A can be divided into:

$$A = D - L - U \tag{6}$$

where $\boldsymbol{A} = \mathrm{diag}(a_{11}, a_{22}, \ldots, a_{n+1n+1})$, $-\boldsymbol{L}$ denotes a strictly lower triangular matrix, whose elements below the principal diagonal are corresponding elements of $\boldsymbol{A}$. $-\boldsymbol{U}$ denotes a strictly upper triangular matrix, whose elements above the principal diagonal are corresponding elements of $\boldsymbol{A}$. Owing to the matrix $\boldsymbol{D}$ being invertible, Equation (5) can be modified to:

$$\boldsymbol{d} = \boldsymbol{D}^{-1}(\boldsymbol{L} + \boldsymbol{U})\boldsymbol{d} + \boldsymbol{D}^{-1}\boldsymbol{C} \tag{7}$$

In this case, corner marks (k) and $(k+1)$ are added in Equation (7) to identify the number of iterations. Then, the elementary iterative scheme of $\boldsymbol{d}$ can be written as:

$$\boldsymbol{d}^{(k+1)} = \boldsymbol{M}_1\boldsymbol{d}^{(k)} + \boldsymbol{D}^{-1}\boldsymbol{C} \tag{8}$$

where $\boldsymbol{M}_1 = \boldsymbol{D}^{-1}(\boldsymbol{L} + \boldsymbol{U})$. The component form of Equation (8) can be expressed as:

$$\boldsymbol{d}_i^{(k+1)} = \frac{1}{a_{ii}}\left(-\sum_{\substack{j=1 \\ j \neq i}}^{n+1} a_{ij}\boldsymbol{d}_j^{(k)} + \boldsymbol{C}_i\right),\ i = 1, 2, \ldots, n+1 \tag{9}$$

which is also known as the Jacobi iteration scheme, noting that, before calculating $\boldsymbol{d}_i^{(k+1)}$, the iterative values of the first $i-1$ components in $\boldsymbol{d}^{(k+1)}$ have been generated, which are more approximate to the true value than the results obtained by the previous iteration. Therefore, $\boldsymbol{d}_1^{(k)}, \boldsymbol{d}_2^{(k)}, \ldots, \boldsymbol{d}_{i-1}^{(k)}$ can be replaced by $\boldsymbol{d}_1^{(k+1)}, \boldsymbol{d}_2^{(k+1)}, \ldots, \boldsymbol{d}_{i-1}^{(k+1)}$ to make $\boldsymbol{d}_i^{(k+1)}$ closer to the true value. To improve the convergence rate of the iteration further, a proper parameter μ is selected to conduct the weighted averaging on the aforementioned iterative scheme:

$$\boldsymbol{d}_i^{(k+1)} = \mu\widetilde{\boldsymbol{d}}_i^{(k+1)} + (1-\mu)\boldsymbol{d}_i^{(k)}, i = 1, 2, \ldots n+1 \tag{10}$$

with:

$$\widetilde{\boldsymbol{d}}_i^{(k+1)} = \frac{1}{a_{ii}}\left(-\sum_{j=1}^{i-1} a_{ij}\boldsymbol{d}_j^{(k+1)} - \sum_{j=i+1}^{n+1} a_{ij}\boldsymbol{d}_j^{(k)} + \boldsymbol{C}_i\right) \tag{11}$$

Substituting Equation (11) into Equation (10) leads to the SOR iteration scheme:

$$a_{ii}\boldsymbol{d}_i^{(k+1)} + \mu\sum_{j=1}^{i-1} a_{ij}\boldsymbol{d}_j^{(k+1)} = a_{ii}\boldsymbol{d}_i^{(k)} - \mu\sum_{j=i}^{n+1} a_{ij}\boldsymbol{d}_j^{(k)} + \mu\boldsymbol{C}_i,\ i = 1, 2, \ldots, n+1 \tag{12}$$

Note that $\boldsymbol{A}$ is a tridiagonal positive definite matrix, so the optimal value μ_{opt} of the relaxation factor can be expressed as [29]:

$$\mu_{opt} = \frac{2}{1 + \sqrt{1 - [\rho(\boldsymbol{M}_1)]^2}} \tag{13}$$

where $\rho(\boldsymbol{M}_1)$ denotes the spectral radius of $\boldsymbol{M}_1$.

It is worth noting that, for a flat element, the aforementioned method can be directly applied to calculate the NURBS trajectory, but for a curved element, the polishing shaft should always lie along the normal direction of the element. In this case, it is necessary to solve two trajectories of end-point and reference-point on the polishing shaft to determine the polishing attitude (for more details, please see [21]).

3. Fairing of the NURBS

Although the constructed cubic NURBS by the above method satisfies the G2 continuity characteristics, i.e., the second-order derivative functions of the trajectory is continuous, the motion stability of the trajectory is still influenced by curvature saltation. In this section, a fairing optimisation method is proposed by adjusting the weight factors of NURBS.

The fairing optimisation can be classified into two categories, i.e., global and local fairing according to the number of the adjusted point on NURBS. Considering the significant computational burden of global fairing caused by the large quantity of NC codes for polishing, it is more reasonable to modify the outlier points with curvature saltation by the use of local fairing, which are selected from all NC codes [30]. A filtering process is needed to eliminate the influence of curvature fluctuations caused by discrete calculation.

In the fairing optimisation, the shear jerk of the outlier point is taken as the evaluation index:

$$D_j = \frac{\kappa_{next1} - \kappa_j}{\|\boldsymbol{C}_{next1} - \boldsymbol{C}_j\|} - \frac{\kappa_j - \kappa_{last1}}{\|\boldsymbol{C}_j - \boldsymbol{C}_{last1}\|} \tag{14}$$

where κ_j denotes the curvature at the jth outlier point $\boldsymbol{C}_j$, κ_{next1} and κ_{last1} denote the curvature of the $\boldsymbol{C}_{next1}$ and $\boldsymbol{C}_{last1}$, which are on two adjacent sides of the outlier point. The index indicates the curvature changes of the adjacent outlier points.

To generate the weight factors of the various outlier points, the objective function is defined as:

$$L = \sum_{j=1}^{k} D_j^2 + (\boldsymbol{C}_j - \boldsymbol{C}_{j,0})^2 \tag{15}$$

where $\boldsymbol{C}_{j,0}$ denotes the jth outlier point before optimisation. It can be seen from Equation (15) that the objective function is composed of two parts: part one is the curvature changes of the outlier point after optimisation, and part two is the adjustment amplitude of outlier points before, and after, optimisation. Thus, the objective function means that fairing optimisation is performed on the premise of modifying the NURBS as little as possible.

According to affine invariant principle of NURBS [31], the four-dimensional (4D) space constructed by the control points and weight factors can be expressed as:

$$\boldsymbol{C}^{\omega}(u) = \sum_{i=1}^{n+1} N_{i,3}(u) \begin{bmatrix} \omega_i \boldsymbol{d}_i \\ \omega_i \end{bmatrix} = \sum_{i=1}^{n+1} N_{i,3}(u) \boldsymbol{d}_i^{\omega} \tag{16}$$

Then, the NURBS defined by Equation (1) can be regarded as the projection of the curve $\boldsymbol{C}^{\omega}(u)$ in 4D space on the centre of the hyperplane $\omega = 1$. Based on Equation (2), it can be seen that:

$$\boldsymbol{C}_j^{\omega} = \frac{1}{6}\boldsymbol{d}_{j,last1}^{\omega} + \frac{2}{3}\boldsymbol{d}_j^{\omega} + \frac{1}{6}\boldsymbol{d}_{j,next1}^{\omega} \tag{17}$$

$$\boldsymbol{C}_{j,last1}^{\omega} = \frac{1}{6}\boldsymbol{d}_{j,last2}^{\omega} + \frac{2}{3}\boldsymbol{d}_{j,last1}^{\omega} + \frac{1}{6}\boldsymbol{d}_j^{\omega} \tag{18}$$

$$\boldsymbol{C}_{j,next1}^{\omega} = \frac{1}{6}\boldsymbol{d}_j^{\omega} + \frac{2}{3}\boldsymbol{d}_{next1}^{\omega} + \frac{1}{6}\boldsymbol{d}_{next2}^{\omega} \tag{19}$$

where $\boldsymbol{d}_j^{\omega}$ denotes the control point corresponding to the jth outlier point and $\boldsymbol{d}_{next1}^{\omega}, \boldsymbol{d}_{next2}^{\omega}$ and $\boldsymbol{d}_{last1}^{\omega}, \boldsymbol{d}_{last2}^{\omega}$ denote the control points on the two adjacent sides of the control point $\boldsymbol{d}_j^{\omega}$, respectively.

Furthermore, Equation (15) can be written in 4D space as:

$$L^{\omega} = \sum_{j=1}^{k} \left(D_j^{\omega}\right)^2 + \left(\boldsymbol{C}_j^{\omega} - \boldsymbol{C}_{j,0}^{\omega}\right)^2 \tag{20}$$

where $\boldsymbol{C}_j^{\omega} - \boldsymbol{C}_{j,0}^{\omega}$ can be equivalently simplified as $\boldsymbol{d}_j^{\omega} - \boldsymbol{d}_{j,0}^{\omega}$ and D_j^{ω} can be approximately expressed as [32]:

$$D_j^{\omega} \approx \frac{\left(\boldsymbol{C}_{j,next1}^{\omega}\right)'' - \left(\boldsymbol{C}_j^{\omega}\right)''}{l_{j,next1}^{\omega}} - \frac{\left(\boldsymbol{C}_j^{\omega}\right)'' - \left(\boldsymbol{C}_{j,last1}^{\omega}\right)''}{l_{j,last1}^{\omega}} \tag{21}$$

where $\left(\boldsymbol{C}_j^{\omega}\right)''$ denotes the second derivative of the NURBS at $\boldsymbol{C}_j^{\omega}$, $l_{j,next1}^{\omega} = \|\boldsymbol{C}_{j,next1,0}^{\omega} - \boldsymbol{C}_{j,0}^{\omega}\|$ and $l_{j,last1}^{\omega} = \|\boldsymbol{C}_{j,0}^{\omega} - \boldsymbol{C}_{j,last1,0}^{\omega}\|$. Substituting Equation (21) into Equation (20) yields:

$$L^{\omega} = \sum_{j=1}^{k}\left(\frac{\left(\boldsymbol{C}_{j,next1}^{\omega}\right)'' - \left(\boldsymbol{C}_j^{\omega}\right)''}{l_{j,next1}^{\omega}} - \frac{\left(\boldsymbol{C}_j^{\omega}\right)'' - \left(\boldsymbol{C}_{j,last1}^{\omega}\right)''}{l_{j,last1}^{\omega}}\right) + \left(\boldsymbol{d}_j^{\omega} - \boldsymbol{d}_{j,0}^{\omega}\right)^2 \tag{22}$$

To calculate the minimum value of L^{ω}, the partial derivative of Equation (22) about $\boldsymbol{d}_j^{\omega}$ is set to 0:

$$\begin{aligned} \frac{\partial L^{\omega}}{\partial \boldsymbol{d}_j^{\omega}} = \sum_{j=1}^{k} 2\left(\frac{\boldsymbol{d}_{j,next2}^{\omega} - 3\boldsymbol{d}_{j,next1}^{\omega} + 3\boldsymbol{d}_j^{\omega} - \boldsymbol{d}_{j,last1}^{\omega}}{l_{j,next1}^{\omega}} - \frac{\boldsymbol{d}_{j,next1}^{\omega} - 3\boldsymbol{d}_j^{\omega} + 3\boldsymbol{d}_{j,last1}^{\omega} - \boldsymbol{d}_{j,last2}^{\omega}}{l_{j,last1}^{\omega}}\right) \\ \cdot\left(\frac{3}{l_{j,next1}^{\omega}} + \frac{3}{l_{j,last1}^{\omega}}\right) + 2\left(\boldsymbol{d}_j^{\omega} - \boldsymbol{d}_{j,0}^{\omega}\right) = 0 \end{aligned} \tag{23}$$

Then, the equations for $\boldsymbol{d}_1^{\omega}, \boldsymbol{d}_2^{\omega}, \ldots, \boldsymbol{d}_k^{\omega}$ can be expressed as:

$$\begin{aligned} \boldsymbol{A}^{\omega}\boldsymbol{d}^{\omega} &= \boldsymbol{C}^{\omega} \\ \boldsymbol{A}^{\omega} &= diag\left(\left(\frac{3}{l_{j,next1}^{\omega}} + \frac{3}{l_{j,last1}^{\omega}}\right)^2 + 1\right) \\ \boldsymbol{d}^{\omega} &= \left(\boldsymbol{d}_1^{\omega}, \boldsymbol{d}_2^{\omega}, \ldots, \boldsymbol{d}_k^{\omega}\right)^{\mathrm{T}} \\ \boldsymbol{C}^{\omega} &= \left(\boldsymbol{C}_1^{\omega}, \boldsymbol{C}_2^{\omega}, \ldots, \boldsymbol{C}_k^{\omega}\right)^{\mathrm{T}} \\ \boldsymbol{C}_j^{\omega} = \boldsymbol{d}_{j,0}^{\omega} - &\left(\frac{\boldsymbol{d}_{j,next2}^{\omega} - 3\boldsymbol{d}_{j,next1}^{\omega} - \boldsymbol{d}_{j,last1}^{\omega}}{l_{j,next1}^{\omega}} - \frac{\boldsymbol{d}_{j,next1}^{\omega} + 3\boldsymbol{d}_{j,last1}^{\omega} - \boldsymbol{d}_{j,last2}^{\omega}}{l_{j,last1}^{\omega}}\right)\left(\frac{3}{l_{j,next1}^{\omega}} + \frac{3}{l_{j,last1}^{\omega}}\right) \end{aligned} \tag{24}$$

According to Equation (24), the optimised control points and weight factors corresponding to the outlier points can then be generated.

4. NURBS Interpolation

The NURBS interpolation is used to discretise the parametric curve to the NC commands based on the planned feed-rate. In this section, an interpolation method for optical polishing is proposed aiming to minimise the realisation error of the trajectory runtime.

4.1. Feed-Rate Planning

Feed-rate planning is the main influencing factor in interpolating the NC commands along the trajectory. For the optical polishing, to guarantee the desired runtime of trajectory, the specific method is displayed as follows:

Step 1: considering that now there is no analytical solution to calculate the length of NURBS, the Simpson formula is used to obtain the estimation of the length through numerical iteration:

$$p = \frac{up - low}{6}\left(f(low) + 4f(mid) + f(up)\right) \tag{25}$$

with:

$$f(u) = \frac{ds}{du} = \sqrt{x'(u) + y'(u) + z'(u)} \tag{26}$$

where $x'(u), y'(u), z'(u)$ respectively denote the first-order derivatives of $x(u), y(u), z(u)$, which are the one-dimensional curves of $\mathbf{C}(u)$ along x, y, z axes. up and low denote the upper and lower boundaries of u, $mid = (up + low)/2$.

Step 2: the length between the two points corresponding to the parameters up and low is calculated:

$$l = \|\mathbf{C}(up) - \mathbf{C}(low)\| \tag{27}$$

Step 3: the error between the aforementioned two lengths is calculated:

$$e = |p - l| \tag{28}$$

The convergence threshold $[e]$ is given. If $e > [e]$, let $up = mid$ and repeat Steps 1 and 2. If $e < [e]$, turn to Step 4.

Step 4: the parameter interval $(0, 1)$ was sectioned by the equivalent distance $up - low$ to generate the knot vector $(u_0, u_1, \ldots, u_n)$. Thus, the length of the NURBS is:

$$s \approx \sum_{i=1}^{n} \|\mathbf{C}(u_i) - \mathbf{C}(u_{i-1})\| \tag{29}$$

Equipped with the length at hand, the S-curve motion law is invoked to guarantee that the feed-rates of the adjacent NURBS segments are changed smoothly. Moreover, the initial sections of the trajectory segments are defined as the feed-rate transition zone. Then, the feed-rates remain constant until the end of the trajectory segments. In this case [33]:

$$s = 2v_{low}t_a + Jt_a^3 + v_{up}t_v \tag{30}$$

$$v_{up} = v_{low} + Jt_a^2 \tag{31}$$

where v_{low} and v_{up} denote the initial and final feed-rates of the trajectory segment. J, $2t_a$ and t_v denote the jerk, acceleration (deceleration) time and the uniform motion time, respectively. Thus, the runtime of the trajectory segment is $t_d = 2t_a + t_v$.

It is worth noting that previous studies have shown that the feed-rates, when limited by the runtime of the trajectory segments, are much lower than the maximum value which is constrained by the chord error and the driving capacity. Therefore, there is no need to check the feed-rate again.

4.2. NURBS Interpolation

The essence of interpolation is to generate the NC command along the trajectory according to the period t_s. As each interpolation point of the NURBS corresponds to one curve parameter, only the curve parameters need to be solved.

Taking u as the function of t, the second-order Taylor expansion can be expressed as:

$$u_{i+1} = u_i + \left.\frac{du}{dt}\right|_{t=t_i} t_s + \frac{1}{2}\left.\frac{d^2u}{dt^2}\right|_{t=t_i} t_s^2 \tag{32}$$

The feed-rate $v(u)$ can be written as:

$$v(u) = \|\frac{d\mathbf{C}(u)}{du}\| = \|\frac{d\mathbf{C}(u)}{du}\frac{du}{dt}\| \tag{33}$$

Furthermore:

$$\frac{du}{dt} = \frac{v(u)}{\|\mathbf{C}'(u)\|} \tag{34}$$

Calculating the derivative of Equation (34):

$$\frac{d^2u}{dt^2} = \frac{\frac{dv(u)}{du}\frac{du}{dt}}{\|\mathbf{C}'(u)\|} - v(u)\frac{\frac{d(\|\mathbf{C}'(u)\|)}{du}\frac{du}{dt}}{\|\mathbf{C}'(u)\|^2} \tag{35}$$

where:

$$\frac{d(\|\mathbf{C}'(u)\|)}{du} = \frac{\mathbf{C}''(u)\cdot\mathbf{C}'(u)}{\|\mathbf{C}'(u)\|}$$

Substituting Equations (33)–(35) into Equation (32) gives:

$$u_{i+1} = u_i + \frac{v_C(u_i)}{\|\mathbf{C}'(u_i)\|}t_s + \frac{1}{2}\left[\frac{v'_C(u_i)}{\|\mathbf{C}'(u_i)\|^2}v_C(u_i) - \frac{\mathbf{C}''(u)\cdot\mathbf{C}'(u)}{\|\mathbf{C}'(u)\|^4}v_C^2(u_i)\right]t_s^2 \tag{36}$$

Substituting u_{i+1} into the curve equation obtained through the fairing optimisation described in Section 3, the next interpolation point can be acquired. Repeating this process until:

$$t_i = \text{ceil}\left(\frac{t_d}{t_s}\right) * t_s \tag{37}$$

where ceil$(\cdot)$ denotes an integer that is rounded up. In this way, interpolation of all trajectory segment can be completed.

5. Experiments

Both simulation and experiments were carried out to validate the effectiveness of the presented method on the prototype of the hybrid polishing robot. As shown in Figure 2, it is mainly composed of a 6-DOF (degrees of freedom) hybrid robot, a polishing effector, a magnetic worktable, a column, and a CNC system. The hybrid robot is composed of a 3-DOF (3U$\underline{\text{P}}$S and UP) parallel mechanism and a 3-DOF wrist. The UP limb and the wrist form a UP$\underline{\text{S}}$ or UP$\underline{\text{RRR}}$ limb. Here, R, U, S, and P represent, respectively, revolute, universal, spherical, and prismatic joints, and the underlined $\underline{\text{P}}$, $\underline{\text{S}}$, and $\underline{\text{R}}$ denote the actuated prismatic, spherical, and revolute joints, respectively. The CNC system is built upon an IPC+PMAC open architecture, consisting of a host control computer responsible for reconstruction of the parametric curves, trajectory interpolation, and NC command generation, and a PMAC motion controller for servo-control of the actuated joints.

1. CNC system 2. Column 3. Hybrid robot 4. Polishing effector 5. Magnetic worktable

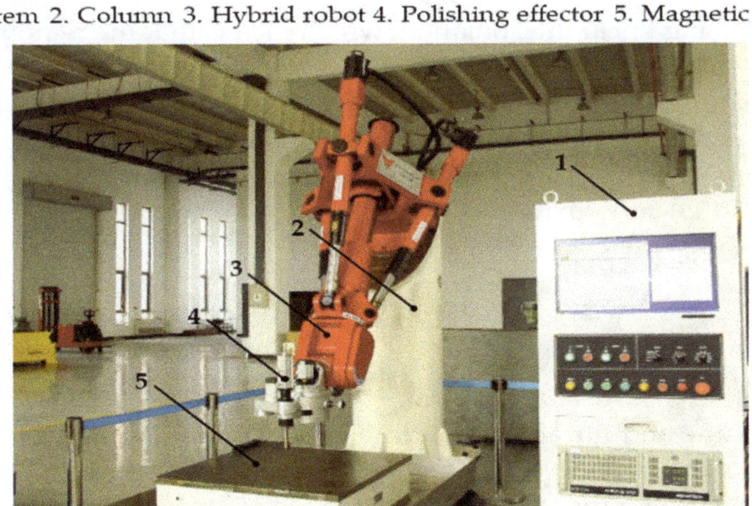

Figure 2. The prototype of the polishing robot.

Without loss of generality, a segment of NC codes was taken from the polishing trajectory to validate the effectiveness of the proposed interpolation method. According to the SOR iterative scheme mentioned in Section 2, Figure 3 shows the two fitted trajectories. Then, the optimisation algorithm described in Section 3 is used to smooth the curvature of the trajectory. The change threshold of curvature for judging outlier points is given as 0.01. As shown in Table 1, the fairness of the two trajectories is both significantly improved. The maximum curvatures of the two trajectories are reduced by as much as 79.7% and 63.3% and the maximum absolute values of shear jerks are decreased by 91.2% and 90.2%, correspondingly. Considering the optimisation and interpolation methods for the two trajectories are same, experimental results of the end-point trajectory are just shown in the following discussion for the sake of simplicity.

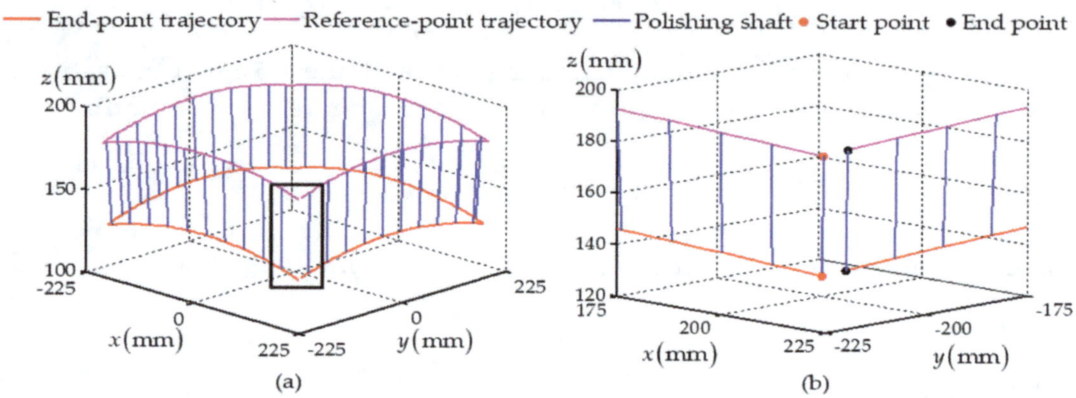

Figure 3. The initial fitted polishing trajectory for fairing optimisation and interpolation (**a**) the fitted polishing trajectory (**b**) the partial enlarged view.

Based on the optimisation trajectory shown in Figure 4, the discrete NC command sequences were generated and sent to the PMAC motion controller, which were mapped into the servo-command of actuated joints through an inverse kinematic model. The PMAC motion controller synchronously gathered the positions and velocities fed back from the servo-motors. The interpolation and sampling periods are 10 and 20 ms, respectively. To validate the effectiveness of the method, a comparison experiment was carried out utilizing the linear interpolation method. Figure 5 shows the experimental result, which is computed by the feedback positions of all the actuated joints. It can be seen from the partial enlarged view that the NC commands generated by the proposed method are closer to the NC codes than those found using linear interpolation, which means that the removal position of polishing process can be reached more accurately. Figures 6 and 7 show the runtime between the adjacent NC nodes. Compared with the desired value, the proposed method is able to realise the runtime more precisely, which indicates that the removal quantity during the polishing process can be more precisely controlled, correspondingly. To evaluate the effect of the interpolation method, the indices are defined as the interpolation error (e_i) and the runtime error of the trajectory (e_t):

$$e_i = \|\mathbf{C}_{i0} - \mathbf{C}_i\|_2,\ i = 1, 2, \ldots, n \tag{38}$$

$$e_{ti} = |t_{i0} - t_i|,\ i = 1, 2, \ldots, n \tag{39}$$

where $\mathbf{C}_{i0}$ and $\mathbf{C}_i$ denote the ith desired NC code and actual NC code after interpolation, t_{i0} and t_i denote the desired runtime and the actual runtime between the $i-1$th and ith NC codes. It can be seen from Table 2 that, compared to the linear interpolation method, the interpolation error and the runtime error generated through the proposed method are an order of magnitude smaller. These data further imply that the proposed interpolation method has more precision than the linear interpolation method in the polishing process.

Table 1. Changes in the two trajectories before, and after, fairing optimisation.

Trajectory	Before		After	
	Maximum Curvature	Maximum Shear Jerk	Maximum Curvature	Maximum Shear Jerk
End-point	1.249	0.285	0.253	0.025
Reference-point	1.380	0.325	0.506	0.032

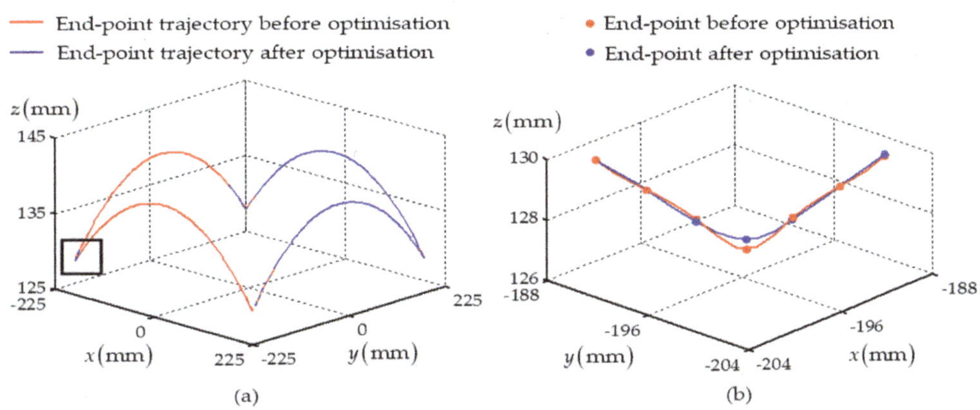

Figure 4. The end-point trajectory before, and after, fairing optimization: (**a**) the end-point trajectory; and (**b**) the partial enlarged view

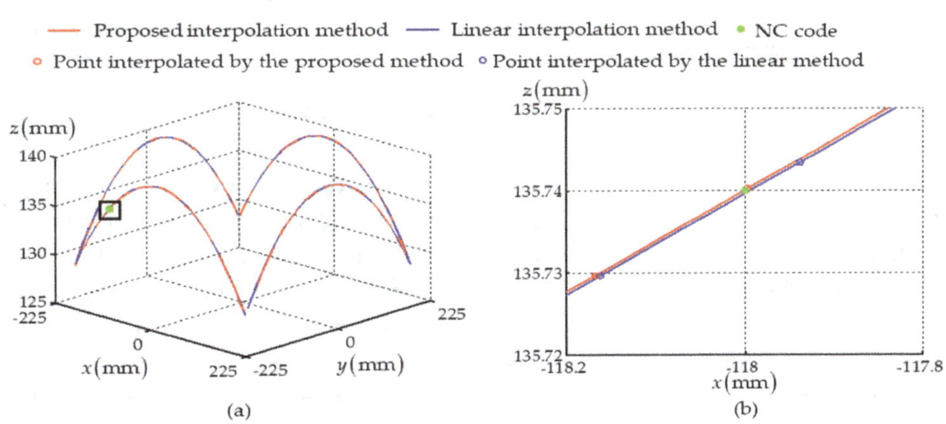

Figure 5. The end-point trajectory interpolated by different methods: (**a**) the end-point trajectory; and (**b**) the partial enlarged view.

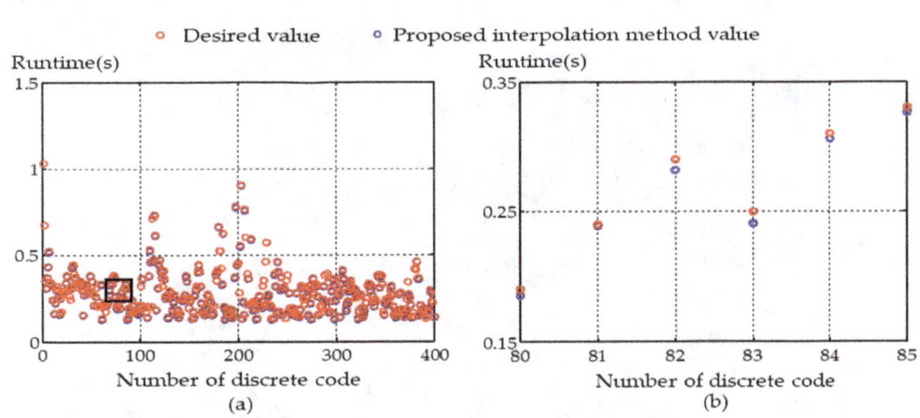

Figure 6. Runtime of the trajectory interpolated by the proposed method: (**a**) runtime of the trajectory; and (**b**) the partial enlarged view.

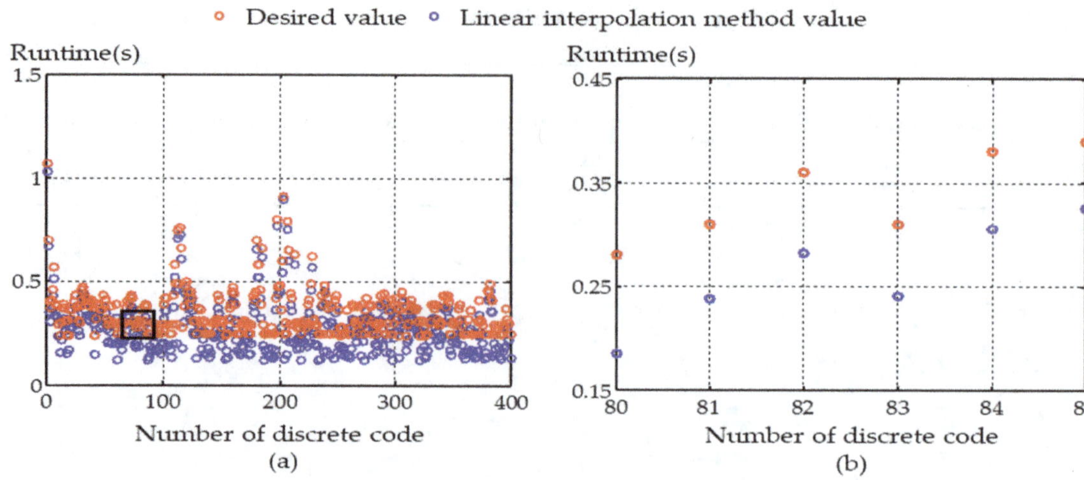

Figure 7. Runtime of the trajectory interpolated by the linear method: (**a**) runtime of the trajectory; and (**b**) the partial enlarged view.

Table 2. Comparison of the effect of the two interpolation methods.

Method	Interpolation Error (mm)		Runtime Error (s)	
	Maximum	Mean	Maximum	Mean
Proposed	0.005	0.002	0.010	0.005
Linear	0.091	0.076	0.130	0.076

To validate the modification effect of the interpolation method on the surface error of optical elements, polishing experiments were carried out on fused silica elements by, respectively, using the linear interpolation method and the proposed interpolation method, as shown in Figure 8. Using the Nanovea contour graph, the polished zone (70 mm $\times$ 70 mm) of the elements was detected. The experimental results obtained using the two interpolation methods are displayed in Figures 9 and 10, respectively. It can be seen, from the figures, that, after conducting linear interpolation-based polishing, the surface error (PV) decreases from $11.894\lambda(\lambda = 633\text{nm})$ to 7.422λ. In contrast, using the proposed interpolation method, the surface error (PV) is reduced from 11.282λ to 6.267λ. The convergence rate of the surface error is increased from 37.59% to 44.44%, which further verifies the effectiveness of the proposed method.

Figure 8. Polishing experiment on fused silica elements.

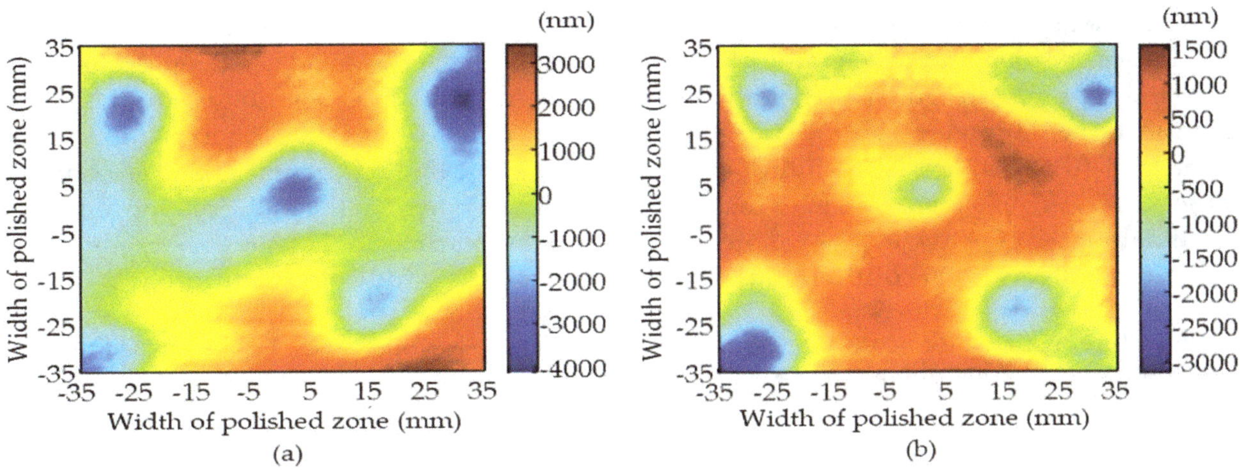

Figure 9. Changes in surface errors (**a**) before and (**b**) after conducting the linear interpolation-based polishing.

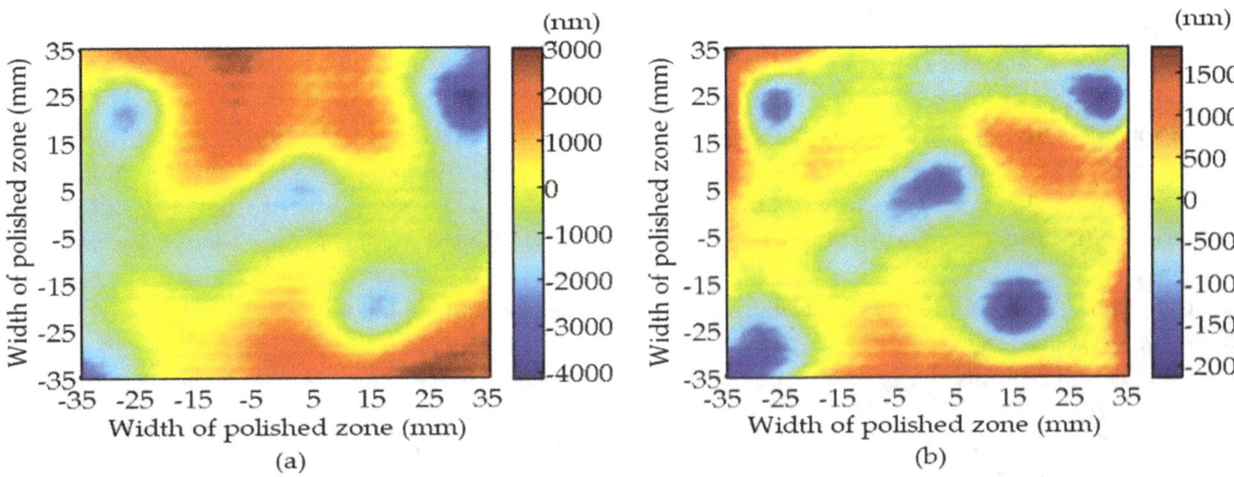

Figure 10. Changes in surface errors (**a**) before and (**b**) after conducting the proposed interpolation-based polishing.

6. Conclusions

A new trajectory planning method for optical polishing is proposed in this paper. It is developed to generate NC commands that can decrease the interpolation error and the runtime error. First, to obtain the NURBS trajectory without loss of the information about the runtime, a global fitting method is presented based on SOR iteration theory to deal with the problem of the computational burden arising from the large quantity of NC codes for polishing. Then, taking the shear jerk of the NURBS as the evaluation index, a fairing optimisation method is carried out to smooth the curvature saltation of the trajectory. Finally, the feed-rate planning method and the path interpolation scheme are proposed to reduce the realisation error of the trajectory runtime.

Simulation results verify that the fairing optimisation proposed in this research can modify the curvature saltation at the expense of trajectory accuracy compared to the given NC codes; however, the curvature saltation only occurs at the turning point of the trajectory and the trajectory error magnitude generated by the optimisation algorithm is consistent with the linear interpolation. Therefore, the trajectory accuracy is not treated as the index with which to evaluate the smoothing effect in Section 5. The effect of the optimisation algorithm on the optical polishing will be investigated in future work.

It can be seen from Figures 5–7, and Table 2, that the runtime error of the proposed interpolation method arises because the trajectory runtime cannot be divided exactly by the interpolation period and is rounded up to an integer. In contrast, the runtime error of the linear interpolation method arises as a result of the acceleration and deceleration that occurs frequently between adjacent short line segments. Similar to the runtime error, the interpolation error of the proposed interpolation method is a result of the numerical calculation error of the trajectory length and that of the linear interpolation method is due to the transition for the discontinuity of the adjacent short line segments. Then the convergence rate of the surface error is increased with the help of the improvement in the interpolation, and runtime, errors, validating the effectiveness of the proposed interpolation process. For the aforementioned error, their individual effects on the optical polishing need to be further indicated in future work.

Author Contributions: D.Z. wrote the manuscript and performed the experiments, H.G. analysed the data and revised the manuscript. All authors discussed the results and commented on the manuscript.

References

1. Rupp, W. Conventional optical polishing techniques. *Int. J. Opt.* **1971**, *18*, 1–16. [CrossRef]
2. Kaltsoukalas, K.; Makris, S.; Chryssolouris, G. On generating the motion of industrial robot manipulators. *Rob. Comput. Integr. Manuf.* **2015**, *32*, 65–71. [CrossRef]
3. Makris, S.; Tsarouchi, P.; Matthaiakis, A. Dual arm robot in cooperation with humans for flexible assembly. *CIRP Ann.* **2017**, *66*, 13–16. [CrossRef]
4. Lei, W.; Wang, S. Robust real-time NURBS path interpolators. *Int. J. Mach. Tools Manuf.* **2009**, *49*, 625–633. [CrossRef]
5. Jahanpour, J.; Alizadeh, M. A novel acc-jerk-limited NURBS interpolation enhanced with an optimized S-shaped quintic feedrate scheduling scheme. *Int. J. Adv. Manuf. Technol.* **2015**, *77*, 1889–1905. [CrossRef]
6. Liu, X.; Peng, J.; Si, L. A novel approach for NURBS interpolation through the integration of acc-jerk-continuous-based control method and look-ahead algorithm. *Int. J. Adv. Manuf. Technol.* **2017**, *88*, 961–969.
7. Zhao, J.; Li, L.; Wang, G. Research of NURBS Curve Real-time Interpolation Feed Speed Planning. *Tool Eng.* **2015**, *49*, 7–11.
8. Lin, M.; Tsai, M.; Yau, H. Development of a dynamics-based NURBS interpolator with real-time look-ahead algorithm. *Int. J. Mach. Tools Manuf.* **2007**, *47*, 2246–2262. [CrossRef]
9. Park, H.; Lee, J. B-spline curve fitting based on adaptive curve refinement using dominant points. *Comput.-Aided Des.* **2007**, *39*, 439–451. [CrossRef]
10. Park, H. B-spline surface fitting based on adaptive knot placement using dominant columns. *Comput.-Aided Des.* **2011**, *43*, 258–264. [CrossRef]
11. Zhou, H.; Wang, Y.; Liu, Z. On non-uniform rational B-splines curve fitting based on the least control points. *J. Xian Jiaotong Univ.* **2008**, *42*, 73–77.
12. Xu, J. B-spline Curve Approximation Based on Feature Points Automatic Recognition. *J. Mech. Eng.* **2009**, *45*, 212–217. [CrossRef]
13. Zhao, H.; Lu, Y.; Zhu, L. Look-ahead interpolation of short line segments using B-spline curve fitting of dominant points. *P. I. Mech. Eng. B-J. Eng.* **2014**, *229*, 1–13. [CrossRef]
14. Yang, X.; Hu, Z.; Zhong, Z. Research on the NURBS Curve Fitting for Tool Path Generation. *China Mech. Eng.* **2009**, *20*, 983–984.
15. Li, W.; Liu, Y.; Yamazaki, K. The design of a NURBS pre-interpolator for five-axis machining. *Int. J. Adv. Manuf. Technol.* **2008**, *36*, 927–935. [CrossRef]
16. Lin, K.; Ueng, W.; Lai, J. CNC codes conversion from linear and circular paths to NURBS curves. *Int. J. Adv. Manuf. Technol.* **2008**, *39*, 760–773. [CrossRef]
17. Tsai, M.; Nien, H. Development of a real-time look-ahead interpolation methodology with spline-fitting technique for high-speed machining. *Int. J. Adv. Manuf. Technol.* **2010**, *47*, 621–638. [CrossRef]
18. Wang, J.; Yau, H. Real-time NURBS interpolator: application to short linear segments. *Int. J. Adv. Manuf. Technol.* **2009**, *41*, 1169–1185. [CrossRef]
19. Timar, S.; Farouki, R. Time-optimal traversal of curved paths by Cartesian CNC machines under both constant and speed-dependent axis acceleration bounds. *Rob. Comput. Integr. Manuf.* **2007**, *23*, 563–579. [CrossRef]

20. Sencer, B. Smooth Trajectory Generation and Precision Control of 5-Axis CNC Machine Tools. Ph.D. Thesis, The University of British Columbia, Vancouver, BC, Canada, October 2009.
21. Boyadjieff, C.; Farouki, R.; Timar, S. *Smoothing of Time-Optimal Feed rates for Cartesian CNC Machines*; Springer: Berlin, Germany, 2005; pp. 84–101.
22. Sencer, B.; Altintas, Y.; Croft, E. Feed optimization for five-axis CNC machine tools with drive constraints. *Int. J. Mach. Tools Manuf.* **2008**, *48*, 733–745. [CrossRef]
23. Lu, L.; Zhang, L.; Ji, S. An offline predictive feedrate scheduling method for parametric interpolation considering the constraints in trajectory and drive systems. *Int. J. Adv. Manuf. Technol.* **2016**, *83*, 2143–2157. [CrossRef]
24. Zhao, H.; Zhu, L.; Ding, H. A parametric interpolator with minimal feed fluctuation for CNC machine tools using arc-length compensation and feedback correction. *Int. J. Mach. Tools Manuf.* **2013**, *75*, 1–8. [CrossRef]
25. Sun, Y.; Zhao, Y.; Bao, Y. A novel adaptive-feedrate interpolation method for NURBS tool path with drive constraints. *Int. J. Mach. Tools Manuf.* **2013**, *77*, 74–81. [CrossRef]
26. Zhang, J.; Zhang, L.; Zhang, K. Double NURBS trajectory generation and synchronous interpolation for five-axis machining based on dual quaternion algorithm. *Int. J. Adv. Manuf. Technol.* **2016**, *83*, 2015–2025. [CrossRef]
27. Peng, J.; Liu, X.; Si, L. A Novel Approach for NURBS Interpolation with Minimal Feed Rate Fluctuation Based on Improved Adams-Moulton Method. *Math. Probl. Eng.* **2017**, *2017*, 1–10. [CrossRef]
28. Bai, Z.; Parlett, B.; Wang, Z. On generalized successive over relaxation methods for augmented linear systems. *Numer. Math.* **2005**, *102*, 1–38. [CrossRef]
29. Huang, J. Another Version of SOR Iteration and Its Generalization. Master's Thesis, Dalian University of Technology, Dalian, China, June 2013.
30. Li, Y. Study and Realization on Construction, Fairing and Smooth Joining Between Adjacent of the NURBS Surface. Master's Thesis, Xi'an University of Technology, Xi'an, China, March 2010.
31. Surhone, L.; Timpledon, M.; Marseken, S. *Non-Uniform Rational B-Spline*; Betascript Publishing: Montana, MT, USA, 2010.
32. Zhang, H.; Jiang, D.; Ding, Y. A weight-based optimal faring algorithm for planar cubic NURBS curves. *J. Southwest Minzu Univ.* **2005**, *31*, 351–355.
33. Li, X.; Wu, Y.; Leng, H. Research on a New S-curve Acceleration and Deceleration Control Method. *Modul. Mach. Tool. Autom. Manuf. Technol.* **2007**, 50–53.

5

Multiellipsoidal Mapping Algorithm

Carlos Villaseñor, Nancy Arana-Daniel *, Alma Y. Alanis, Carlos Lopez-Franco and Javier Gomez-Avila

Centro Universitario de Ciencias Exactas e Ingenierías, Universidad de Guadalajara, Blvd Marcelino García Barragán 1421, Guadalajara 44430, Mexico; cavp@outlook.com (C.V.); almayalanis@gmail.com (A.Y.A.); clzfranco@gmail.com (C.L.-F.); javier.ega@hotmail.com (J.G.-A.)

* Correspondence: nancyaranad@gmail.com.

Abstract: The robotic mapping problem, which consists in providing a spatial model of the environment to a robot, is a research topic with a wide range of applications. One important challenge of this problem is to obtain a map that is information-rich (i.e., a map that preserves main structures of the environment and object shapes) yet still has a low memory cost. Point clouds offer a highly descriptive and information-rich environmental representation; accordingly, many algorithms have been developed to approximate point clouds and lower the memory cost. In recent years, approaches using basic and "simple" (i.e., using only planes or spheres) geometric entities for approximating point clouds have been shown to provide accurate representations at low memory cost. However, a better approximation can be implemented if more complex geometric entities are used. In the present paper, a new object-mapping algorithm is introduced for approximating point clouds with multiple ellipsoids and other quadratic surfaces. We show that this algorithm creates maps that are rich in information yet low in memory cost and have features suitable for other robotics problems such as navigation and pose estimation.

Keywords: object mapping; Geometric Algebra; Differential Evolution

---

1. Introduction

Many algorithms have been developed for the robotic mapping problem [1], which arises when a robot is provided with a spatial model of the environment. Unlike many 3D reconstruction algorithms that use interpolation to obtain the best object rendering, the principal aim of robotic mapping is to create maps that are rich in information and low in memory cost. The environment and objects could be acquired through 3D sensor like Laser Rangefinder or multiples camera view with algorithms like Structure from Motion [2]. With these techniques we obtain dense point clouds.

On the one hand, there exist 3D mapping algorithms, e.g., Simultaneous Localization and Mapping (SLAM), that model key point maps. The principal features of such algorithms are representations with low memory cost and ease of use. However, the maps are not information-rich; consequently, such algorithms cannot be used in many applications, such as object manipulation and aerial navigation. On the other hand, many algorithms create extensive object representations as 3D reconstructions based on the Radial Basis Function (RBF) [3] or other functions [4], resulting in richer information maps, albeit at a high memory cost; additionally, such algorithms are difficult to use. Finding a good balance between representation and memory cost is an important current research topic.

In the last decade, algorithms based on object mapping with geometric entities have become very popular for many applications [1], such as urban and office environment mapping. Object mapping refers to algorithms that approximate the volume shape of a point cloud (it is to say approximate a point cloud) with multiply geometric entities, this is analogous to use curve fitting in a regression problem, but we use geometric entities to describe and preserve the volume shapes instead of functions because there is not dependent variables. The quality of the approximation could be measured using the mean distance of the points to the geometric entity.

Geometric entities have been used before to approximate point clouds, for instance, we observe the use of multiplanar representations [5] and spherical and linear representations [6–9]. Each of these approaches fits the parameters of their geometric entities to obtain the best approximation of sensor data. However, if more complex entities are used, we can obtain approximations of point clouds with better accuracy, and using fewer entities than those obtained by using muliltiplanar, spherical or linear representations.

In this paper, a new object-mapping algorithm is presented. This algorithm is capable of approximating the point cloud with multiple ellipsoids and other quadratic surfaces such as spheres, pairs of planes, and pseudocylinders, which allows us to reduce the number of geometric entities used to represent the objects and complete scenes of the environment, this feature is shown in a controlled experiment, but to automate this process a hierarchical clustering must be implemented, as discussed in the future work section.

This algorithm can also be related to Hyperellipsoidal Neurons (HNs) [10] based on Geometric Algebra (GA) [11], where every neuron is trained with k-means++ [12] and Differential Evolution (DE) [13] for adapting the point cloud implicit surface.

In the following sections, we show that the multiellipsoidal mapping algorithm is capable of creating information-rich maps with low memory cost and that the entities obtained with the algorithm are even capable of deforming into other quadratic surfaces due to their representation in GA as multivectors. The entities are described in GA; then, the map is suitable for developing new algorithms that work in this algebra, such as path planning [14], pose estimation [15,16], and other tasks [17].

The paper is organized as follows. In Section 2, we introduce several mathematical and algorithmic tools used in this paper. Section 3 presents the adaptation strategy of the ellipsoidal surfaces to a point cloud. In Section 4, we explore various experiments to show the performance of the algorithm. In Section 5, the differences from object-mapping algorithms based on function approximations are discussed. Finally, conclusions are presented in Section 6, and in Section 7 we discuss the future work.

2. Mathematical Background

In this section, we introduce the mathematical framework of GA to represent the geometric entities. We also introduce the basic notions of k-means++ and DE algorithms to optimize the representation. The notation shown here will be used in the development of the proposed algorithm.

2.1. Geometric Algebra

GAs are Clifford Algebras [11,18] constructed over a bilinear form of the vector space $\mathbb{R}^{p,q,r}$, where (p, q, r) is the algebraic signature. This GA is denoted by $\mathbb{G}_{p,q,r}$ and it is an equivalent notation for $C\ell_{p,q,r}(\mathbb{R})$. The elements that belong to this algebra are called multivectors. Let us denote by "$\circ$" the Clifford product, by "$\bullet$" the inner product and by "$\wedge$" the outer product. Then, for two basis vectors, (1) states the Clifford product behavior, where $e_i \wedge e_j$ is a bivector or a 2-vector element; all vectors in $\mathbb{R}^{p,q,r}$ are included in $\mathbb{G}_{p,q,r}$ as 1-vectors (a k-vector describe a multivector span by k basis vectors). In addition, the properties of the bilinear form shown in (2) are present, where "$\cdot$" is the common dot product used in linear algebra, thus $\mathbb{R}^{p,q,r} \subset \mathbb{G}_{p,q,r}$.

$$e_i \circ e_j = \begin{cases} e_i \wedge e_j & \text{if } i \neq j \\ 1 & \text{if } 1 \leq i = j \leq p \\ -1 & \text{if } p < i = j \leq q \\ 0 & \text{if } q < i = j \leq r \end{cases} \tag{1}$$

$$a \circ a = a \bullet a = a \cdot a, \quad a \in \mathbb{R}^{p,q,r} \tag{2}$$

GAs are associative and anticommutative algebras; additionally, due to the generality of the Clifford product, many properties of other mathematical frameworks are present [11] such as complex numbers, quaternions, and other Cayley-Dickson algebras, in addition to Pauli matrices and spinor algebras.

Nevertheless, GAs are not only useful for integrating multiple mathematical frameworks but also present attractive features for geometric entity representation. In a GA, multivectors by themselves could represent geometric entities in their inner or outer product representation instead of the traditional geometric locus. The importance of this feature is that the geometric entities, as well as their operators, are represented as elements of the same algebra (both are represented as multivectors).

2.2. Hyperconformal Geometric Algebra

The most used GA is perhaps the Conformal Geometric Algebra (CGA) for the 3D case $\mathbb{G}_{4,1}$, where the algebraic signature is $(4,1,0)$, and its extension to n-dimensions $\mathbb{G}_{n+1,1}$ to represent the $\mathbb{R}^n$ vector space. In this algebra, it is possible to represent points, pairs of points, lines, planes, circles and spheres. Many algorithms for robotics and machine vision [19], e.g., finding path planning algorithms [14], pose estimation [15], structure extraction from motion [16], geometric entity detection [6,7], and robotic mapping algorithms [8,9], have been developed with this algebra.

However, to use more complex geometric entities as algebra elements, other algebras must be used. The GA $\mathbb{G}_{6,3}$ [20] is a generalization of $\mathbb{G}_{4,1}$, in which such geometric entities like ellipsoids, planes, pair of planes, pseudocylinders, spheres and others deformed quadratic surfaces can be represented as multivectors. In [10], $\mathbb{G}_{6,3}$ was extended to any dimension in the so-called Hyperconformal GA (HGA) with a notation $\mathbb{G}_{2n,n}$ for representing the vector space $\mathbb{R}^n$. This algebra is constructed by using a homogeneous stereographic projection in (3) over every coordinate. Figure 1 describes a graphical representation of the projection.

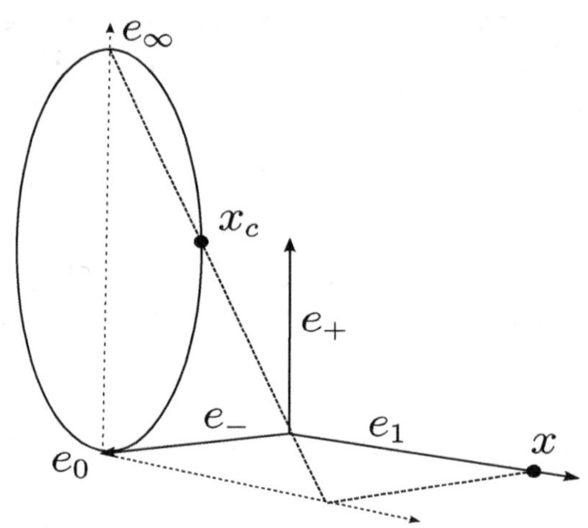

Figure 1. Homogeneous stereographic projection.

The null basis are related to (3) by $e_0 = e_+ + e_-$ and $e_\infty = e_+ - e_-$ and they have the property $e_0^2 = e_\infty^2 = 0$. Given a three-dimensional space with basis (e_x, e_y, e_z), the previous notation must be extended. Consider that the stereographic projection is applied in every axis, e.g., the null basis of the coordinate e_x are denoted with e_{0x} and $e_{\infty x}$.

$$x_c = 2\frac{x}{x^2+1}e_1 + \frac{x^2-1}{x^2+1}e_+ + e_- = xe_1 + \frac{1}{2}x^2 e_\infty + e_0 \tag{3}$$

Hence, the entities in $\mathbb{G}_{6,3}$ can be calculated using the null basis. Let $x = p_x e_x + p_y e_y + p_z e_z$ be a point in $\mathbb{R}^3$, the its representation in $\mathbb{G}_{6,3}$ is denoted by X as is shown in (4), where $e_{i\infty}$ is the point at infinity in the i homogeneous stereographic projection and $e_0 = 1/3(e_{0x} + e_{0y} + e_{0z})$ is the point at zero.

$$X = p_x e_x + p_y e_y + p_z e_z + \frac{1}{2}(p_x^2 e_{\infty x} + p_y^2 e_{\infty y} + p_z^2 e_{\infty z}) + e_0 \tag{4}$$

In $\mathbb{G}_{6,3}$, H denotes a 1-vector that represents an fixed-axes ellipsoid with center (c_x, c_y, c_z) and semiaxis (r_x, r_y, r_z), as can be seen in (5), where $e_\infty = e_{\infty z} + e_{\infty y} + e_{\infty z}$.

$$E = \frac{c_x}{r_x^2} e_x + \frac{c_y}{r_y^2} e_y + \frac{c_z}{r_z^2} e_z + \frac{1}{2}\left(\frac{c_x^2}{r_x^2} + \frac{c_y^2}{r_y^2} + \frac{c_z^2}{r_z^2} - 1\right) e_\infty + \frac{1}{r_x^2} e_{0x} + \frac{1}{r_y^2} e_{0y} + \frac{1}{r_z^2} e_{0z} \tag{5}$$

These notations are different from the ones shown in [20], but already presented in [10]. Other geometric entities are derived from the ellipsoid, as shown in Table 1. In Section 3, this theory is used to develop the optimization of the ellipsoidal surfaces.

Table 1. Geometric entities that can be represented by a deformed ellipsoid.

Entity	Representation
Sphere	$S = E$ if $r_x = r_y = r_z$
Pseudocylinder	$C = \lim_{r_i \to \infty} E, i \in \{x, y, z\}$
Pair of planes	$P_p = \lim_{r_i, r_j \to \infty} E, i, j \in \{x, y, z\}$ and $i \neq j$

2.3. k-Means Algorithm

The k-means algorithm is a popular algorithm for clustering and is frequently used in unsupervised techniques. For a set of points $P = \{p_1, p_2, \cdots, p_n\}$, the k-means algorithm aims to find the partition $S = \{S_1, S_2, \cdots, S_k\}$ with $k \leq n$, as shown in (6), where c_i is the centroid of the cluster S_i.

$$S = \underset{S}{\operatorname{argmin}} \sum_{i=1}^{k} \sum_{p \in S_i} p - c_i^2 \tag{6}$$

For this paper, this optimization problem is solved using Lloyd's algorithm with a variant of initialization known as k-means++ [12].

2.4. Differential Evolution

DE is an evolutionary algorithm for multivariate functions optimization with many highly successful applications [13]. DE proposes a set of Candidate Solutions (CS) that compete for the best performance in the objective function $f(\cdot)$. In Figure 2, the basic DE scheme is described.

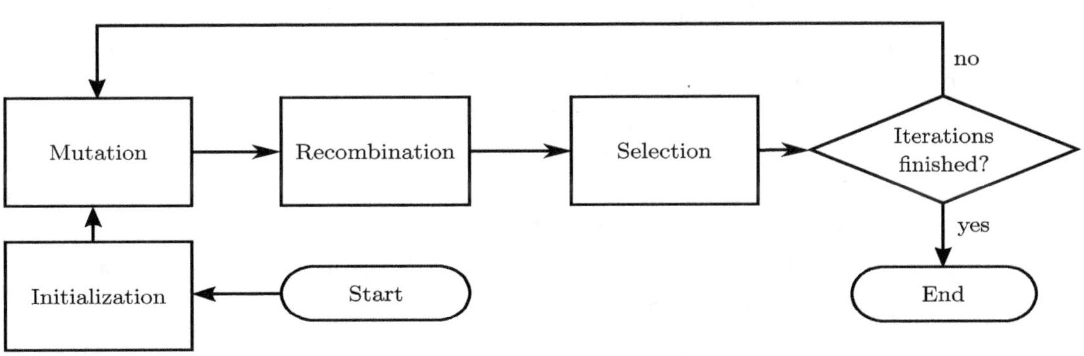

Figure 2. Differential Evolution scheme.

A CS is a vector whose elements match the variables in $f(\cdot)$. The population is randomly initialized inside predefined bounds that limit the search space. Afterwards in the mutation process, a donor vector v_i is calculated for every CS x_i, by randomly selecting three different CSs, $\{r_1, r_2, r_3\}$ and applying (7), where F is the mutation factor $F \in [0, 2]$.

$$v_i = r_1 + F(r_2 - r_3) \tag{7}$$

In the recombination process, the CS x_i is recombined with the donor vector v in every dimension j to obtain a new CS u_i as is shown in (8), where $rand() \sim U[0, 1]$, and CR is the crossover rate.

$$u_i(j) = \begin{cases} v_i(j) & \text{if } rand() \leq CR \\ x_i(j) & \text{if } rand() > CR \end{cases} \tag{8}$$

Finally, in the selection process the new CS u_i is compared to the actual CS x_i, and use (9) (for a minimization problem) to choose whether to keep the actual solution or replace it with the new one.

$$x_i = \begin{cases} u_i & \text{if } f(u_i) \leq f(x_i) \\ x_i & \text{otherwise} \end{cases} \tag{9}$$

The mutation, recombination, and selection function for a certain number of iterations is continuously performed, and ultimately, the CS with the best performance in $f(\cdot)$ is returned.

3. Ellipsoidal Surfaces Optimization

In 2017, we presented the Hyperellipsoidal Neuron (HN) [10], where the neuron represents an hyperellipsoidal decision surface. The HN is capable of deforming the decision surface into geometric entities, such as a pair of planes, and spheres and pseudocylinders. In this paper, we use the same propagation of the HN for representing the ellipsoidal surfaces.

We use the parametrization functions $\psi_1(\cdot)$ and $\psi_2(\cdot)$ defined in (10) and (11) respectively.

$$\psi_1(x) = \left[x_1, \cdots, x_n, 1, -\frac{1}{2}x_1^2, \cdots, -\frac{1}{2}x_n^2\right]^T \tag{10}$$

$$\psi_2(E) = \left[\frac{c_1}{r_1^2}, \cdots, \frac{c_n}{r_n^2}, -\frac{1}{2}\left(\frac{c_1^2}{r_1^2} + \cdots + \frac{c_n^2}{r_n^2} - 1\right), \frac{1}{r_1^2}, \cdots, \frac{1}{r_n^2}\right]^T \tag{11}$$

Note that the product $\psi_1(X)^T \psi_2(E)$ is a parameterization in $\mathbb{R}^{2n+1}$ of the the inner product in the algebra $\mathbb{G}_{2n,n}$, and by the definition of an ellipsoid in the inner product null space, it is possible to ensure that a point lies on the ellipsoid surface if $\psi_1(X)^T \psi_2(E) = 0$.

In [10] was presented a training algorithm of HNs for classification. However, the aim of this paper is not to classify points but to approximate surfaces of cloud points with ellipsoids. Hence, in what follows, the development of a new training algorithm to solve the problem of obtaining object maps of environments is presented.

Training Algorithm for 3D Mapping

Two sets of parameters, the center $\{c_x, c_y, c_z\}$ and the semi-axes $\{r_x, r_y, r_z\}$, are adapted to represent an ellipsoid. Similar to the case of RBF networks, the ellipsoid center is trained with k-means++, taking the centroid of the cluster as the center of the ellipsoid. Then, for a point cloud with n points $\{X_1, X_2, \cdots, X_n\}$, k clusters $\{S_1, S_2, \cdots, S_k\}$ and k centroids $\{c_1, c_2, \cdots, c_k\}$ for the ellipsoids are found.

For training the semiaxes, the inner product of each point and ellipsoid $\psi_1(X)^T \psi_2(E)$ is minimized, consequently the distance between the points and the ellipsoid surface is minimized. Additionally, the volume of the ellipsoid, defined as $V = \frac{4}{3}\pi r_x r_y r_z$, must be penalized to avoid trivial solutions, e.g., an ellipsoid contained all of the point cloud or a ellipsoid approximating just one point.

Finally, the fitness function for the DE algorithm in (12) is designed, where every cluster S_i, calculated by k-means++, is used for adapting the semiaxes $\{r_x, r_y, r_z\}$ of an ellipsoid with center c_i. The first term penalizes the distance from every point in the cluster to the ellipsoid surface (outside or inside) by applying the root mean square error (RMSE), and the second term penalizes the density of each cluster S_i by computing the ratio of the volume of the entity to the number of points contained in S_i, i.e., S_i.

$$(r_x, r_y, r_z) = \underset{(r_x, r_y, r_z)}{\arg\min} \quad \alpha \sqrt{\frac{1}{S_i} \sum_{X \in S_i} [\psi_1(X)^T \psi_2(E)]^2} + \frac{4(1-\alpha)}{3S_i} \pi r_x r_y r_z \tag{12}$$

The free parameter α controls how much the ellipsoid can grow; α and the parameter k control the granularity of the ellipsoidal map. The granularity refers to the number and size of the ellipsoids that represent an object.

4. Experiments

To show the capabilities of the proposed algorithm, in this section, the results of experiments are presented. For all of the following experiments, DE is used with parameter settings: $F = 1.2$ and $CR = 0.7$ and a fixed population of 10 particles and 50 iterations. The objective function has the parameter $\alpha = 0.8$, that was chosen heuristically for a good granularity balance. The point clouds are provided by an SRI-500 Laser Rangefinder from Acuity Technologies capable of scanning 800,000 points per second at distances of up to 500 feet (150 meters approximately).

4.1. Object Mapping

In experiment 1, as shown in Figure 3, the point cloud (left) is composed of 10,916 three-dimensional points and the multiellipsoidal map (right) contains 150 ellipsoids.

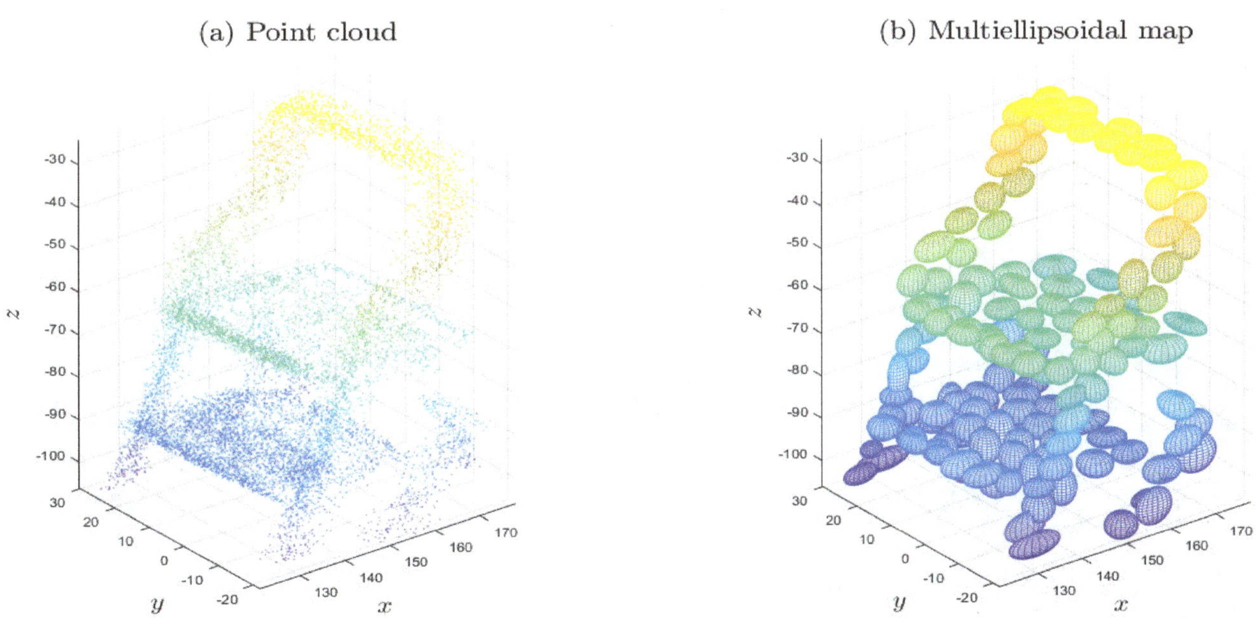

Figure 3. Experiment 1.

Similarly, in experiment 2, shown in Figure 4, a map of a tree of 31,049 points is adapted with 400 ellipsoids. Notably, the ellipsoids that are on the floor of the approximation are projected onto the 2D plane defined by the x and y axes of the figure.

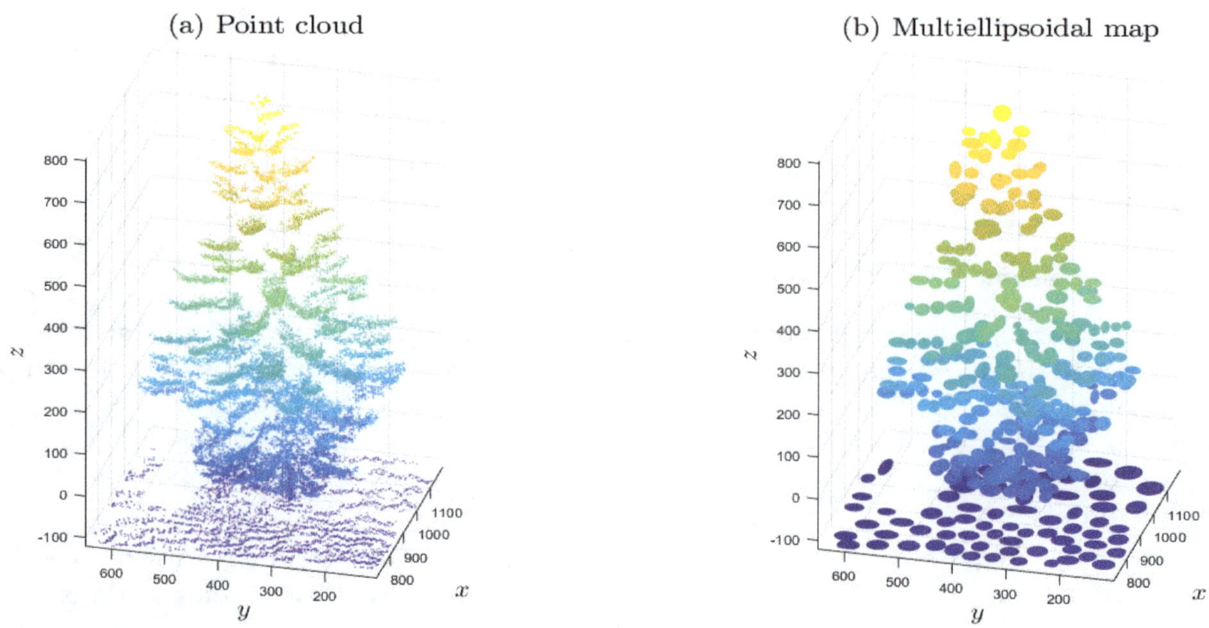

Figure 4. Experiment 2.

In Figure 5, the results of experiment 3 are shown. This is an example of a person with opened arms. As can be seen, concave areas are represented accurately by several ellipsoids. The point cloud has 40,883 points, and the multiellipsoidal map contains only 350 ellipsoids.

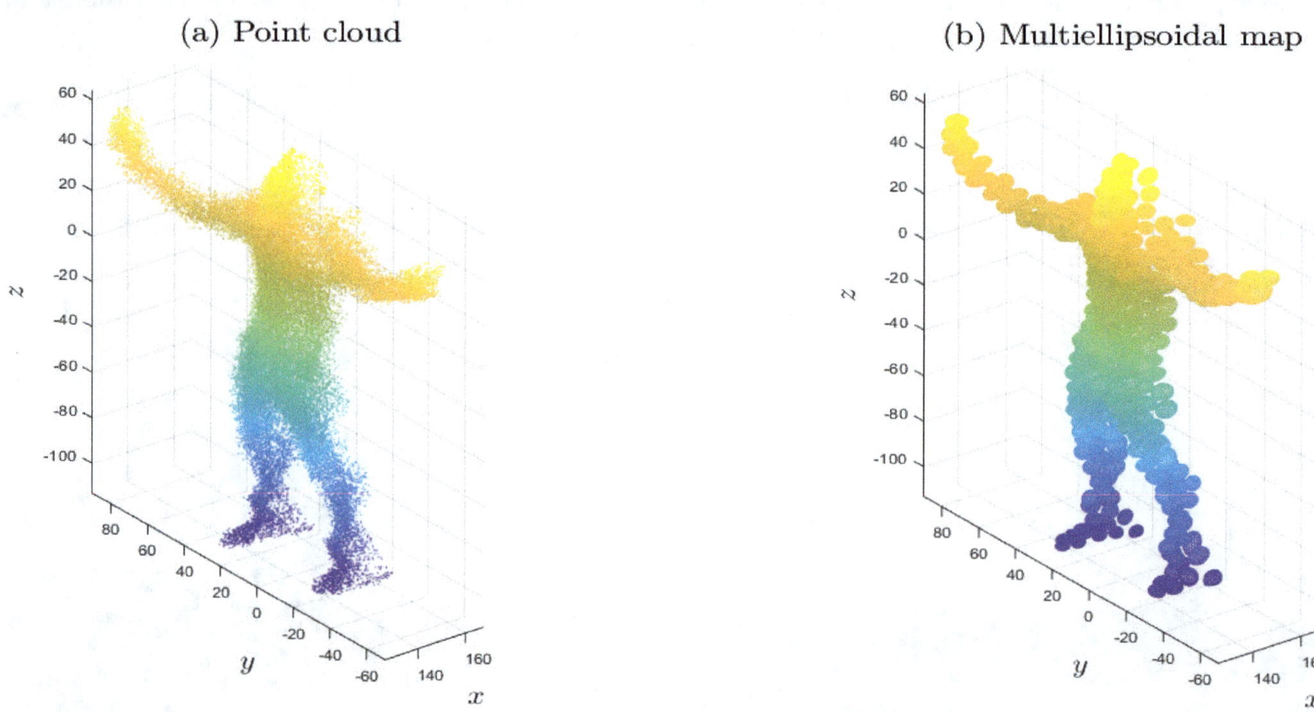

Figure 5. Experiment 3.

In experiment 4 in Figure 6, a man is standing with his back toward the observer. The point cloud is formed by 37,142 points, and the multiellipsoidal map contains 350 ellipsoids.

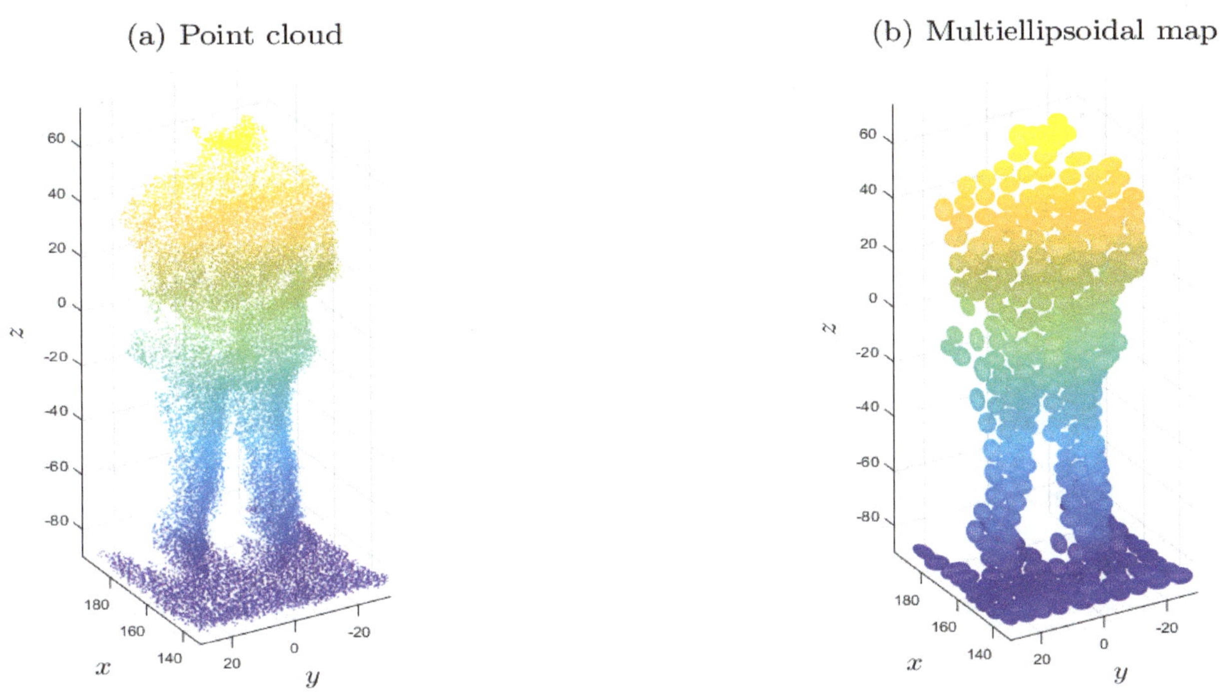

Figure 6. Experiment 4.

In Figure 7, another human figure is considered, in this case, a man with open arms sitting in a chair. The point cloud is formed by 42,272 points, and the multiellipsoidal map has 350 ellipsoids.

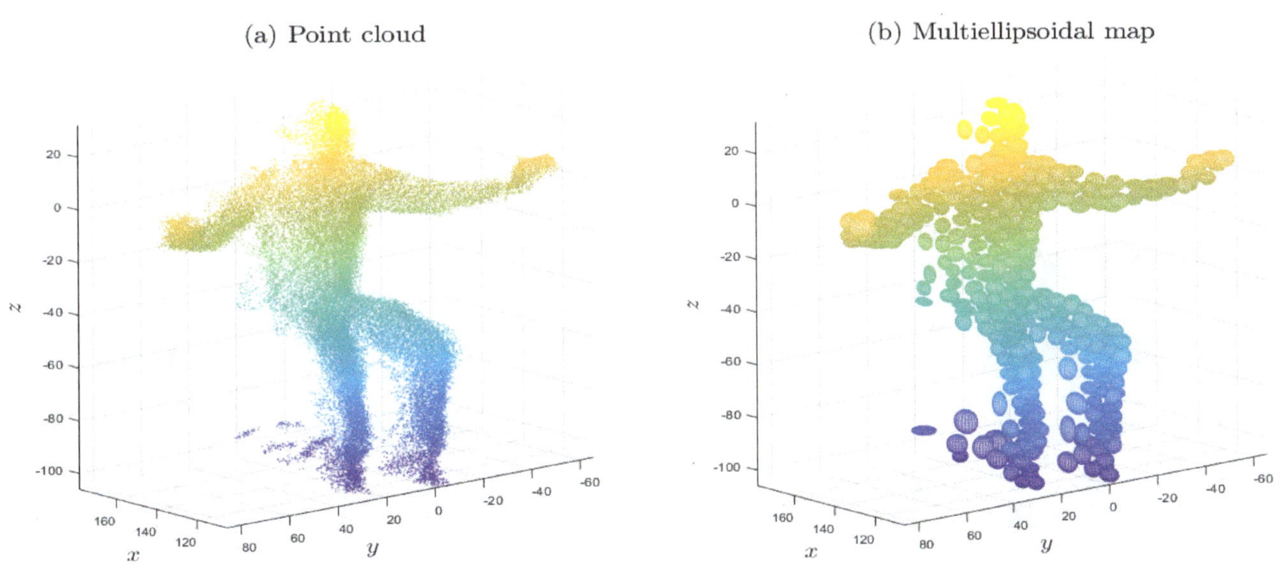

Figure 7. Experiment 5.

Summarizing these five experiments, it is observed that a good approximation and representation of the objects are obtained even when a small number of ellipsoids is used. Table 2 shows the memory cost. Consider a four-byte floating-point number; then, the memory cost of the cloud point (three floating-point numbers for each point) and of the multiellipsoidal map (six floating-point numbers for each ellipsoid) is calculated. Finally, a percentage cost of each map is shown, where it is easy to observe that an information-rich map with low memory cost has been obtained.

Table 2. Memory cost of the representation—part 1.

Experiment	Point Cloud		Multiellipsoidal Map		Percentage Cost
	Points	Bytes	Ellipsoids	Bytes	
Experiment 1	10,916	130,992	150	3600	2.7482%
Experiment 2	31,049	372,588	400	9600	2.5765%
Experiment 3	40,883	490,596	350	8400	1.7122%
Experiment 4	37,142	445,704	350	8400	1.8846%
Experiment 5	42,272	507,264	350	8400	1.6559%

4.2. Varying the Number of Ellipsoids

The percentage cost shown in Table 2 depends on the chosen number of ellipsoids (k). To show how the representation changes with k, consider experiment 6 in Figure 8, where a point cloud of an office chair is approximated with various numbers of ellipsoids. The number of ellipsoids that are shown depends on the application.

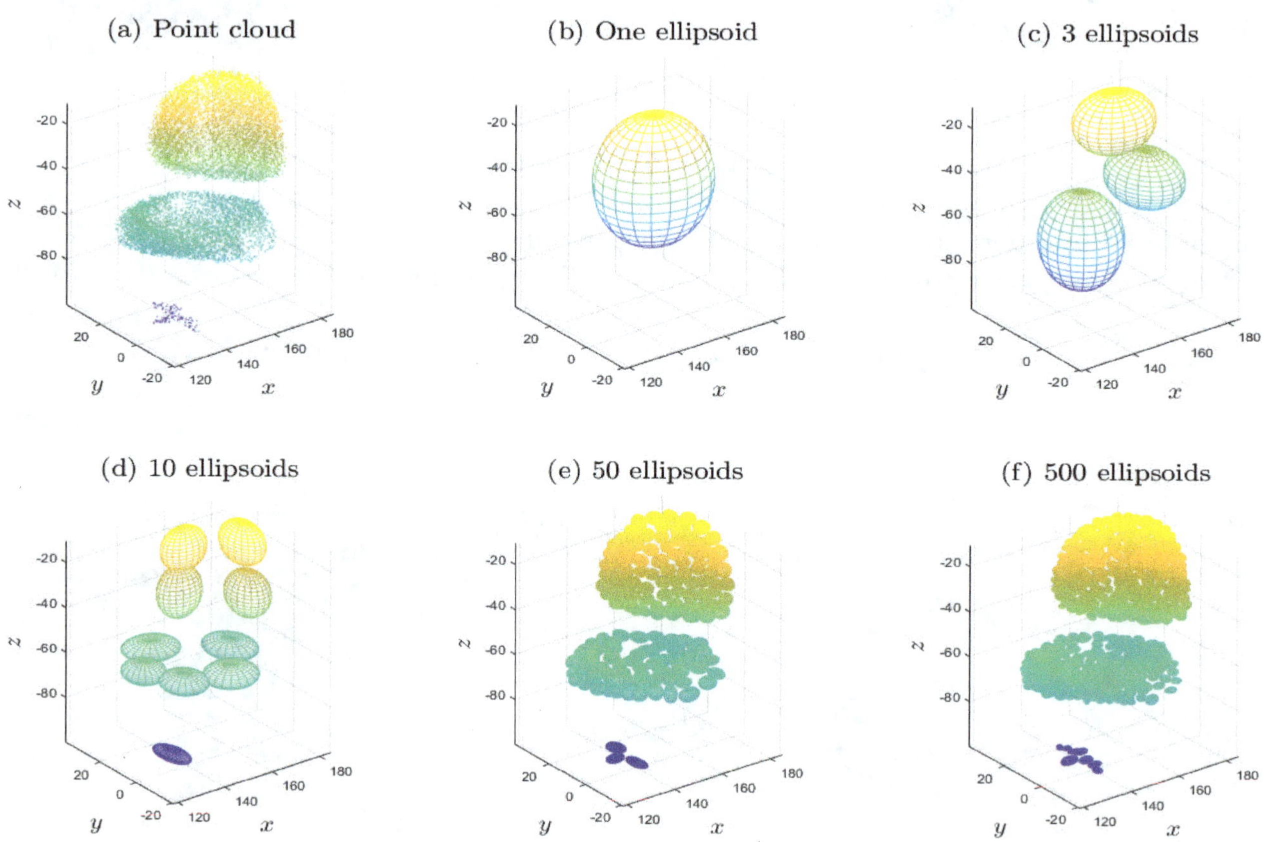

Figure 8. Experiment 6.

4.3. Ellipsoid Deformation

Table 1 presents the geometric entities that can be obtained when an ellipsoid is deformed using the GA framework. Figure 9 shows the results of experiment 7, where a trash can is approximated with two ellipsoids. We use a threshold of four meters for deforming the ellipsoids. It can be observed that an ellipsoid is deformed into a pseudocylinder for adapting all of the points of the sides of the trash can; additionally, to represent the floor, another ellipsoid is deformed into a pair of planes. This representation is very useful for robotic navigation; however, because a partition clustering technique is used, the ellipsoid size is also controlled by the number of ellipsoids. For a more general way of segmenting, it is always possible to choose to change to a hierarchical clustering algorithm which would allow us to obtain the mentioned deformations of the ellipsoids into pseudo cylinders or pair of planes.

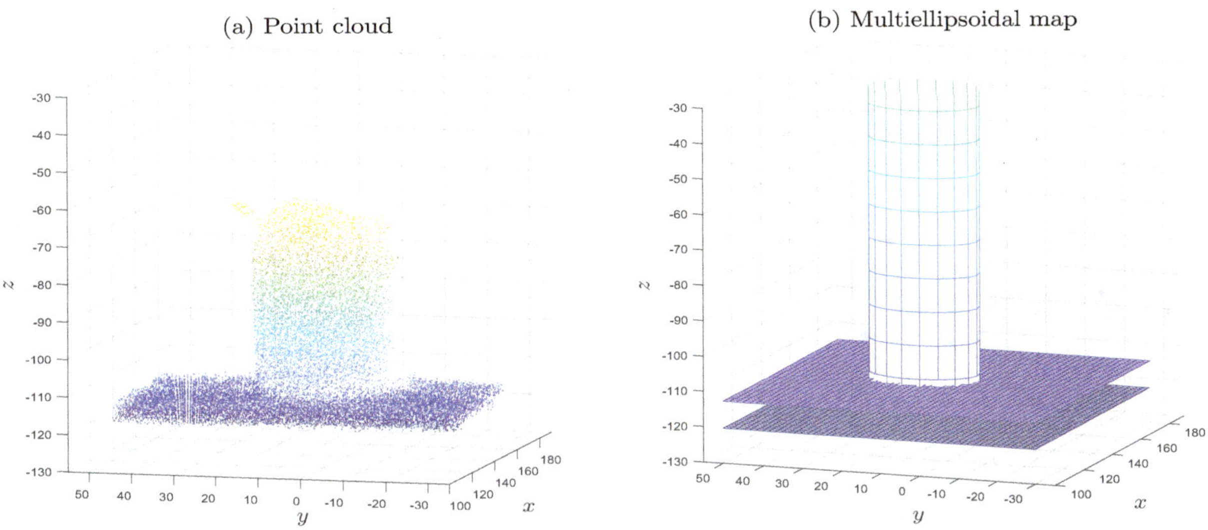

Figure 9. Experiment 7.

4.4. Environment Mapping

The multiellipsoidal mapping algorithm presented is useful for mapping not only for objects but also environments. Figures 10 and 11 presents two experiments that show the proposed algorithm's capabilities for mapping environments. In Table 3, the memory cost of representing these environments is shown.

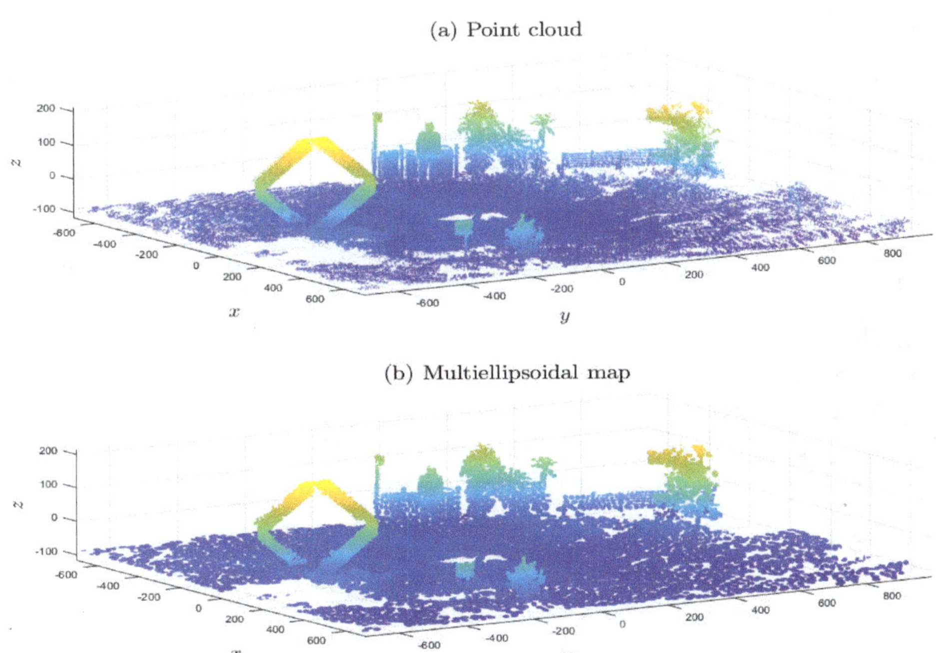

Figure 10. Experiment 8.

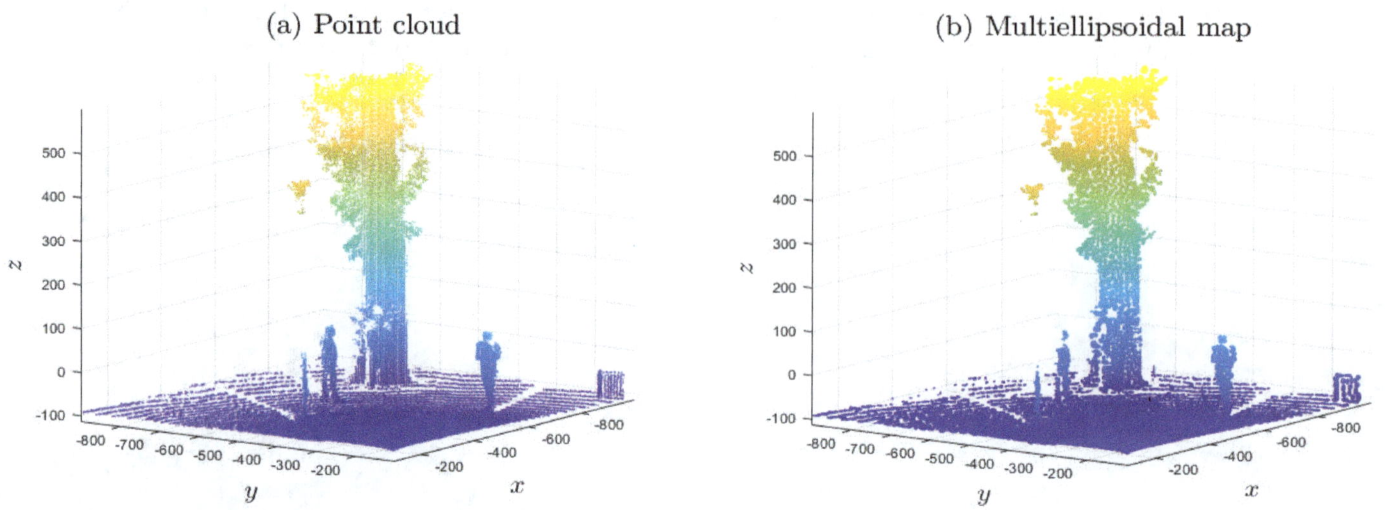

Figure 11. Experiment 9.

Table 3. Memory cost of the representation—part 2.

Experiment	Point Cloud		Multiellipsoidal Map		Percentage Cost
	Points	Bytes	Ellipsoids	Bytes	
Experiment 8	714,446	8,573,352	8880	213,120	2.4858%
Experiment 9	70,447	845,364	5920	142,080	16.8069%

4.5. Comparison to Spherical Mapping

Multiplanar mapping algorithms have been used for many applications such as urban mapping and office-like environments representation; however, these algorithms are not suitable to mapping free form objects. To solve this problem, dense multiplanar representations have been developed [5]; however, another problem then arises because, in such cases, the planes boundaries must be defined.

Spherical mapping [8,9] solves both problems by approximating the cloud point with spheres. Because the proposed approach is an extension of the spherical mapping, both approaches are empirically compared to show that ellipsoids can more accurately represent a point cloud.

Let us consider the model error for the point cloud of experiment 1. Let $x \in S_i$ represent the points in the cluster S_i and d be the closest to x three-dimensional points on the spherical or ellipsoidal surface. Then, the RMSE defined by (13) is shown, where i is the number of the cluster. Figure 12 shows the ellipsoidal and spherical approximations with $k = 100$. The spherical mapping is created in the exactly same way as ellipsoidal, except for adapting only one radius.

$$e_i = \sqrt{\frac{\sum_{x \in S_i}(d(x) - x)^2}{S_i}} \tag{13}$$

In Figure 13, we present the histograms of error of the ellipsoidal and the spherical approaches. In the left, the error e_i calculated using Equation (13) of each ellipsoid is shown. In the same way, in the right we have the calculated error for each sphere. In Table 4, their statistical measures are provided.

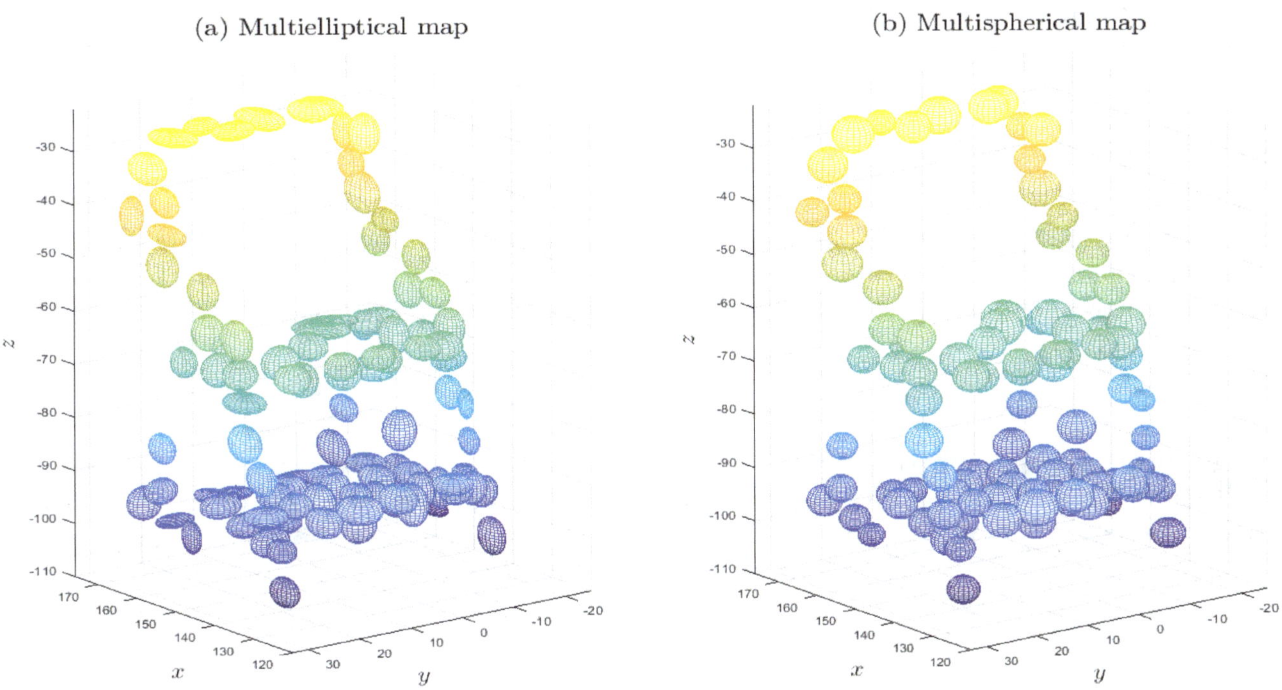

Figure 12. Maps of experiment 10.

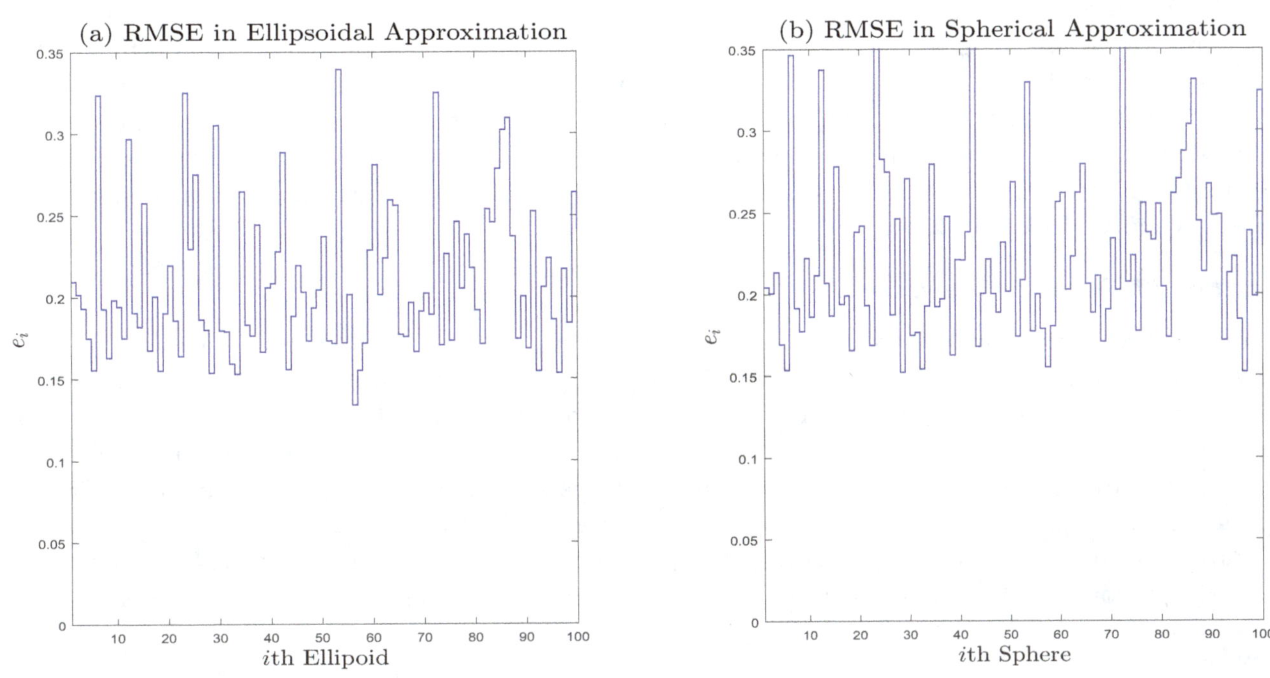

Figure 13. Errors of experiment 10.

Table 4. Errors of ellipsoidal and spherical approximations.

Approximation	Mean	STD	Maximum	Minimum	Total
Ellipsoidal	0.2079	0.0451	0.3604	0.1351	20.7936
Spherical	0.2258	0.0540	0.4647	0.1475	22.5769

Note that the multiellipsoidal mapping has a smaller error than the spherical representation due to the degrees of freedom of the ellipsoid. Hence, in this experiment, it has been empirically shown that the ellipsoidal mapping represents an improvement in the spherical mapping algorithm.

5. Discussion of Other Mapping Algorithms

In this section, the differences between mapping algorithms based on the approximation of geometric entities and those based on function approximation are discussed; in particular, the difference between the approximation of geometric entities represented in GA as multivectors [6–9] and approximations of functions such as the Gaussian approximation with RBF networks [3] or generalized distance functions [4].

5.1. *The Best Approximation*

Regarding the best approximation, the RBF is a successful technique because, in similarity with the Multi-layer Perceptron, the universal approximation theorem [21] shows that RBF can be used for approximating any free form object. However, there is no theorem that shows that planes, spheres, and ellipsoids are in fact universal approximators. Consequently, it is possible to claim that the approximation of functions will result in the best fit of the point cloud. This is a good property if the application is point cloud interpolation.

5.2. *Compactness and Heuristics*

As can be seen, function approximation results in the best approximation; however, because functions with the domain $(-\infty, \infty)$ are considered, such functions have to be bounded, and a heuristic is needed to determine the plane position (and the points transformation) where the 3D function is defined. This problem implies that more parameters for the approximation must be used. In contrast, using geometric entities of GA, self-bounded entities such as spheres and ellipsoids and infinite entities such as pseudocylinders, planes, and pairs of planes are obtained. Additionally, because all the entities belong to the same algebra and the geometric locus is defined by their null space, heuristics to fix the representation are not required. Hence, we conclude that geometric entities are useful for easy implementation and compactness. We can also argue that the best information compression can be obtained as shown in Tables 2–4.

5.3. *The Best Mathematical Framework*

For most robotic applications, obtaining the map is merely an initial step. Hence, it is important that the map be suitable for subsequent tasks. In the case of functions, it is computationally expensive to apply a rigid transformation as is needed in many applications. In contrast, geometric entities represented by multivectors are suitable for such transformations, because the same operators can be used in every entity. These features make GA suitable to pose estimation [15], movement estimation [16], navigation over rough terrain [14], computation of inverse kinematics [22], object manipulation [23], 3D reconstruction of buildings [6], and other applications in computer graphics [24]. In conclusion, the mapping algorithms based on GA offer a suitable framework for algorithms development.

6. Conclusions

In this paper, a new mapping algorithm based in $\mathbb{G}_{6,3}$ GA has been presented, capable of adapting multiples ellipsoids to obtain an object representation that is more compact than the point cloud. This multiellipsoidal mapping offers information-rich maps suitable for representation and approximation at low memory cost. Furthermore, the use of the GA framework allows us to work with an algebraic representation of ellipsoids that is capable of deforming the ellipsoids by themselves into other quadratic surfaces, such as spheres, pair of planes and pseudocylinders; this feature is valuable for robotic mapping.

As our results show, compared to other object-mapping algorithms like multiplanar mapping [5] and spherical volume registration [8,9], ellipsoids could adapt better free form objects. In the Section 5, we discussed that in contrast to mapping algorithms based in RBF [3] or other functions [4], the ellipsoid is an element of GA $\mathbb{G}_{6,3}$, and is hence easier to manipulate than functions. This property will provide a better framework for other robotic tasks such as path planning and navigation.

7. Future Work

The proposed algorithm is strongly related to the clustering algorithms that segment the cloud point to get other quadratic surfaces; a hierarchical clustering methodology is necessary. This improvement will reduce the number of geometric entities. Consequently, no k parameter should be found.

The presented algorithm can also be abstracted as a one-layer neural network consisting in HNs. The extension of this neural network could be trained for on-line capabilities (e.g., trained with Extended Kalman Filter), this could lead to a dynamic multielliptical mapping algorithm. Furthermore, the parameters k and α were selected heuristically; for a better understanding of how these parameters affect the granularity of the mapping algorithm, a new study has to be done.

Author Contributions: Conceptualization, C.V. and N.A.-D.; Formal analysis, C.V. and J.G.-A.; Funding acquisition, A.Y.A. and C.L.-F.; Investigation, C.V., N.A.-D.; Methodology, C.V.; Project administration, N.A.-D.; Software, C.V. and J.G.-A.; Supervision, N.A.-D. and C.L.-F.; Validation, J.G.-A.; Writing—original draft, C.V.; Writing—review and editing, N.A.-D. and A.Y.A.

References

1. Thrun, S. Robotic mapping: A survey. *Explor. Artif. Intell. New Millenn.* **2002**, *1*, 1–35.
2. Szeliski, R. *Computer Vision: Algorithms and Applications*; Springer Science & Business Media: Berlin, Germany, 2010.
3. Carr, J.C.; Beatson, R.K.; Cherrie, J.B.; Mitchell, T.J.; Fright, W.R.; McCallum, B.C.; Evans, T.R. Reconstruction and representation of 3D objects with radial basis functions. In Proceedings of the 28th Annual Conference on Computer Graphics and Interactive Techniques, Los Angeles, CA, USA, 12–17 August 2001; ACM: New York, NY, USA, 2001; pp. 67–76.
4. Poranne, R.; Gotsman, C.; Keren, D. *3D Surface Reconstruction Using a Generalized Distance Function. Computer Graphics Forum*; Wiley Online Library: Hoboken, NJ, USA, 2010; Volume 29, pp. 2479–2491.
5. Argiles, A.; Civera, J.; Montesano, L. Dense Multi-Planar Scene Estimation From a Sparse Set of Images. In Proceedings of the 2011 IEEE/RSJ International Conference on Intelligent Robots and Systems, San Francisco, CA, USA, 25–30 September 2011; pp. 4448–4454.
6. Bayro-Corrochano, E.; Bernal-Marin, M. Generalized Hough transform and conformal geometric algebra to detect lines and planes for building 3D maps and robot navigation. In Proceedings of the 2010 IEEE/RSJ International Conference on Intelligent Robots and Systems, Taipei, Taiwan, 18–22 October 2010; pp. 810–815.
7. López-González, G.; Arana-Daniel, N.; Bayro-Corrochano, E. *Conformal Hough Transform for 2D and 3D Cloud Points. Iberoamerican Congress on Pattern Recognition*; Springer: Berlin, Germany, 2013; pp. 73–83.
8. Rivera-Rovelo, J.; Bayro-Corrochano, E. Segmentation and volume representation based on spheres for non-rigid registration. In *International Workshop on Computer Vision for Biomedical Image Applications*; Springer: Berlin, Germany, 2005; pp. 449–458.
9. Rivera-Rovelo, J.; Bayro-Corrochano, E.; Dillmann, R. Geometric neural computing for 2d contour and 3d surface reconstruction. In *Geometric Algebra Computing*; Springer: Berlin, Germany, 2010; pp. 191–209.
10. Villaseñor, C.; Arana-Daniel, N.; Alanís, A.Y.; López-Franco, C. Hyperellipsoidal Neuron. In Proceedings of the International Joint Conference on Neural Networks (IJCNN), Anchorage, Alaska, 14–19 May 2017; pp. 788–794.

11. Hestenes, D.; Sobczyk, G. *Clifford Algebra to Geometric Calculus: A Unified Language for Mathematics and Physics*; Springer Science & Business Media: Berlin, Germany, 2012; Volume 50.
12. Arthur, D.; Vassilvitskii, S. k-means++: The advantages of careful seeding. In Proceedings of the Eighteenth Annual ACM-SIAM Symposium on Discrete Algorithms, New Orleans, LA, USA, 7–9 January 2007; Society for Industrial and Applied Mathematics: Philadelphia, PA, USA, 2007; pp. 1027–1035.
13. Das, S.; Suganthan, P.N. Differential evolution: A survey of the state-of-the-art. *IEEE Trans. Evolut. Comput.* **2011**, *15*, 4–31. [CrossRef]
14. Valencia-Murillo, R.; Arana-Daniel, N.; López-Franco, C.; Alanís, A.Y. Rough Terrain Perception Through Geometric Entities for Robot Navigation. In Proceedings of the 2nd International Conference on Advances in Computer Science and Engineering, Los Angeles, CA, USA, 1–2 July 2013.
15. Rosenhahn, B.; Sommer, G. Pose estimation in conformal geometric algebra part i: The stratification of mathematical spaces. *J. Math. Imaging Vis.* **2005**, *22*, 27–48. [CrossRef]
16. Arana-Daniel, N.; Villaseñor, C.; López-Franco, C.; Alanís, A.Y. Bio-inspired aging model-particle swarm optimization and geometric algebra for structure from motion. In *Iberoamerican Congress on Pattern Recognition*; Springer: Berlin, Germany, 2014; pp. 762–769.
17. Bayro-Corrochano, E. *Geometric Computing: For Wavelet Transforms, Robot Vision, Learning, Control and Action*; Springer Publishing Company: Berlin, Germany, 2010.
18. Dorst, L.; Fontijne, D.; Mann, S. *Geometric Algebra for Computer Science: An Object-Oriented Approach to Geometry*; Morgan Kaufmann Publishers Inc.: Burlington, MA, USA, 2009.
19. Bayro-Corrochano, E.; Reyes-Lozano, L.; Zamora-Esquivel, J. Conformal geometric algebra for robotic vision. *J. Math. Imaging Vis.* **2006**, *24*, 55–81. [CrossRef]
20. Zamora-Esquivel, J. G6,3 geometric algebra; description and implementation. *Adv. Appl. Clifford Algebras* **2014**, *24*, 493–514. [CrossRef]
21. Hornik, K. Approximation capabilities of multilayer feedforward networks. *Neural Netw.* **1991**, *4*, 251–257. [CrossRef]
22. Hildenbrand, D.; Zamora, J.; Bayro-Corrochano, E. Inverse kinematics computation in computer graphics and robotics using conformal geometric algebra. *Adv. Appl. Clifford Algebras* **2008**, *18*, 699–713. [CrossRef]
23. Hildenbrand, D.; Bayro-Corrochano, E.; Zamora, J. Advanced geometric approach for graphics and visual guided robot object manipulation. In Proceedings of the 2005 IEEE International Conference on Robotics and Automation, Barcelona, Spain, 18–22 April 2005; pp. 4727–4732.
24. Vince, J. *Geometric Algebra for Computer Graphics*; Springer Science & Business Media: Berlin, Germany, 2008.

6

Predictable Trajectory Planning of Industrial Robots with Constraints

Youdong Chen * and Ling Li

School of Mechanical Engineering and Automation, Beihang University, Beijing 100191, China; liuyunmuzi@126.com
* Correspondence: chenyd@buaa.edu.cn.

Abstract: Trajectory prediction is currently attracting considerable attention. This paper proposes geodesic trajectory planning with end-effector and joint constraints to predict the trajectory properties of the end-effector, such as velocities, accelerations, and smoothness. The prediction of the trajectory properties is independent of the joint trajectories. The prediction makes it possible to adjust the trajectory properties in line with a light computational burden. To demonstrate the effectiveness of the proposed method, experiments were conducted using the Efort robot. The experiments show that the proposed method can predict the properties of the trajectory and modify the trajectory to meet the constraints.

Keywords: predictable trajectory planning; geodesic; constrained motion

---

1. Introduction

There are growing interests in trajectory prediction. Trajectory prediction can be classified into two kinds: trajectory prediction for future time and trajectory prediction for future tasks. Future actions have been predicted for obstacle avoiding and trajectory re-planning [1,2]. Appropriate trajectories for future tasks have been predicted by using machine learning [3,4]. Fast motion planning has been obtained from experience. The transfer of previously optimized trajectories to a new situation cannot be made in the joint space. Little attention has been paid to Cartesian trajectory prediction that predicts the motion performed by the robot end-effector under joint velocity/acceleration constraints.

Many practical tasks impose constraints on robot motions, such as machining, welding, and cutting and so on. RRT (rapidly-exploring random tree)/PRM (probabilistic roadmap)-based approaches were used to plan the path with end-effector constraints [5–9]. The Cartesian positions of the end-effector at the via-points were achieved, but the motions of the end-effector between via points are not predictable, in view of the nonlinear effects introduced by the direct kinematics.

There are limits of joint positions, velocities, and accelerations. To obtain smooth motions, the joint constraints should be bounded. Joint constraints are often met by optimizing appropriate objective functions, such as keeping the joints close to their range centers [10], using the joint ranges in a weighted pseudo-inversion [11], or defining an infinity norm to be minimized at the velocity levels [12]. However, this method did not track the assigned paths. Inverse kinematics was used to enable joint velocity and acceleration not exceeding their limits [13], which was achieved by scaling the task time. However, these methods cannot meet joint constraints.

Generally, there are three questions. (1) The motions of the end-effector between via points are not predictable in trajectory planning with end-effector constraints. (2) There is no guarantee that the end-effector will track the assigned paths in trajectory planning with joint constraints. (3) End-effector constraints and joint constraints have seldom been considered simultaneously in the previous studies. In this paper, we attempt to solve these problems.

Geodesic is an important concept in Riemannian geometry and can be used in robot trajectory planning [14,15] and machine tool interpolation [16]. By selecting suitable Riemannian metrics and local coordinates, the direct output of the proposed method is joint trajectories.

In this paper, we present a predictable trajectory planning with constraints using geodesics. The end-effector position, velocities, and accelerations can be predicted without forward kinematics. The prediction of end-effector motion properties can be obtained before the calculation of the geodesic method. While tracking the constrained end-effector paths, the joint velocity/acceleration constraints can be met by scaling the motion time. The trajectory based on this method is predictable and has a light computational burden. Furthermore, the end-effector motions are of high precision.

This paper is organized as follows. The predictable trajectory planning with constraints is presented in Section 2. Using null constraints and a virtual end-effector constraint, predictable linear and circular trajectory planning are described in Section 3. Experiments using the Efort robot are illustrated in Section 4. Finally, the conclusion is given.

2. Predictable Trajectory Planning with Constraints

2.1. Predictable Trajectory Planning with End-Effector Constraints

The robot end-effector moves under some constraints to accomplish some tasks in practical applications. The end-effector position constraint H and orientation constraint F can be described by

$$\begin{cases} H(p) = 0 \\ F(n, o, a) = 0 \end{cases} \tag{1}$$

where p and $\left\{ \begin{array}{ccc} n & o & a \end{array} \right\}$ are the end-effector position and orientation respectively.

The end-effector trajectories in Cartesian space can be predicted without forward kinematics. The prediction is independent of the joint trajectories and can be made before the calculation of the proposed geodesic method. The prediction can adjust the end-effector motion properties and has light computational burdens.

First of all, the corresponding Cartesian paths meet the end-effector constraints defined in Equation (1). In Cartesian space, the end-effector moves the shortest paths that are usually not linear paths. Because the linear paths may not satisfy the end-effector constraints the shapes of the Cartesian paths are determined by the constraints.

Secondly, the Cartesian velocity is constant. Because the velocity of a geodesic is constant, the velocity always equal to the initial velocity. The velocity of a geodesic trajectory v is

$$v = \frac{L_d}{t_w} = v_0, \tag{2}$$

where L_d, t_w and v_0 are the path length, the motion time, and the initial velocity, respectively.

Because the velocity of a geodesic is constant, the end-effector performs a uniform curved motion. The tangential acceleration is zero. The centripetal acceleration a is

$$a = \frac{v^2}{r}, \tag{3}$$

where r is the radius of the curvature of a geodesic path.

Finally, the joint and Cartesian trajectories are both smooth. Because the solutions of the geodesic equations are C^3 continuous [10], the joint trajectories are smooth as are the Cartesian trajectories.

The joint trajectories are calculated by the geodesic method. The position and orientation are the functions of the joint variables,

$$\begin{aligned} p &= p(q_1 \cdots q_k) \\ n &= n(q_{k+1} \cdots q_n) \\ o &= o(q_{k+1} \cdots q_n) \\ a &= a(q_{k+1} \cdots q_n) \end{aligned} \tag{4}$$

where q_i are joint variables. By the implicit function theorem [17], there are q_{i_0} and q_{j_0} which are derived from Equation (1)

$$\begin{aligned} q_{i_0} &= q_{i_0}(q_1, \cdots q_{i_0-1}, q_{i_0+1}, \cdots q_{k_0}) \\ q_{j_0} &= q_{j_0}(q_{k_0+1}, \cdots q_{j_0-1}, q_{j_0+1}, \cdots q_n) \end{aligned} \tag{5}$$

The position and orientation vectors can be described as,

$$\begin{aligned} &p = p(q_1, \cdots q_{i_0-1}, q_{i_0+1}, \cdots q_{k_0}) \\ &\mathrm{O} \cong n + o + a = \mathrm{O}(q_{k_0+1}, \cdots q_{j_0-1}, q_{j_0+1}, \cdots q_n) \end{aligned} \tag{6}$$

In this way, the end-effector constraints are implicit in the position and orientation vector. For simplicity, they can be denoted as,

$$\begin{aligned} u_i &= \begin{cases} q_i, i \le i_0 - 1 \\ q_{i+1}, k_0 \ge i \ge i_0 + 1 \end{cases} \\ v_i &= \begin{cases} q_{k+i}, i \le j_0 - k_0 - 1 \\ q_{k+i+1}, n \ge i \ge j_0 + 1 \end{cases} \\ p &= (p_1, p_2, p_3) \\ \mathrm{O} &= (\mathrm{O}_1, \mathrm{O}_2, \mathrm{O}_3) \end{aligned} \tag{7}$$

Define Riemannian metrics as

$$\begin{aligned} {}^d g &= \sum_{\alpha,i,j} \frac{\partial p_\alpha}{\partial u_i}\frac{\partial p_\alpha}{\partial u_j} du_i du_j = \sum_{i,j} {}^d g_{ij} du_i du_j \\ {}^o g &= \sum_{\alpha,i,j} \frac{\partial \mathrm{O}_\alpha}{\partial v_i}\frac{\partial \mathrm{O}_\alpha}{\partial v_j} dv_i dv_j = \sum_{i,j} {}^o g_{ij} dv_i dv_j \end{aligned} \tag{8}$$

where du_i is the differential of u_i, and $\Sigma_{\alpha,i,j}$ is the summation notation with the index α, i, j. The quantities relevant position and orientation with superscript are d and o, respectively. The geodesic equations are

$$\begin{cases} \frac{d^2 u_i}{dt^2} + \sum\limits_{k,j} {}^d\Gamma^i_{kj} \frac{du_k}{dt}\frac{du_j}{dt} = 0 \\ \frac{d^2 v_l}{dt^2} + \sum\limits_{m,b} {}^o\Gamma^l_{bm} \frac{dv_m}{dt}\frac{dv_b}{dt} = 0 \end{cases} \tag{9}$$

The Christoffel symbols ${}^d\Gamma^i_{kj}$ and ${}^o\Gamma^l_{mb}$ are given as

$$\begin{cases} {}^d\Gamma^i_{kj} = \frac{1}{2}\sum\limits_m {}^d g^{ei} \left(\frac{\partial^d g_{ke}}{\partial u_j} + \frac{\partial^d g_{je}}{\partial u_k} - \frac{\partial^d g_{kj}}{\partial u_e} \right) \\ {}^o\Gamma^l_{bm} = \frac{1}{2}\sum\limits_f {}^o g^{fl} \left(\frac{\partial^o g_{nf}}{\partial v_m} + \frac{\partial^o g_{mf}}{\partial v_b} - \frac{\partial^o g_{bm}}{\partial v_f} \right) \end{cases} \tag{10}$$

where ${}^d g^{ei}$ and ${}^o g^{fl}$ are the elements of inverse Riemannian metric coefficient matrices ${}^d G^{-1} = \left({}^d g_{ij}\right)^{-1}$ and ${}^o G^{-1} = ({}^o g_{mn})^{-1}$ respectively [16]. ∂ is the notation of partial differential. The Christoffel symbols are derived from the position and orientation vectors with end-effector constraints. The joint trajectories are obtained by Equation (9) together with Equations (5) and (7) with boundary conditions

$$\begin{cases} u_i(t_0) = {}^0u_i \\ v_l(t_0) = {}^0v_l \\ u_i(t_0 + t_w) = {}^fu_i \\ v_l(t_0 + t_w) = {}^fv_l \end{cases} \tag{11}$$

where t_0, $\{{}^0u_i, {}^0v_l\}$ and $\left\{{}^fu_i, {}^fv_l\right\}$ are the start time, the start and end joint positions, respectively. In other words, the direct output of the geodesic method is joint trajectories.

As we can see above, trajectory planning with the end-effector constraints is given. The prediction of the end-effector motion properties can be made before the calculation of the geodesic method. The prediction is demonstrated in Section 3.1 in a simple way. The reason why the trajectories can be predicted is that they are geodesic trajectories. Geodesic has many merits, such as the shortest path, constant velocity.

2.2. *Joint Velocity/Acceleration Limits Avoidance*

The joint velocity and acceleration limits must be taken into account. Assume that the trajectory planned by the proposed method is,

$$\left\{t_w, q_i, \dot{q}_i, \ddot{q}_i (i = 1, \cdots, n)\right\} \tag{12}$$

Introducing a set of joint velocity/acceleration limits as follows,

$$\begin{cases} \left|\dot{q}_i\right| \le \dot{q}_{i,\text{lim}} \\ \left|\ddot{q}_i\right| \le \ddot{q}_{i,\text{lim}} \end{cases} \tag{13}$$

where $\dot{q}_{i,\text{lim}}$ and $\ddot{q}_{i,\text{lim}}$ are the velocity and acceleration limits of the ith joint. Define a scaling factor

$$\alpha = \max\left\{\frac{\left|\dot{q}_i\right|}{\dot{q}_{i,\text{lim}}}, \frac{\left|\ddot{q}_i\right|}{\ddot{q}_{i,\text{lim}}}, i = 1, \cdots, n\right\} \tag{14}$$

If $\alpha \ge 1$, there is at least one of the joint velocities or accelerations out of its limit. Update the motion time by the scaling factor,

$$t_{new} = \alpha t_w \tag{15}$$

The end-effector motions are predictable. The end-effector moves along the desired paths. The end-effector velocity is renewed as,

$$v_{new} = \frac{1}{\alpha} v \tag{16}$$

Along the geodesic path, the tangential acceleration is still zero. The centripetal acceleration is

$$a_{new} = \frac{{v_{new}}^2}{r} = \frac{1}{\alpha^2} a \tag{17}$$

The new joint trajectories are obtained by solving the geodesic equations, Equation (9) under the updated motion time with the boundary conditions,

$$\begin{cases} u_i(t_0) = {}^0u_i \\ v_l(t_0) = {}^0v_l \\ u_i(t_0 + t_{new}) = {}^fu_i \\ v_l(t_0 + t_{new}) = {}^fv_l \end{cases} \quad (18)$$

Actually, the joint velocity and acceleration limits avoidance is achieved by scaling the motion time. Our method can track the assigned paths, and the velocities while the accelerations can be predicted in advance.

2.3. Acceleration/Deceleration Profile

Because the geodesic is constant-speed, the planning geodesic trajectory is constant speed. The acceleration/deceleration profile should be considered at the endpoints in the practical applications.

Generally, the acceleration/deceleration profile is given as

$$\frac{d^2q_i}{dt^2} + \sum_{k,j} \Gamma^i_{kj} \frac{dq_k}{dt}\frac{dq_j}{dt} = \frac{acc}{\sqrt{\sum_j (\frac{dq_j}{dt})^2}} \frac{dq_i}{dt} \quad (19)$$

where Γ^i_{kj} and acc are the Christoffel symbols and the acceleration respectively [10]. In fact, acc is the tangential acceleration. For linear motions, the centripetal acceleration equals zero. The tangential acceleration is the acceleration acc. The acceleration of curved motions acc is actually the tangential acceleration.

3. Predictable Linear and Circular Trajectory Planning

3.1. Predictable Linear Trajectory Planning

If the constraint defined in Equation (1) is zero, the method illustrated in Section 2 degenerates into linear trajectory planning. Because the shortest path in the Cartesian space is a linear path, the Riemannian metrics ${}^dg$ and ${}^og$ become [9],

$$\begin{cases} {}^dg = (dp)^2 \\ {}^og = (dn)^2 + (do)^2 + (da)^2 \end{cases} \quad (20)$$

The geodesic equations are

$$\begin{cases} \frac{d^2q_i}{dt^2} + \sum_{k,j} {}^d\Gamma^i_{kj} \frac{dq_k}{dt}\frac{dq_j}{dt} = 0 \\ \frac{d^2q_l}{dt^2} + \sum_{m,b} {}^o\Gamma^l_{bm} \frac{dq_b}{dt}\frac{dq_n}{dt} = 0 \end{cases} \quad (21)$$

We predict that the end-effector performs linear paths with the joint trajectories derived by Equation (19). The velocity and acceleration are constant and zero respectively. The Cartesian trajectories are smooth. These predictions are validated as follows.

The joint variables $\theta_1, \cdots, \theta_{k_0}$ are regarded as local coordinates of the position space. The components of the position vector p_x, p_y and p_z can also be chosen as coordinates. The Riemannian metric defined in Equation (19) can be rewritten as

$${}^dg = (dp)^2 = (\ dp_x \quad dp_y \quad dp_z\ )I(\ dp_x \quad dp_y \quad dp_z\ )^T, \quad (22)$$

where I is a 3×3 identity matrix. The geodesic equations become

$$\begin{cases} \frac{d^2p_x}{dt^2} = 0 \\ \frac{d^2p_y}{dt^2} = 0 \\ \frac{d^2p_z}{dt^2} = 0 \end{cases} \tag{23}$$

which are equations of the linear path. Since geodesic has no relation to the coordinates, Equations (19) and (21) represent the same geodesic. The robot end-effector will move linearly under the joint trajectories determined by Equation (19). The velocity and acceleration are constant and zero respectively from Equation (21). The end-effector motion properties can be made before the calculation of the geodesic method. The Cartesian trajectories are smooth by the theory of the ordinary differential equation.

3.2. Predictable Circular Trajectory Planning

A geodesic of a spherical surface is a big circle. To obtain circular geodesic trajectory planning, we can impose a virtual end-effector constraint as

$$(p - p_0)^2 = R^2, (R > 0) \tag{24}$$

where p_0 and R are the center and the radius of a spherical surface respectively. The corresponding Riemannian metrics and geodesic equations can be obtained as in Section 2.

It is predicted that the end-effector performs circular paths with joint trajectories derived by the geodesic equations. The velocity, tangential acceleration, and centripetal acceleration are constant v, zero, and $\frac{v^2}{R}$ respectively. The Cartesian trajectories are smooth. Similar validation can be obtained.

4. Experiment

The proposed predictable trajectory planning method and joint velocity/acceleration limits avoidance schemes were validated in the following experiment. The experiment used a 6-DOF industrial robot Efort robot (ER3A-3C (HD)) produced by Anhui Efort intelligent equipment Co., Ltd. The link parameters of the Efort robot are shown in Table 1. The joint velocity limits given by the manufacture are presented in Table 2. The joint acceleration limit is 2×10^4 (rad/s^2).

Table 1. Link parameters of the Efort robot.

i	α_{i-1}	a_{i-1} (mm)	d_i(mm)	θ_i
1	0	0	0	θ_1
2	$-\pi/2$	50	0	θ_2
3	0	270	0	θ_3
4	$-\pi/2$	70	299	θ_4
5	$\pi/2$	0	0	θ_5
6	$-\pi/2$	0	0	θ_6

Table 2. The joint velocity limits of the Efort robot.

Joint	1	2	3	4	5	6
limits(rad/s)	4.01	4.01	4.36	5.59	5.59	7.33

An actual machining process is performed on the Efort robot. The coordinates and time stamps of the via points on the machined path are listed in Table 3. The paths are shown in Figure 1. The desired Cartesian velocity is 800 mm/s. To reduce the vibration, the Cartesian velocities at the corner points are set at 200 mm/s.

Before the experiment, we can predict the trajectories. The robot will move along the path shown in Figure 1 with the end-effector constraints. There will be an acceleration zone, a constant-velocity zone and a deceleration zone in a normal block. In a short block, there will be only an acceleration/deceleration zone or a constant-velocity zone. After avoiding the joint limits, the end-effector will move the desired path. The motion time and velocities will be scaled.

In the constant-velocity zones, the trajectories are generated by Equation (19). The trajectories in the acceleration and deceleration zones are generated by Equation (17). The corresponding results are shown in Figure 2. In the first acceleration zone, the velocity creeps up from 0 to 200 mm/s in 0.2 s. The velocity runs to 800 mm/s within 0.04 s. The total motion time grows from 6.18 s to 6.49 s with the joint avoidance scheme. Corresponding velocities are also changed.

The joint trajectories are shown in Figure 3. As shown in Table 4, the maximum velocity and the maximum acceleration of joint 3 are 5.06 rad/s and 2.32 $\times$ 104 rad/s2, respectively. Both of the values exceed their limits. The method of joint velocity/acceleration limits avoidance introduced in Section 2.2 is employed to adjust the joint velocity and acceleration trajectories. Actually, there are three areas where the joint velocities exceed their limits. These trajectories can be modified to comply with the velocity and acceleration constraints by scaling the motion time. Scaling the neighboring zones using the scaling factor defined in Equation (12), the motion time is updated by the scaling factor as in Equation (13). The maximum velocity and the maximum acceleration of joint 3 reduced to 4.36 rad/s and 1.85 $\times$ 104 rad/s2. The maximum joint velocities and accelerations are shown in Table 4. Table 4 shows that most of the values are reduced and all values are in the allowable range. The modified joint trajectories are shown in Figure 3. To see the details clearly, the first modified zone is taken as an example, as shown in Figure 4. In this zone, the motion time (0.48–0.76) (s) is scaled to (0.48–0.85) (s). The velocity of the constant-velocity zone is reduced from 800 mm/s to 592.57 mm/s, as shown in Figure 2. The Cartesian paths marked red in Figure 1 are the joint modified zones. The Cartesian path marked green in Figure 1 is above the plane. As is shown, the predicted Cartesian path coincides with the modified test one. The Cartesian velocity shown in Figure 2 is consistent with the predictions that are acceleration, constant velocity, and deceleration zones. It can be observed that the proposed method allows us to keep the joint velocities and accelerations below their limits while precisely tracking the desired paths. The machined workpiece is shown in Figure 5. The computational time of each interpolation point is about 1ms, which meets the real time demands. The Cartesian trajectories move as is predicted.

Table 3. The coordinates and time stamps of the via points.

	(px, py, pz) (mm)	Time (s)		(px, py, pz) (mm)	Time (s)
1	(500, −260, 100)	0	14	(310, −70, 80)	2.89
2	(500, −260, 80)	0.2	15	(310, −70, 100)	2.99
3	(500, −60, 80)	0.48	16	(355, −115, 100)	3.31
4	(300, −60, 80)	0.76	17	(355, −115, 80)	3.41
5	(300, −260, 80)	1.04	18	(355, −205, 80)	3.86
6	(500, −260, 80)	1.32	19	(310, −160, 80)	4.18
7	(500, −260, 100)	1.42	20	(310, −160, 100)	4.28
8	(490, −250, 80)	1.49	21	(445, −205, 100)	4.99
9	(490, −250, 100)	1.59	22	(445, −205, 80)	5.09
10	(490, −70, 80)	1.83	23	(400, −250, 80)	5.41
11	(310, −70, 80)	2.07	24	(355, −205, 80)	5.72
12	(490, −250, 80)	2.41	25	(490, −70, 80)	5.98
13	(310, −250, 80)	2.65	26	(490, −70, 100)	6.18

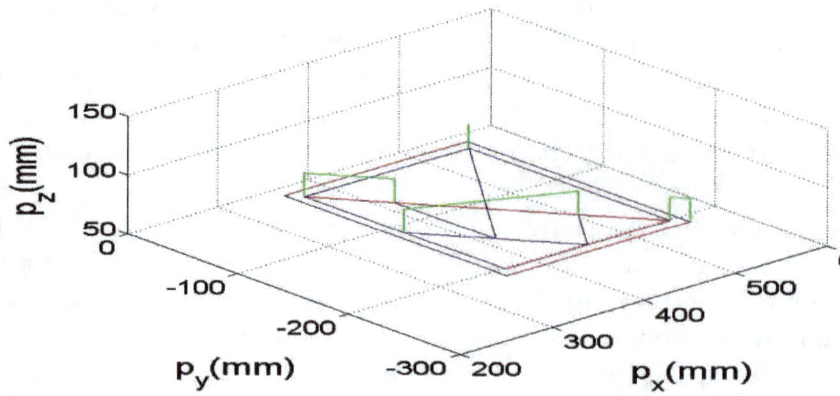

Figure 1. The Cartesian path of the Efort robot.

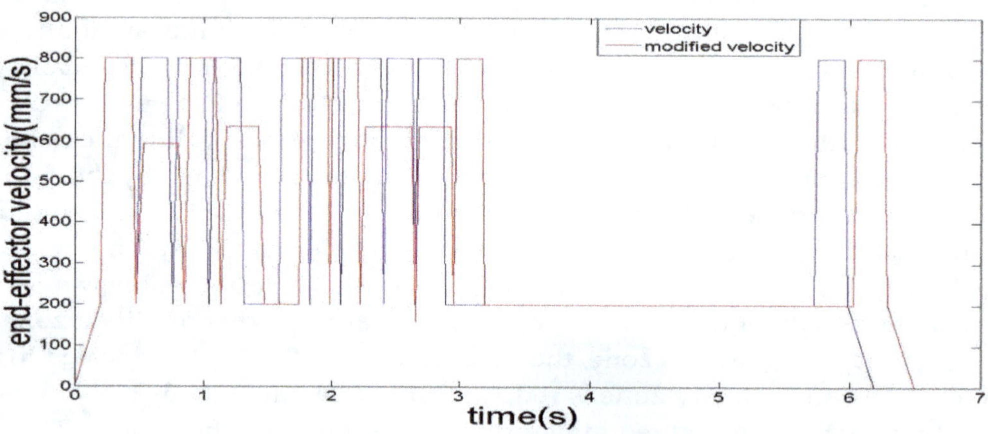

Figure 2. The Cartesian velocity of the Efort robot.

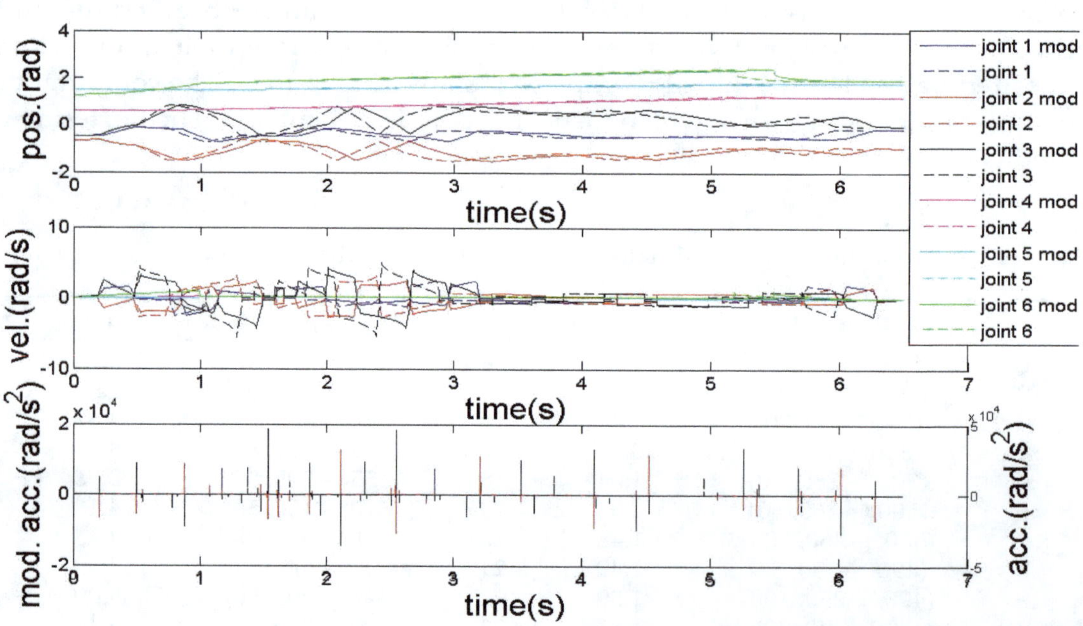

Figure 3. The joint trajectories (Here, pos., vel., acc., mod. are the abbreviation of position, velocity, acceleration and modified, respectively).

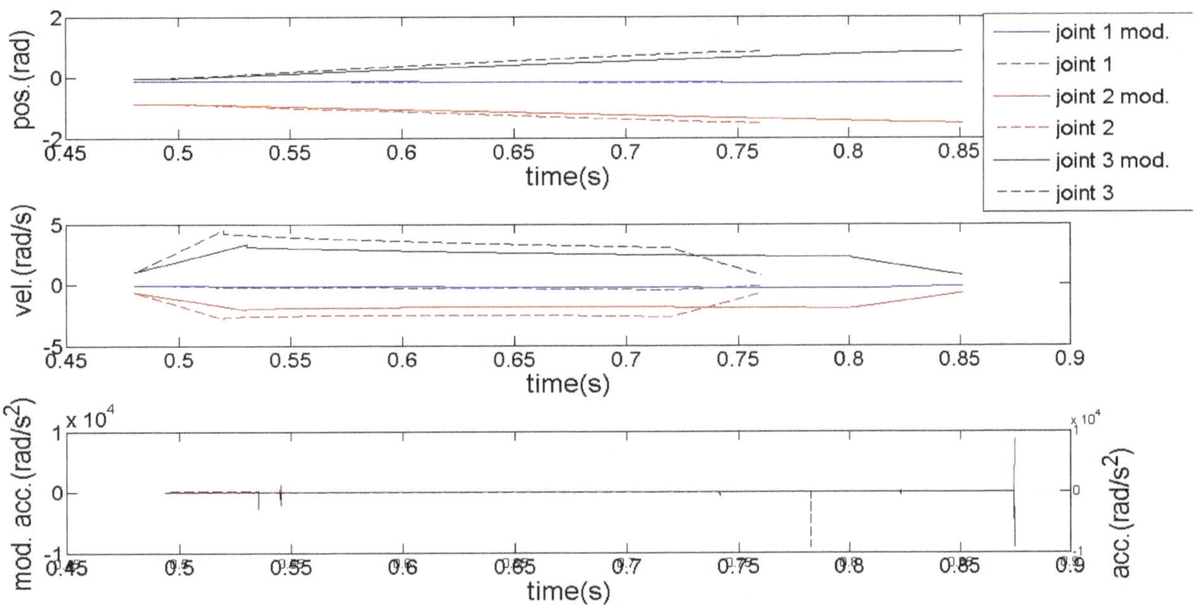

Figure 4. The local joint trajectories of first three joints (Here, pos., vel., acc., mod. are the abbreviation of position, velocity, acceleration and modified, respectively.).

Table 4. Comparisons of joint velocities and accelerations.

	Max. Vel. (rad/s)	Max. Modified Vel. (rad/s)	Max. Acc. (rad/s^2)	Max. Modified Acc. (rad/s^2)
joint 1	2.49	2.49	7.12 × 103	7.10 × 103
joint 2	3.21	2.73	1.37 × 104	1.28 × 104
joint 3	5.51	4.36	2.32 × 104	1.85 × 104
joint 4	0.22	0.22	450.47	214.25
joint 5	0.1	0.1	3.25 × 103	3.16 × 103
joint 6	0.1	0.1	1.67 × 104	1.61 × 104

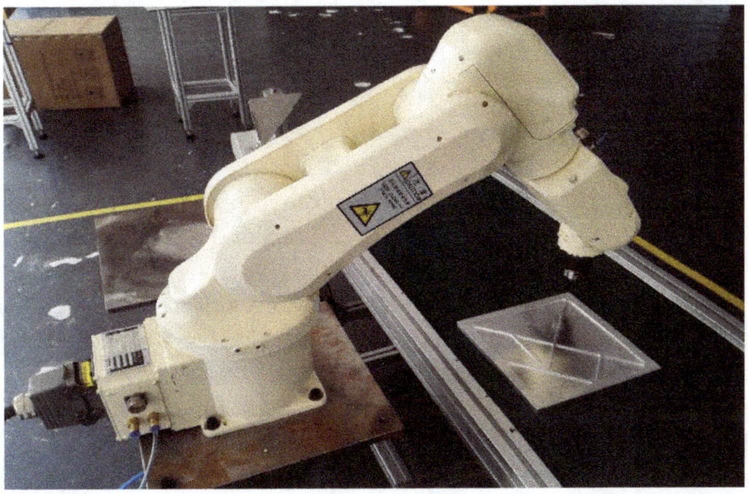

Figure 5. A machined workpiece.

5. Conclusions

In this paper, a predictable trajectory planning of industrial robots with constraints is presented. The end-effector and joint velocity/acceleration constraints are all considered. The end-effector motions, velocities, and accelerations can be predicted. The prediction is embedded in two aspects. (1) The direct output of the method is the joint trajectories while the Cartesian trajectories even between via points can be predicted. (2) The end-effector paths can be predicted with velocity and acceleration constraints. Experiments with this method using the Efort robot are given. The results show that the trajectory can be predicted and modified on-line to meet the robot constraint.

There are also some limitations. In practice, the end-effector constraints may not be described easily. In the paper, we give two specific end-effector constraints—linear and circular end-effector motions. There are two ways to deal with complicated end-effector constraints. One is to explore the equations of end-effector constraints as in Equation (1). The other is to fit complex end-effector motions by linear and circular motions.

There are also questions to be studied, such as the accuracy of the prediction. Related problems include the size, influence factor, and improvements of the accuracy of the prediction, and so on.

Author Contributions: Y.C. and L.L. conceived and designed the experiments, L.L. performed the experiments, Y.C. and L.L. analyzed the data, L.L. contributed materials/analysis tools, L.L. wrote the original draft preparation, Y.C.; wrote the review and editing, Y.C. acquired the funding.

References

1. Vendrell, E.; Mellado, M.; Crespo, A. Robot planning and re-planning using decomposition, abstraction, deduction, and prediction. *Eng. Appl. Artif. Intell.* **2001**, *14*, 505–518. [CrossRef]
2. Mora, M.C.; Tornero, J. Predictive and multirate densor-based planning under uncertainty. *IEEE Trans. Intell. Transp. Syst.* **2015**, *16*, 1493–1504. [CrossRef]
3. Jetchev, N.; Toussaint, M. Fast motion planning from experience: Trajectory prediction for speeding up movement generation. *Auton. Robots* **2013**, *34*, 111–127. [CrossRef]
4. Jetchev, N.; Toussaint, M. Trajectory prediction in cluttered voxel environments. In Proceedings of the 2010 IEEE International Conference on Robotics and Automation (ICRA), Anchorage, AK, USA, 3–7 May 2010; pp. 2523–2528.
5. Yao, Z.; Gupta, K. Path planning with general end-effector constraints. *Robot. Auton. Syst.* **2007**, *55*, 316–327. [CrossRef]
6. Jaillet, L.; Porta, J.M. Path planning under kinematic constraints by rapidly exploring manifolds. *IEEE Trans. Robot.* **2013**, *29*, 105–117. [CrossRef]
7. Berenson, D.; Srinivasa, S.S.; Ferguson, D.; Kuffner, J.J. Manipulation planning on constraint manifolds. In Proceedings of the 2010 IEEE International Conference on Robotics and Automation (ICRA), Kobe, Japan, 12–17 May 2009; pp. 625–632.
8. Suh, C.; Taewoong, T.; Kim, B.; Noh, H.; Kim, M.; Park, F.C. Tangent space RRT: A randomized planning algorithm on constraint manifolds. In Proceedings of the 2010 IEEE International Conference on Robotics and Automation (ICRA), Shanghai, China, 9–13 May 2011; pp. 4968–4973.
9. Berenson, D.; Srinivasa, S.S. Probabilistically complete planning with end-effector pose constraints. In Proceedings of the 2010 IEEE International Conference on Robotics and Automation (ICRA), Anchorage, AK, USA, 3–7 May 2010; pp. 2724–2730.
10. Samson, C.; Borgne, M.L.; Espiau, B. *Robot Control: The Task Function Approach*; Clarendon: Oxford, UK, 1991.
11. Chanand, T.; Dubey, R. A weighted least-norm solution based scheme for avoiding joint limits for redundant joint manipulators. *IEEE Trans. Robot. Autom.* **1995**, *11*, 286–292.
12. Deo, A.S.; Walker, I.D. Minimum effort inverse kinematics for redundant manipulators. *IEEE Trans. Robot. Autom.* **1997**, *13*, 767–775. [CrossRef]
13. Antonelli, G.; Chiaverini, S.; Fusco, G. A new on-line algorithm for inverse kinematics of robot manipulators ensuring path tracking capability under joint limits. *IEEE Trans. Robot. Autom.* **2003**, *19*, 162–167. [CrossRef]
14. Chen, Y.; Li, L.; Ji, X. Smooth and Accurate Trajectory Planning for Industrial Robots. *Adv. Mech. Eng.* **2014**, *6*, 342137. [CrossRef]
15. Zefran, M.; Kumar, V.; Croke, C.B. On the generation of smooth three-dimensional rigid body motions. *IEEE Trans. Robot. Autom.* **1998**, *14*, 576–589. [CrossRef]
16. Chen, Y.; Li, L. Smooth geodesic interpolation for five-axis machine tools. *IEEE/ASME Trans. Mechatron.* **2016**, *21*, 1592–1603. [CrossRef]
17. Rudin, W. *Principles of Mathematical Analysis*, 3rd ed.; McGraw-Hill Companies, Inc.: New York, NY, USA, 1976.

7

Robust Adaptive Path Following Control of an Unmanned Surface Vessel Subject to Input Saturation and Uncertainties

Yunsheng Fan *, Hongyun Huang and Yuanyuan Tan

School of Marine Electrical Engineering, Dalian Maritime University, Dalian 116026, China; iamhongyun@gmail.com (H.H.); tanyuanyuan@dlmu.edu.cn (Y.T.)

* Correspondence: yunsheng@dlmu.edu.cn

Abstract: This paper investigates the path following control problem of an unmanned surface vessel (USV) subject to input saturation and uncertainties including model parameters uncertainties and unknown time-varying external disturbances. A nonlinear robust adaptive control scheme is proposed to address the issue, more specifically, steering a USV to follow the desired path at a certain velocity assignment despite the involved disturbances, by utilizing the finite-time currents observer based line-of-sight (LOS) guidance and radial basis function neural networks (RBFNN). Backstepping and Lyapunov's direct method are the main design frameworks. Based on the finite-time currents observer and adaptive control technique, an improved LOS guidance law is proposed to obtain the desired approaching angle to the desired path, making compensations for the effects of unknown time-varying ocean currents. Then, a kinetic controller with the capability of uncertainties estimation and disturbances rejection is proposed based on the RBFNNs, where the adaptive laws including leakage terms estimate the approximation error and the unknown time-varying disturbances. Subsequently, sophisticated auxiliary control systems are employed to handle input saturation constraints of actuators. All error signals of the closed-loop system are proved to be locally uniformly ultimately bounded (UUB). Numerical simulations demonstrated the effectiveness and robustness of the proposed path following control method.

Keywords: unmanned surface vessel; path following; integral line-of-sight; finite-time currents observer; radial basis function neural networks; input saturation

---

1. Introduction

Unmanned surface vessel (USV) as an intelligent and autonomous marine equipment has received more and more attention from the control community, for broad application in the cluttered ocean environment, especially in cases where human intervention is not possible [1]. Generally speaking, three different types of control technologies play a crucial role in the development of USVs: path following control, trajectory tracking control and set-point control [2]. Many researchers propose lots of relevant control strategies and address the issues to a various extent. This study continues along the works and contributions of the predecessors. This paper aims at the path following control, mainly discussing the guidance and control of a USV. Path following is usually defined as steering a vessel to follow the desired path at a certain speed, which is not specified with temporal constraint [3]. Although there are considerable theoretical studies regarding the path following and practical engineering achievements, practical studies of the path following control for USVs have progressed haltingly amid great difficulties. It is essential to develop a highly accurate and robust path following controller for a USV when executing various vital missions. Therefore, under the circumstance of severe sea state, the safe operation and mission execution can be guaranteed.

Two aspects play crucial roles in the path following control scheme: guidance and control [4]. Guidance can refer to the popular and effective line-of-sight (LOS) guidance, refining missile guidance approach or marine guidance [5,6]. This special guidance law exploits the geometry relationships to generate a yaw angle known as the approaching angle, which is fed into the control system. In other words, the control system tracks the reference yaw angle signal together with the specified velocity tracking. Hence, the performance of the path following heavily depends on the guidance system. It turns out that classic LOS guidance is simple and effective [7], albeit with limitations in the case of being exposed to the complex ocean environment induced by waves, wind, and ocean currents. Moreover, the traditional LOS guidance will cause large cross-tracking error when the marine surface vehicles are in steady state, which strictly depends on the path curvature and the magnitude of the drift force. Therefore, the traditional LOS was extended to various forms such as integral LOS (ILOS) guidance and adaptive LOS (ALOS) guidance by Fossen. In [8], an ILOS guidance with integral action is proposed to handle the constant and irrotational ocean currents and other environmental disturbances including wind or waves. In [9], a new ILOS with time-varying lookahead distance is presented with the capability of canceling the effects of constant environmental disturbances, like constant ocean currents. In [10], the direct and indirect adaptive ILOS based path following controller is proposed to deal with the ocean currents. Another way to improve the LOS guidance is to estimate the sideslip angle caused by the external disturbance as an observer-based strategy. In [11], a novel adaptive LOS (ALOS) guidance with small computation footprint is proposed where the adaptive laws dominate the sideslip angle compensation rate. The sideslip angle is treated as an unknown constant, which significantly limits the application of ALOS. In [12], a reduced-order extended state observer (ESO) based LOS guidance is proposed to deal with the time-varying sideslip angle, which is appropriate for straight-line and curved path following. In [13], the magnitude and convergence speed of the sideslip angle is considered, which can be identified by the constructed finite-time sideslip angle observer. However, on the one hand, it should be noted that the aforementioned literature (e.g., [8–10]) merely solve the problem in the kinematic level, i.e., ignoring the along-tracking error. On the other hand, the method in [10] only adds the ocean currents in the kinematic model in terms of the relative surge and sway velocities, whereas the along-tracking error is omitted.

The dynamics control system as the execution system of a USV is another crucial constituent in the path following scheme. In general, USVs do not have an independent actuator in their sway direction. This nonholonomic constraint and underactuated nature makes the control system design much more challenging [14]. To satisfy the successful execution of missions and achieve expected performance and robustness in the full range of work space from calm ocean environment to severe ocean environment along with inaccuracy system parameters, various nonlinear control methods have been proposed [15–22]. In [15], external disturbances rejection method based on the reduced-order linear ESO is proposed. In [16], Do proposed a global path-tracking controller for underactuated ships under deterministic and stochastic loads where weak nonlinearly and strong Lyapunov design method is introduced and the estimations of disturbances are updated by projection algorithms. In [17,18], nonlinear disturbance observers are utilized to estimate the ubiquitous external disturbances in the dynamics model of marine vehicles, where the disturbance estimation errors are proved to be UUB. In [19], a nonlinear adaptive PI sliding mode tracking controller is proposed to solve the environmental disturbance problems by relaxing the assumption of knowing the upper bound of the disturbance. In [20,21], a novel adaptive switching-gain-based control method is proposed for a general uncertain Euler–Lagrange system, which is not only insensitive to the nature of uncertainties but can also alleviate the overestimation–underestimation problem. Moreover, intelligent control such as fuzzy and neural control has been applied to deal with the uncertainties and disturbances of underactuated marine surface ships. In [22], Wang proposed an adaptive online constructive fuzzy controller where the fuzzy approximator is used to approximate the unknown disturbances. Besides, robust radial basis function neural network control laws and iterative neural network control laws are proposed in [23,24], respectively, which are devoted to identifying and compensating the dynamical uncertainties and external disturbances. In addition, it should be noted that few of the aforementioned path following control schemes take into account input saturation. In fact, the practical constraint of actuators determined by the maximum forces and moments would degrade the performance of the control system or even make it unstable. Therefore, it is essential to implement the emulate of the constraint on the control laws design process for reliability and robustness.

Motivated by the observations and considerations mentioned above, a finite-time currents observer based ILOS guidance is proposed to deal with unknown time-varying ocean currents, which is suitable for any desired parametric path with high accuracy control performance. Subsequently, adaptive control laws based on the RBFNN are designed, which solve input saturation with sophisticated auxiliary systems simultaneously. A robust adaptive controller is developed to address the path following problem, which is verified to be effective via numerical simulations. The main contributions of the paper are summarized as follows.

(1) A finite-time currents observer based LOS guidance is presented to obtain the desired yaw angle and estimate the unknown time-varying ocean currents precisely, which significantly influences the performance of the control subsystem.
(2) The RBF neural networks are incorporated into the kinetic controller to solve the uncertainties, which does not require any prior knowledge of the dynamics of the USV and disturbances, and the adaptive laws are designed to estimate the compound bounds of approximating errors and external time-varying disturbances.
(3) The input constraint effect is analyzed with auxiliary systems and the states of auxiliary systems are utilized to make compensations for input saturation, which attenuates the challenge of the actuators.

The rest sections are organized as follows. Section 2 presents the preliminaries and problem formulation. Section 3 provides the design details of the guidance, the kinetic controller, and the stability analysis. Simulation results are presented and analyzed in Section 4. Finally, Section 5 concludes the paper.

2. Preliminaries and Problem Formulation

2.1. RBFNN Approximation

Consider an unknown smooth nonlinear function $f(\mathbf{x}) : \mathbb{R}^m \to \mathbb{R}$ can be approximated on a compact set $\mathbf{\Omega} \in \mathbb{R}^m$ by the following RBFNN:

$$f(\mathbf{x}) = \mathbf{W}^{*T}\boldsymbol{\varphi}(\mathbf{x}) + \varepsilon \tag{1}$$

where $\mathbf{x} \in \mathbf{\Omega}$ is the input vector, ε is the approximation error and satisfies $|\varepsilon| \leqslant \bar{\varepsilon}$ and $\bar{\varepsilon}$ is a constant, and the node number of the NN is $l > 1$. $\mathbf{W}^* \in \mathbb{R}^l$ represents the optimal weight vector, which is defined by

$$\mathbf{W}^* = \arg \min_{\hat{\mathbf{W}} \in R^l} \left\{ \sup_{\mathbf{x} \in \mathbf{\Omega}} \left| f(\mathbf{x}) - \hat{\mathbf{W}}^T\boldsymbol{\varphi}(\mathbf{x}) \right| \right\} \tag{2}$$

where $\hat{\mathbf{W}}$ is the estimation of $\mathbf{W}^*$. $\boldsymbol{\varphi}(\mathbf{x}) = [\varphi_1, \varphi_2, \ldots, \varphi_l]^T : \mathbf{\Omega} \to \mathbb{R}^l$ represents the radial function vector, the element of which is chosen as the Gaussian function:

$$\varphi_i(\mathbf{x}) = \exp\left(-\frac{(x - b_i)^T (x - b_i)}{c_i^2} \right) (i = 1, 2, \ldots, l) \tag{3}$$

where $\mathbf{b} = [b_1, b_2, \ldots, b_l]^T$ and $\mathbf{c} = [c_1, c_2, \ldots, c_l]^T$ are the centers of receptive field and spread of the Gaussian function, respectively.

2.2. USV Model

As illustrated in Figure 1, the position and orientation describe the horizontal plane motion of a USV neglecting roll, pitch, and heave, where $\{i\}$ and $\{b\}$ represent the inertial frame and body fixed frame, respectively. The desired continuous path $P_d(\theta) = [x_d(\theta), y_d(\theta)]^T$ is parameterized by a scalar variable θ and $P = [x, y]^T$ is the position coordinate. The kinematic equations of a USV can be described by relative velocities as follows [23]

$$\begin{cases} \dot{x} = u_r \cos\psi - v_r \sin\psi + V_x \\ \dot{y} = u_r \sin\psi + v_r \cos\psi + V_y \\ \dot{\psi} = r \end{cases} \tag{4}$$

where u_r and v_r are the relative surge and sway velocities; x, y, and ψ express the position and orientation in $\{i\}$; and V_x and V_y are the ocean currents represented in $\{b\}$. The dynamics of a USV is expressed as follows [25]

$$\begin{cases} \dot{u}_r = \underbrace{\frac{m_{22}}{m_{11}} v_r r - \frac{d_{11}}{m_{11}} u_r - \sum_{i=2}^{3} \frac{d_{ui}}{m_{11}} |u_r|^{i-1} u_r}_{f_u} + \frac{1}{m_{11}} \tau_u + \frac{1}{m_{11}} \tau_{wu} \\ \dot{v}_r = \underbrace{-\frac{m_{11}}{m_{22}} u_r r - \frac{d_{22}}{m_{22}} v_r - \sum_{i=2}^{3} \frac{d_{vi}}{m_{11}} |v_r|^{i-1} v_r}_{f_v} + \frac{1}{m_{22}} \tau_{wv} \\ \dot{r} = \underbrace{\frac{(m_{11} - m_{22})}{m_{33}} u_r v_r - \frac{d_{33}}{m_{33}} r - \sum_{i=2}^{3} \frac{d_{ri}}{m_{11}} |r|^{i-1} r}_{f_r} + \frac{1}{m_{33}} \tau_r + \frac{1}{m_{33}} \tau_{wr} \end{cases} \tag{5}$$

where the positive constant parameters $m_{jj} (j = 1, 2, 3)$ are the inertia including added mass; d_{ii}, d_{ui}, d_{vi}, and $d_{ri} (i = 2, 3)$ are the linear and quadratic hydrodynamic damping in surge, sway, and yaw; τ_{wu}, τ_{wv}, and τ_{wr} are the unknown time-varying environmental disturbances; and τ_u and τ_r are the available control inputs of the surge force and the yaw moment thereby viewing it as the underactuated control problem. Since the model parameters are directly related to the operation conditions [26], the parameters of the model is uncertain. Actually, due to the physical constraint, the control inputs are subject to nonlinear saturations, which is shown as follows

$$\tau_i = \begin{cases} \tau_{i\max}, & \tau_{i0} > \tau_{i\max} \\ \tau_{i0}, & \tau_{i\min} \leqslant \tau_{i0} \leqslant \tau_{i\max} \\ \tau_{i\min}, & \tau_{i0} < \tau_{i\min} \end{cases} \tag{6}$$

where $\tau_{i\min}$ and $\tau_{i\max}$ $(i = u, r)$ are the minimum and maximum control inputs produced by the actuators, referring to actual constraints of the motor's rotational speed and rudder deflection; and τ_{i0} is the command control input of the path following controller.

Assumption 1. *Assume that all states of a USV are measurable.*

Assumption 2. *The time-varying ocean currents $\boldsymbol{v} = [V_x, V_y]^T$ are assumed to be irrotational and bounded, and there exist a positive constant M, such that $\|\dot{\boldsymbol{v}}\| \leqslant M, M > 0$. The disturbances $\tau_{wi} (i = u, v, r)$ are unknown time-varying and bounded, and the first derivative of them are also bounded such that $|\dot{\tau}_{wi}| \leqslant \bar{\tau}_w$, where $\bar{\tau}_w$ is unknown constant.*

Remark 1. *Note that in (5) the off-diagonal terms of the inertia and damping are ignored. No matter a large scale surface vessel or a highly maneuverable unmanned surface vessel, these terms are relatively small than the main diagonal terms. Therefore, it is reasonable to omit these terms.*

Remark 2. *The external disturbances τ_{wu}, τ_{wv}, and τ_{wr} in Equation (5) represent the compound disturbances of the wind and wave disturbances. The ocean currents as the form of hydrodynamic terms with relative velocities are represented in Equation (4).*

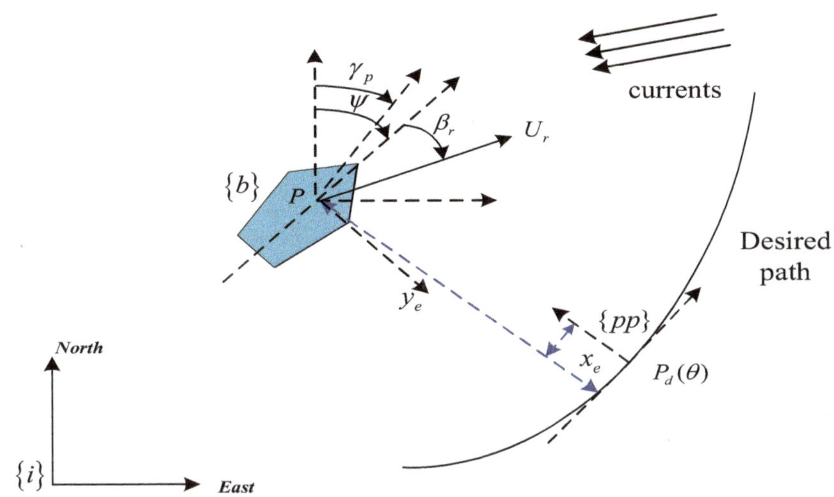

Figure 1. USV path following guidance information illustration.

2.3. Control Objective

In Figure 1, a local path parallel reference frame is denoted as $\{pp\}$. To arrive at$\{pp\}$, $\{i\}$ should be rotated an angle $\gamma_p(\theta) = \text{atan2}\,(y'_d(\theta), x'_d(\theta))$, where the notation $y'_d(\theta) = dy_d(\theta)/d\theta$ is used. Therefore, the position errors can be given as follows

$$\begin{bmatrix} x_e \\ y_e \end{bmatrix} = \underbrace{\begin{bmatrix} \cos\gamma_p & -\sin\gamma_p \\ \sin\gamma_p & \cos\gamma_p \end{bmatrix}^T}_{\mathbf{R}^{\mathbf{T}}(\gamma_p)} (P - P_d(\theta)) \tag{7}$$

where x_e and y_e are the along-tracking error and cross-tracking error, respectively, and $\mathbf{R}(\gamma_p)$ is the rotation matrix. Meanwhile, we also have

$$\dot{P}_d(\theta) = \mathbf{R}(\gamma_p)[U_{pp}, 0]^T \tag{8}$$

where $[U_{pp}, 0]^T$ is the velocity of $\{pp\}$ with respect to $\{i\}$, represented in $\{pp\}$.

The path following control problem is concerned with designing control laws to reach and then keep following the desired path. Once the path is reached, the vessel can maintain a desired surge velocity assignment of u_d. It is worth noting that a constant speed profile is frequently chosen in many cases. Therefore, the control objective is as follows:

$$\sup_{t>0}|x_e| \leqslant \delta_1, \sup_{t>0}|y_e| \leqslant \delta_2, \sup_{t>0}|u_r - u_d| \leqslant \delta_3 \tag{9}$$

where δ_1, δ_2, and δ_3 are small positive constants. Meanwhile, it is guaranteed that all error signals of the closed-loop system are locally UUB.

Assumption 3. *The desired path should be sufficiently smooth such that its first derivative $\dot{P}_d$ is bounded. In addition, the desired speed assignment u_d and its first derivative are bounded.*

3. Main Results

This section presents the design details of the path following controller to satisfy the control objective that concluded in Section 2. The whole controller consists of two parts: the modified ILOS guidance module and the kinetic controller, which is shown in Figure 2. The ILOS guidance is utilized to calculate the desired yaw angle, where the constructed finite-time currents observer can provide the fast and precise estimation of the unknown time-varying ocean currents. By incorporating the RBFNNs into the backstepping design method, the kinetic controller are developed. Finally, the stability analysis is presented to validate the feasibility of the proposed control approach.

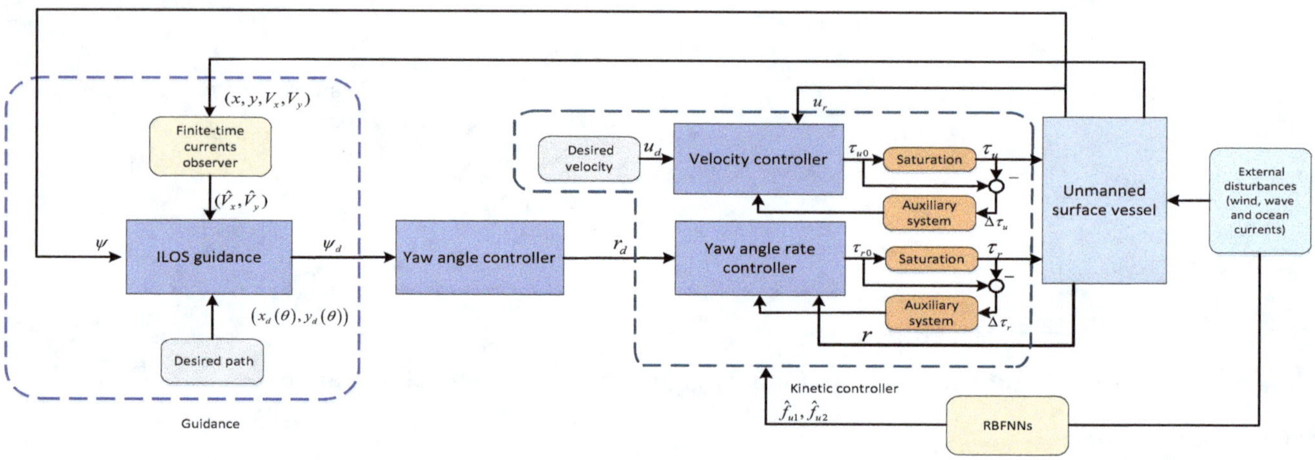

Figure 2. The structure diagram of the path following controller.

3.1. Guidance

This section is devoted to designing the ILOS guidance to calculate the approaching angle with respect to the desired path. Then, the time derivative of Equation (7) is taken to obtain the position errors dynamics

$$\begin{cases} \dot{x}_e = u_r\cos(\psi-\gamma_p) - v_r\sin(\psi-\gamma_p) + y_e\dot{\gamma}_p + \theta_x - U_{pp} \\ \dot{y}_e = u_r\sin(\psi-\gamma_p) + v_r\cos(\psi-\gamma_p) - x_e\dot{\gamma}_p + \theta_y \end{cases} \tag{10}$$

where $\theta_x = V_c\cos(\beta_c-\gamma_p)$ and $\theta_y = V_c\sin(\beta_c-\gamma_p)$; $V_c = \sqrt{V_x^2+V_y^2} \leqslant V_{c\max}$ and $\beta_c = \text{atan2}(V_y,V_x)$. By using the kinematic relationship given by Equation (8), we have $\dot{\theta} = U_{pp}/\sqrt{\dot{x}_d^2+\dot{y}_d^2}$. In addition, the virtual target velocity U_{pp} can be regarded as the extra degree-of-freedom to stabilize the cross-tracking error [5]. To prescribe the desired yaw angle ψ_d for ψ, the following form of the guidance law is taken

$$\psi_d = \gamma_p - \beta_r + \text{atan2}\left(-\left(y_e+\alpha_y\right),\Delta\right) \tag{11}$$

where $\beta_r = \text{atan2}\left(v_r,u_r\right)$; Δ is the specific look-ahead distance; and α_y is the virtual control input. It should be noted that α_y is introduced to shape the dynamics of the ILOS guidance by inherently adding an integral action.

3.1.1. Estimations of Ocean Currents

Before designing the kinematic controller, the unknown ocean currents need be identified precisely. The finite-time currents observer for V_x and V_y are designed as follows:

$$\begin{cases} \dot{\hat{\boldsymbol{\varepsilon}}}_e = \boldsymbol{\Lambda} + \mathbf{g} \\ \boldsymbol{\Lambda} = -\rho_1 \mathbf{L}^{1/2} sig^{1/2}(\hat{\boldsymbol{\varepsilon}}_e - \boldsymbol{\varepsilon}_e) + \hat{\boldsymbol{v}} \\ \dot{\hat{\boldsymbol{v}}} = -\rho_2 \mathbf{L} sig(\hat{\boldsymbol{v}} - \boldsymbol{\Lambda}) \end{cases} \tag{12}$$

where $sig^{\alpha}(\bullet) = |\bullet|^{\alpha}\text{sign}(\bullet)$; $\hat{\boldsymbol{\varepsilon}}_e = [\hat{x}, \hat{y}]^T$; $\mathbf{g} = [U_r\cos(\psi+\beta_r), U_r\sin(\psi+\beta_r)]^T$, $U_r = \sqrt{u_r^2+v_r^2}$; $\hat{\boldsymbol{v}} = [\hat{V}_x, \hat{V}_y]^T$; $\mathbf{L} = diag(l_1, l_2) > 0$; $\rho_1 > 0, \rho_2 > 0$.

Theorem 1. *Considering the proposed finite-time currents observer in Equation (12), the unknown time-varying ocean currents $\boldsymbol{v}$ can be precisely estimated within a finite time.*

Proof. According to Equation (4), we have

$$\dot{\boldsymbol{\varepsilon}}_e = \mathbf{g} + \boldsymbol{v} \tag{13}$$

Together with the finite-time currents observer, we have

$$\begin{cases} \dot{\tilde{\boldsymbol{\varepsilon}}}_e = -\rho_1 \mathbf{L}^{1/2} sig^{1/2}(\hat{\boldsymbol{\varepsilon}}_e - \boldsymbol{\varepsilon}_e) + \tilde{\boldsymbol{v}} \\ \dot{\tilde{\boldsymbol{v}}} = -\rho_2 \mathbf{L} sig(\hat{\boldsymbol{v}} - \boldsymbol{\Lambda}) - \dot{\boldsymbol{v}} \\ \quad \in -\rho_2 \mathbf{L} sig(\hat{\boldsymbol{v}} - \boldsymbol{\Lambda}) + [-M, M] \end{cases} \tag{14}$$

where $\tilde{\boldsymbol{\varepsilon}}_e = \hat{\boldsymbol{\varepsilon}}_e - \boldsymbol{\varepsilon}_e$ and $\tilde{\boldsymbol{v}} = \hat{\boldsymbol{v}} - \boldsymbol{v}$.

In light of Lemma 2 in [27], one can immediately have observer errors $\tilde{\boldsymbol{\varepsilon}}_e$ and $\tilde{\boldsymbol{v}}$ converge to zero within a finite time. It implies that there exists a finite time T_0 and finite constants ε_b and v_b such that

$$\begin{cases} \|\tilde{\boldsymbol{\varepsilon}}_e\| < \varepsilon_b, \|\tilde{\boldsymbol{v}}\| < v_b, \ \forall t < T_0 \\ \tilde{\boldsymbol{\varepsilon}}_e \equiv 0, \tilde{\boldsymbol{v}} \equiv 0, \ \forall t \geqslant T_0 \end{cases} \tag{15}$$

This concludes the proof. □

Therefore, the parametric currents estimation errors $\tilde{\theta}_x = \theta_x - \hat{\theta}_x$ and $\tilde{\theta}_y = \theta_y - \hat{\theta}_y$ satisfy the following inequality

$$\begin{cases} \tilde{\theta}_k < \sqrt{2(V_{c\max}^2 + v_b^2)} := \bar{\theta}, \forall t < T_0 \\ \tilde{\theta}_k \equiv 0, \forall t \geqslant T_0 \end{cases} \quad (k = x, y) \tag{16}$$

Remark 3. *With the availability of $\hat{V}_x$ and $\hat{V}_y$ of the finite-time currents observer, the estimations of parametric currents $\hat{\theta}_x$ and $\hat{\theta}_y$ can be reliably obtained.*

3.1.2. Design of Kinematic Controller

In this section, the kinematic control laws including the virtual control law and path variable update law are presented.

Substituting Equation (11) into Equation (10) results in

$$\begin{cases} \dot{x}_e = U_r\cos(\psi - \gamma_p + \beta_r) + \theta_x - U_{pp} + y_e\dot{\gamma}_p \\ \dot{y}_e = -\dfrac{U_r(y_e + \alpha_y)}{\sqrt{(y_e+\alpha_y)^2 + \Delta^2}} + U_r\phi(y_e, \tilde{\psi})\tilde{\psi} + \theta_y - x_e\dot{\gamma}_p \end{cases} \tag{17}$$

where $\tilde{\psi} = \psi - \psi_d$ and $\phi(y_e, \tilde{\psi}) = \frac{\sin\tilde{\psi}}{\tilde{\psi}}\frac{\Delta}{\sqrt{(y_e+\alpha_y)^2+\Delta^2}} - \frac{\cos\tilde{\psi}-1}{\tilde{\psi}}\frac{(y_e+\alpha_y)}{\sqrt{(y_e+\alpha_y)^2+\Delta^2}}$. Note that the upper bound of $\phi(y_e, \tilde{\psi})$ is 1.73.

The virtual control α_y is designed to cancel θ_y asymptotically [10]

$$\frac{U_r\alpha_y}{\sqrt{(y_e+\alpha_y)^2+\Delta^2}}=\hat{\theta}_y \tag{18}$$

Solving for α_y provides a credible solution (the negative root) as follows

$$\alpha_y=\frac{y_e(\hat{\theta}_y/U_r)^2-\hat{\theta}_y/U_r\sqrt{\Delta^2\left(1-(\hat{\theta}_y/U_r)^2\right)+y_e^2}}{1-(\hat{\theta}_y/U_r)^2} \tag{19}$$

In this case, the boundedness of $\hat{\theta}_y/U_r$ must be ensured such that $|\hat{\theta}_y/U_r| < 1$. Thus, the boundedness of the parametric ocean currents is defined such that $\theta_y < M_\theta < U_r$.

The desire target velocity U_{pp} is chosen to stabilize x_e as follows

$$U_{pp}=U_r\cos(\psi-\gamma_p+\beta_r)+k_1x_e+\hat{\theta}_x \tag{20}$$

where k_1 is a positive design parameter. Thus, the path variable update law can be obtained according to Equation (8):

$$\dot{\theta}=\frac{U_{pp}}{\sqrt{\dot{x}_d^2+\dot{y}_d^2}} \tag{21}$$

Therefore, substituting Equations (18) and (19) into Equation (17), we have

$$\begin{cases}\dot{x}_e=-k_1x_e+\tilde{\theta}_x+y_e\dot{\gamma}_p\\ \dot{y}_e=-\dfrac{U_ry_e}{\sqrt{(y_e+\alpha_y)^2+\Delta^2}}+U_r\phi(y_e,\tilde{\psi})\tilde{\psi}+\tilde{\theta}_y-x_e\dot{\gamma}_p\end{cases} \tag{22}$$

3.2. Design of Kinetic Controller

In this section, a kinetic controller is designed to track the desired approaching angle and the desired surge velocity. Let $\tilde{r}=r-r_d$ and $\tilde{u}_r=u_r-u_d$ be the attitude tracking error and surge velocity tracking error, respectively, where r_d is the virtual control law and u_d is the desired surge velocity.

According to Equation (4), the time derivative of $\tilde{\psi}$ is given by

$$\dot{\tilde{\psi}}=r-\dot{\psi}_d \tag{23}$$

To stabilize Equation (23), the desired intermediate control law for r_d is chosen as"

$$r_d=-k_2\tilde{\psi}+\dot{\psi}_d-U_r\phi y_e \tag{24}$$

where k_2 is a positive design parameter.

From Equation (5), the dynamics of $\tilde{r}$ and $\tilde{u}_r$ are given as follows

$$\begin{cases}m_{33}\dot{\tilde{r}}=m_{33}f_r+\tau_r+\tau_{wr}-m_{33}\dot{r}_d\\ m_{11}\dot{\tilde{u}}_r=m_{11}f_u+\tau_u+\tau_{wu}-m_{11}\dot{u}_d\end{cases} \tag{25}$$

Since the inertia and damping parameters are unknown, RBFNNs are employed to handle the unknown parts

$$\begin{cases}m_{33}f_r-m_{33}\dot{r}_d=\mathbf{W}_1^{*T}\boldsymbol{\varphi}(\mathbf{z})+\varepsilon_1\\ m_{11}f_u-m_{11}\dot{u}_d=\mathbf{W}_2^{*T}\boldsymbol{\varphi}(\mathbf{z})+\varepsilon_2\end{cases} \tag{26}$$

where $\mathbf{W}_i^*(i=1,2)$ is the ideal constant weight matrix satisfying $\|\mathbf{W}_i{}^*\| \leqslant W_{iM}$; $\boldsymbol{\varphi}(\mathbf{z})$ is the radial basis function with $\mathbf{z}=[u_r, v_r, r]^T$ being the input vector to the RBFNNs; and $\varepsilon_i(i=1,2)$ is the approximation error with unknown constant upper bound such that $|\varepsilon_i| \leqslant \bar{\varepsilon}_i$. Denote the unknown parts as $m_{33}f_r - m_{33}\dot{r}_d = f_{u1}$ and $m_{11}f_u - m_{11}\dot{u}_d = f_{u2}$. Furthermore, there exists bounded functions δ_1 and δ_2 such that $|\varepsilon_1| + |\tau_{wr}| \leqslant \delta_1$, and $|\varepsilon_2| + |\tau_{wu}| \leqslant \delta_2$.

Therefore, the nominal control inputs are chosen as follows by considering the input saturation

$$\begin{cases} \tau_{r0} = -k_3\tilde{r} - \tilde{\psi} - \hat{\mathbf{W}}_1\boldsymbol{\varphi}(\mathbf{z}) - \hat{\delta}_1 h(\tilde{r}) + k_{\varsigma 1}\sigma_1 \\ \tau_{u0} = -k_4\tilde{u}_r - \hat{\mathbf{W}}_2\boldsymbol{\varphi}(\mathbf{z}) - \hat{\delta}_2 h(\tilde{u}_r) + k_{\varsigma 2}\sigma_2 \end{cases} \tag{27}$$

with update laws

$$\begin{cases} \dot{\hat{\mathbf{W}}}_1 = \Gamma_1\left(\boldsymbol{\varphi}(\mathbf{z})\tilde{r} - \iota_1\hat{\mathbf{W}}_1\right) \\ \dot{\hat{\delta}}_1 = \xi_1\left(\tilde{r}h(\tilde{r}) - \lambda_1(\hat{\delta}_1 - \delta_1^0)\right) \end{cases} \tag{28}$$

$$\begin{cases} \dot{\hat{\mathbf{W}}}_2 = \Gamma_2\left(\boldsymbol{\varphi}(\mathbf{z})\tilde{u}_r - \iota_2\hat{\mathbf{W}}_2\right) \\ \dot{\hat{\delta}}_2 = \xi_2\left(\tilde{u}_r h(\tilde{u}_r) - \lambda_2(\hat{\delta}_2 - \delta_2^0)\right) \end{cases} \tag{29}$$

where k_3 and k_4 are positive design parameters; $k_{\varsigma 1}$ and $k_{\varsigma 2}$ are positive design parameters; $\hat{\delta}_i(i=1,2)$ is the estimation of δ_i; $h(\tilde{r}) = \tanh(\tilde{r}/\chi_r)$, $h(\tilde{u}_r) = \tanh(\tilde{u}_r/\chi_u)$, $\chi_j(j=r,u)$ is a positive constant; $\boldsymbol{\Gamma}_i(i=1,2)$ is a positive define design matrix; $\lambda_i(i=1,2)$ is a small design parameter; $\delta_j^0(j=r,u)$ is the prior estimation of δ_j; and $\sigma_i(i=1,2)$ is the state of the auxiliary system.

To compensate for the constraint effects of input saturation, auxiliary dynamic systems [28] are given as follows

$$\dot{\sigma}_1 = \begin{cases} -k_{\sigma_1}\sigma_1 - \dfrac{|\tilde{r}\Delta\tau_r| + 0.5\Delta\tau_r{}^2}{\sigma_1^2}\sigma_1 + \Delta\tau_r, & |\sigma_1| \geqslant \mu_1 \\ 0, & |\sigma_1| < \mu_1 \end{cases} \tag{30}$$

and

$$\dot{\sigma}_2 = \begin{cases} -k_{\sigma 2}\sigma_2 - \dfrac{|\tilde{u}_r\Delta\tau_u| + 0.5\Delta\tau_u^2}{\sigma_2^2}\sigma_2 + \Delta\tau_u, & |\sigma_2| \geqslant \mu_2 \\ 0, & |\sigma_2| < \mu_2 \end{cases} \tag{31}$$

where $k_{\sigma_i}(i=1,2)$ are positive design parameters; $\mu_i(i=1,2)$ are small positive constants; $\Delta\tau_r = \tau_r - \tau_{r0}$; $\Delta\tau_u = \tau_u - \tau_{u0}$.

Therefore, the closed-loop attitude and surge velocities tracking errors dynamics become by virtue of Equations (25) to (27)

$$\begin{cases} m_{33}\dot{\tilde{r}} = -k_3\tilde{r} - \tilde{\psi} - \tilde{\mathbf{W}}_1\boldsymbol{\varphi}(\mathbf{z}) - \tilde{\delta}_1 h(\tilde{r}) + k_{\varsigma 1}\sigma_1 + \varepsilon_1 + \tau_{wr} + \Delta\tau_r. \\ m_{11}\dot{u}_e = -k_4 u_e - \tilde{\mathbf{W}}_2\boldsymbol{\varphi}(\mathbf{z}) - \tilde{\delta}_2 h(u_e) + k_{\varsigma 2}\sigma_2 + \varepsilon_2 + \tau_{wu} + \Delta\tau_u. \end{cases} \tag{32}$$

where $\tilde{\mathbf{W}}_i = \hat{\mathbf{W}}_i - \mathbf{W}_i^*(i=1,2)$ are weight matrix estimation errors; and $\tilde{\delta}_i = \hat{\delta}_i - \delta_i(i=1,2)$ are the adaptive terms estimation errors.

3.3. Stability Analysis

In this section, the main theorem of the path following controller is presented.

Theorem 2. *Consider the USV model in Equations (4) and (5) in the presence of uncertainties and unknown external time-varying disturbances under input saturation, and suppose that Assumptions 1 and 2 are satisfied, under the guidance law in Equation (11) along with the finite-time currents observer in Equation (12). The given path is parameterized by θ with the update laws in Equation (21), and the desired velocity u_d is given as well. The control laws in Equation (27) together with the adaptive laws in Equations (28) and (29) and the auxiliary* *systems in Equations (30) and (31) are incorporated to assist in handling input saturation, guaranteeing that all tracking error signals are locally UUB.*

Proof. Consider the following Lyapunov function

$$V = \frac{1}{2}x_e^2 + \frac{1}{2}y_e^2 + \frac{1}{2}\tilde{\psi}^2 + \frac{1}{2}m_{33}\tilde{r}^2 + \frac{1}{2}m_{11}\tilde{u}_r^2 + \frac{1}{2}\sum_{i=1}^{2}\sigma_i^2 + \frac{1}{2}\sum_{i=1}^{2}\tilde{\mathbf{W}}_i^T\boldsymbol{\Gamma}_i^{-1}\tilde{\mathbf{W}}_i + \frac{1}{2\xi_i}\sum_{i=1}^{2}\tilde{\delta}_i^2 \tag{33}$$

Taking the time derivative of Equation (33) along with Equations (22)–(24) and (32) yields

$$\begin{aligned}\dot{V} \leqslant & -k_1x_e^2 - \frac{U_r}{\sqrt{(y_e+\alpha_y)^2+\Delta^2}}y_e^2 - k_2\tilde{\psi}^2 - k_3\tilde{r}^2 - \tilde{\mathbf{W}}_1\boldsymbol{\varphi}(\mathbf{z})\tilde{r} - \hat{\delta}_1h(\tilde{r})\tilde{r} \\ & + k_{\zeta1}\sigma_1\tilde{r} + \delta_1\tilde{r} + \Delta\tau_r\tilde{r} - k_4\tilde{u}_r^2 - \tilde{\mathbf{W}}_2\boldsymbol{\varphi}(\mathbf{z})\tilde{u}_r - \hat{\delta}_2h(\tilde{u}_r)\tilde{u}_r + k_{\zeta2}\sigma_2\tilde{u}_r + \delta_2\tilde{u}_r + \Delta\tau_u\tilde{u}_r \\ & + \tilde{\theta}_xx_e + \tilde{\theta}_yy_e + \sum_{i=1}^{2}\sigma_i\dot{\sigma}_i + \sum_{i=1}^{2}\tilde{\mathbf{W}}_i^T\boldsymbol{\Gamma}_i^{-1}\dot{\hat{\mathbf{W}}}_i + \frac{1}{\xi_i}\sum_{i=1}^{2}\tilde{\delta}_i\dot{\hat{\delta}}_i\end{aligned} \tag{34}$$

(1) When $|\sigma_i| \geqslant \mu_i(i=1,2)$, according to Equations (27) to (30) and Young's equalities $k_{\zeta1}\sigma_1\tilde{r} \leqslant \frac{1}{2}\tilde{r}^2 + \frac{1}{2}k_{\zeta1}^2\sigma_1^2$, $\sigma_1\Delta\tau_r \leqslant \frac{1}{2}\sigma_1^2 + \frac{1}{2}\Delta\tau_r^2$, $k_{\zeta2}\sigma_2\tilde{u}_r \leqslant \frac{1}{2}\tilde{u}_r^2 + \frac{1}{2}k_{\zeta2}^2\sigma_2^2$, $\sigma_2\Delta\tau_u \leqslant \frac{1}{2}\sigma_2^2 + \frac{1}{2}\Delta\tau_u^2$, $\tilde{\theta}_xx_e \leqslant \frac{1}{2}\bar{\theta}^2 + \frac{1}{2}x_e^2$, and $\tilde{\theta}_yy_e \leqslant \frac{k_1}{2}\bar{\theta}^2 + \frac{1}{2k_1}y_e^2$, we have

$$\begin{aligned}\dot{V} \leqslant & -\left(k_1 - \frac{1}{2}\right)x_e^2 - \left(\frac{U_r}{\sqrt{y_e^2+\Delta^2}} - \frac{1}{2k_1}\right)y_e^2 - k_2\tilde{\psi}^2 - \left(k_{\sigma1} - \frac{1}{2} - \frac{1}{2}k_{\zeta1}^2\right)\sigma_1^2 - \left(k_3 - \frac{1}{2}\right)\tilde{r}^2 \\ & - \left(k_{\sigma2} - \frac{1}{2} - \frac{1}{2}k_{\zeta2}^2\right)\sigma_2^2 - \left(k_4 - \frac{1}{2}\right)\tilde{u}_r^2 - \sum_{i=1}^{2}\iota_i\tilde{\mathbf{W}}_i^T\hat{\mathbf{W}}_i - \sum_{i=1}^{2}\lambda_i\tilde{\delta}_i\left(\hat{\delta}_i - \delta_i^0\right) \\ & + \tilde{r}\left(\delta_1 - \hat{\delta}_1h\left(\tilde{r}\right)\right) + \tilde{\delta}_1h\left(\tilde{r}\right)\tilde{r} + \tilde{u}_r\left(\delta_2 - \hat{\delta}_2h\left(\tilde{u}_r\right)\right) + \tilde{\delta}_2h\left(\tilde{u}_r\right)\tilde{u}_r + \frac{1+k_1}{2}\bar{\theta}^2\end{aligned} \tag{35}$$

Consider the following inequality of the hyperbolic tangent function holds for any $\chi > 0$ and for any $\varpi \in \mathbb{R}$ [29]

$$0 \leqslant |\varpi| - \varpi\tan\left(\frac{\varpi}{\chi}\right) \leqslant \kappa_\chi\chi \tag{36}$$

where κ_χ is a constant that satisfies $\kappa_\chi = e^{-(\kappa_\chi+1)}$, i.e., $\kappa_\chi = 0.2785$.

It is worth noting that the following equalities hold

$$-\sum_{i=1}^{2}\iota_i\tilde{\mathbf{W}}_i^T\hat{\mathbf{W}}_i \leqslant -\sum_{i=1}^{2}\frac{\iota_i}{2}\tilde{\mathbf{W}}_i^T\tilde{\mathbf{W}}_\mathbf{i} + \sum_{i=1}^{2}\frac{\iota_i}{2}\|\mathbf{W}_i^*\|^2 \tag{37}$$

$$-\sum_{i=1}^{2}\lambda_i\tilde{\delta}_i\left(\hat{\delta}_i - \delta_i^0\right) \leqslant -\sum_{i=1}^{2}\frac{\lambda_i}{2}\tilde{\delta}_i^2 + \sum_{i=1}^{2}\frac{\lambda_i}{2}\left(\delta_i - \delta_i^0\right)^2 \tag{38}$$

$$\tilde{r}\left(\delta_1 - \hat{\delta}_1\tan\left(\frac{\tilde{r}}{\chi_r}\right)\right) + \tilde{\delta}_1\tan\left(\frac{\tilde{r}}{\chi_r}\right)\tilde{r} \leqslant \delta_1\left(|\tilde{r}| - \tilde{r}\tan\left(\frac{\tilde{r}}{\chi_r}\right)\right) \leqslant 0.2785\chi_r\delta_1 \tag{39}$$

$$\tilde{u}_r\left(\delta_2 - \hat{\delta}_2\tan\left(\frac{\tilde{u}_r}{\chi_u}\right)\right) + \tilde{\delta}_2\tan\left(\frac{\tilde{u}_r}{\chi_u}\right)\tilde{u}_r \leqslant \delta_2\left(|\tilde{u}_r| - \tilde{u}_r\tan\left(\frac{\tilde{u}_r}{\chi_u}\right)\right) \leqslant 0.2785\chi_u\delta_2 \tag{40}$$

Substituting Equations (36)–(39) into Equation (35), we have

$$\begin{aligned}\dot{V} \leqslant & -\left(k_1-\frac{1}{2}\right)x_e^2-\left(\frac{U_r}{\sqrt{y_e^2+\Delta^2}}-\frac{1}{2k_1}\right)y_e^2-k_2\tilde{\psi}^2-\left(k_3-\frac{1}{2}\right)\tilde{r}^2-\left(k_4-\frac{1}{2}\right)\tilde{u}_r^2\\ & -\sum_{i=1}^{2}\left(k_{\sigma i}-\frac{1}{2}-\frac{1}{2}k_{\zeta i}^2\right)\sigma_i^2-\sum_{i=1}^{2}\frac{\iota_i}{2}\tilde{\mathbf{W}}_i^T\tilde{\mathbf{W}}_\mathbf{i}-\sum_{i=1}^{2}\frac{\lambda_i}{2}\tilde{\delta}_i^2+\sum_{i=1}^{2}\frac{1}{2}\left(\delta_i-\delta_i^0\right)^2\\ & +\sum_{i=1}^{2}\frac{\iota_i}{2}W_{iM}{}^2+0.2785\left(\chi_r\delta_1+\chi_u\delta_2\right)+\frac{1+k_1}{2}\bar{\theta}^2\\ \leqslant & -\kappa_1 V+\vartheta_1\end{aligned} \tag{41}$$

where $\kappa_1=\min\left\{2k_1-1, \frac{2U_r}{\sqrt{y_e^2+\Delta^2}}-\frac{1}{k_1}, 2k_2, \frac{2k_3-1}{m_{33}}, \frac{2k_4-1}{m_{11}}, \min_{i=1,2}\left(2k_{\sigma i}-1-k_{\zeta i}^2\right), \min_{i=1,2}\left(\iota_i\lambda_{\min}(\mathbf{\Gamma}_i)\right), \min_{i=1,2}\left(\lambda_i\xi_i\right)\right\}$, $k_1 = \max\left\{\frac{1}{2}, \frac{\sqrt{y_e^2+\Delta^2}}{2U_r}\right\}$, $\lambda_{\min}(\bullet)$ denotes the minimum eigenvalue of a matrix, and $\vartheta_1 = \sum_{i=1}^{2}\frac{1}{2}\left(\delta_i-\delta_i^0\right)^2+\sum_{i=1}^{2}\frac{\iota_i}{2}W_{iM}{}^2+0.2785\left(\chi_r\delta_1+\chi_u\delta_2\right)+\frac{1+k_1}{2}\bar{\theta}^2$.

(2) When $|\sigma_i|<\mu_i (i=1,2)$, we have $\sum_{i=1}^{2}\sigma_i\dot{\sigma}_i=0$. According to Equations (30) and (31) and the inequalities $\sum_{i=1}^{2}\frac{1}{2}k_{\zeta i}^2\sigma_i^2\leqslant-\sum_{i=1}^{2}\frac{1}{2}k_{\zeta i}^2\sigma_i^2+\sum_{i=1}^{2}k_{\zeta i}^2\mu_i^2$, $\tilde{r}\Delta\tau_r\leqslant\frac{1}{2}\tilde{r}^2+\frac{1}{2}\Delta\tau_r^2$, and $\tilde{u}_r\Delta\tau_u\leqslant\frac{1}{2}\tilde{u}_r^2+\frac{1}{2}\Delta\tau_u^2$, Equation (34) becomes

$$\begin{aligned}\dot{V} \leqslant & -\left(k_1-\frac{1}{2}\right)x_e^2-\left(\frac{U_r}{\sqrt{y_e^2+\Delta^2}}-\frac{1}{2k_1}\right)y_e^2-k_2\tilde{\psi}^2-\left(k_3-1\right)\tilde{r}^2-\left(k_4-1\right)\tilde{u}_r^2\\ & -\sum_{i=1}^{2}\frac{k_{\zeta i}^2}{2}\sigma_i^2-\sum_{i=1}^{2}\frac{\iota_i}{2}\tilde{\mathbf{W}}_i^T\tilde{\mathbf{W}}_i-\sum_{i=1}^{2}\frac{\lambda_i}{2}\tilde{\delta}_i^2+\sum_{i=1}^{2}\frac{1}{2}\left(\delta_i-\delta_i^0\right)^2+\sum_{i=1}^{2}\frac{\iota_i}{2}W_{iM}{}^2\\ & +0.2785\left(\chi_r\delta_1+\chi_u\delta_2\right)+\sum_{i=1}^{2}k_{\zeta i}^2\mu_i^2+\frac{1}{2}\Delta\tau_r^2+\frac{1}{2}\Delta\tau_u^2+\frac{1+k_1}{2}\bar{\theta}^2\\ \leqslant & -\kappa_2 V+\vartheta_2\end{aligned} \tag{42}$$

where $\kappa_2 = \min\left\{2k_1-1, \frac{2U_r}{\sqrt{y_e^2+\Delta^2}}-\frac{1}{k_1}, 2k_2, \frac{2k_3-2}{m_{33}}, \frac{2k_4-2}{m_{11}}, \min_{i=1,2}\left(k_{\zeta i}^2\right), \min_{i=1,2}\left(\iota_i\lambda_{\min}(\mathbf{\Gamma}_i)\right), \min_{i=1,2}\left(\lambda_i\xi_i\right)\right\}$, $k_1=\max\left\{\frac{1}{2}, \frac{\sqrt{y_e^2+\Delta^2}}{2U_r}\right\}$; $\vartheta_2=\sum_{i=1}^{2}\frac{1}{2}\left(\delta_i-\delta_i^0\right)^2+\sum_{i=1}^{2}\frac{\iota_i}{2}W_{iM}{}^2+0.2785\left(\chi_r\delta_1+\chi_u\delta_2\right)+\sum_{i=1}^{2}k_{\zeta i}^2\mu_i^2+\frac{1}{2}\Delta\tau_r^2+\frac{1}{2}\Delta\tau_u^2+\frac{1+k_1}{2}\bar{\theta}^2$.

Synthesizing Equations (41) and (42), we have

$$\dot{V}\leqslant-\kappa V+\vartheta \tag{43}$$

where $\kappa=\min\{\kappa_1,\kappa_2\}$ and $\vartheta=\max\{\vartheta_1,\vartheta_2\}$ with the design parameters satisfying the conditions: $k_1>\max\left(\frac{1}{2}, \frac{\sqrt{y_e^2+\Delta^2}}{2U_r}\right)$, $k_2>0, k_3>1, k_4>1, k_{\zeta 1}>0, k_{\zeta 2}>0, k_{\sigma 1}>\frac{1}{2}k_{\zeta 1}^2+\frac{1}{2}, k_{\sigma 2}>\frac{1}{2}k_{\zeta 2}^2+\frac{1}{2}$. Then, the following inequality can be obtained

$$0\leqslant V\leqslant\left(V(0)-\frac{\vartheta}{\kappa}\right)e^{-\kappa t}+\frac{\vartheta}{\kappa} \tag{44}$$

In conclusion, it follows the definition of V that the tracking error signals $\eta_e = [x_e, y_e, \tilde{\psi}, \tilde{r}, \tilde{u}_r]^T$ are locally UUB, which ultimately converges to the compact sets $\mathbf{\Omega}_1 = \left\{\eta_e\in\mathbb{R}^5 \mid \|\eta_e\|\leqslant\sqrt{2\left(V(0)-\frac{\vartheta}{\kappa}\right)e^{-\kappa t}+\frac{2\vartheta}{\kappa}}\right\}$. The ultimate compact set can be easily tuned by adjusting the design parameters. Meanwhile, the parameters estimation errors $\tilde{\mathbf{W}}_1, \tilde{\mathbf{W}}_2, \tilde{\delta}_1$, and $\tilde{\delta}_2$ are locally UUB. Theorem 2 is thus proved. □

3.4. Sway Dynamics

For the sway velocity dynamics, considering the Lyapunov function $V_v = \frac{1}{2}v_r^2$ and taking the time derivative of it based on Equation (5) yields

$$\begin{aligned} \dot{V}_v &= v_r \dot{v}_r \\ &= -\frac{m_{11}}{m_{22}} u_r r v_r - \frac{d_{22}}{m_{22}} v_r^2 - \frac{d_{v2}}{m_{11}} |v_r| v_r^2 - \frac{d_{v3}}{m_{11}} v_r^4 + \frac{1}{m_{22}} \tau_{wv} \\ &\leqslant -\chi v_r + \frac{1}{m_{22}} \bar{\tau}_w \end{aligned} \tag{45}$$

where $\chi = \frac{m_{11}}{m_{22}} u_r r$. Based on the above analyses, the boundedness of $\tilde{u}_r$ and $\tilde{r}$ is guaranteed, thus χ is bounded. It should be noted that d_{vi} and $m_{ii} (i = 2, 3)$ are positive constants. According to Krstic, M.; Kanellakopoulos, I.; Kokotovic, P. V. [30], we have

$$\begin{aligned} V_v &\leqslant V_v(t_0) e^{-\chi(t-t_0)} + \bar{\tau}_w \frac{1 - e^{-\chi(t-t_0)}}{m_{22}\chi} \\ &\leqslant V_v(t_0) + \frac{\bar{\tau}_w}{m_{22}\chi} \end{aligned} \tag{46}$$

Therefore, the boundedness of the sway velocity v_r is guaranteed.

4. Simulations

The effectiveness and robustness of the proposed path following control method were evaluated based on the platform of MATLAB. The Cybership II [31] was taken as the control object whose parameters were as follows: $m_{11} = 25.8, m_{22} = 33.8, m_{33} = 2.76, d_{11} = 0.9257, d_{22} = 2.8909, d_{33} = 0.5$. For simplicity, we ignored the off-diagonal terms of the inertia and damping. The maximum actuated force and moment were 2 N and 1.5 Nm. From the beginning of the simulations, the ocean currents with time-varying speed were given as $V_x = 0.08 \sin(0.1t)$ m/s and $V_y = 0.04 \sin(0.1t)$ m/s. The time-varying external disturbances were generated with the first-order Markov process $\dot{\tau}_{wu} + \varsigma_1 \tau_{wu} = w_1, \dot{\tau}_{wv} + \varsigma_2 \tau_{wv} = w_2, \dot{\tau}_{wr} + \varsigma_3 \tau_{wr} = w_3$, where w_i and $\varsigma_i (i = 1, 2, 3)$ are zero-mean Gaussian white noise and constants, respectively [32]. The parameters of the controller are listed in Table 1. The node numbers and widths of RBFNNs were chosen as: node number $l = 21$ and the widths $b_i = 3(i = 1, 2, \ldots, l)$. The neural active region was chosen as $[|u|, |v|, |r|] \in [[0, 2], [0, 1.5], [0, 1.5]]^T$. Performance comparisons between the proposed finite-time currents observer based ILOS guidance with adaptive RBFNN (FCONN) controller and the indirect adaptive observer based [10] ILOS guidance with adaptive RBFNN (IAONN) controller are presented in the following two control scenarios.

Table 1. Parameters of the path following controller.

Notation	Value	Natation	Value	Natation	Value	Natation	Value	Natation	Value
k_1	1	l_2	4	$\Gamma_1 ii(i = 21)$	500	ξ_2	20	χ_r	0.01
k_2	2	ρ_1	0.01	$\Gamma_2 ii(i = 21)$	50	λ_1	0.01	χ_u	1
k_3	4	ρ_2	0.03	ι_1	0.05	λ_2	0.01	Δ	2.51
k_4	5	k_{ζ_1}	1.2	ι_2	0.05	δ_1^0	0.1		
l_1	100	k_{ζ_2}	1.2	ξ_1	50	δ_1^0	0.1		

Case 1: The desired path and speed assignment were chosen as $\boldsymbol{P}_d = [\theta, \theta]^T$ and $u_d = 0.5$ m/s. The initial states were given as $[x(0), y(0), u_r(0), v_r(0), r]^T = [0 \text{ m}, 2 \text{ m}, 0.01 \text{ m/s}, 0 \text{ m/s}, 0 \text{ rad/s}]^T$.

The simulation results are shown in Figure 3a–g. Table 2 summarizes the performance indices based on the integrated absolute error (IAE) and the time integrated absolute error (ITAE), which were used to evaluate the transient performance and steady-state performance. The proposed FCONN control method could drive the USV following the desired path with a high-precision and better transient process (Figure 3a). Their detailed distinction is more clearly shown in Figure 3b,c, specifically the smaller along- and cross-tracking errors and the smaller heading and surge velocity tracking errors. Meanwhile, the lower IAE and ITAE metrical values of the cross-tracking error revealed the better transient and steady-state performance.

Figure 3d shows that the proposed finite-time currents observer could identify the time-varying currents accurately, whereas the IAONN control scheme had an obvious oscillation during the transient process, as well as lower accuracy in the steady state.

Figure 3d demonstrates that the RBFNNs could capture the unknown dynamical uncertainties precisely and Figure 3e presents the compound bounds and their estimation, which played decisive roles in driving the dynamics state r and u_r to their real value. Figure 3f depicts the profile of the control inputs where the input saturation (IS) problem Was effectively compensated by the auxiliary system. The control inputs of the proposed method were in the specified region.

Hence, in the case, the proposed FCONN path following control method was more effective and robust according to these simulation results.

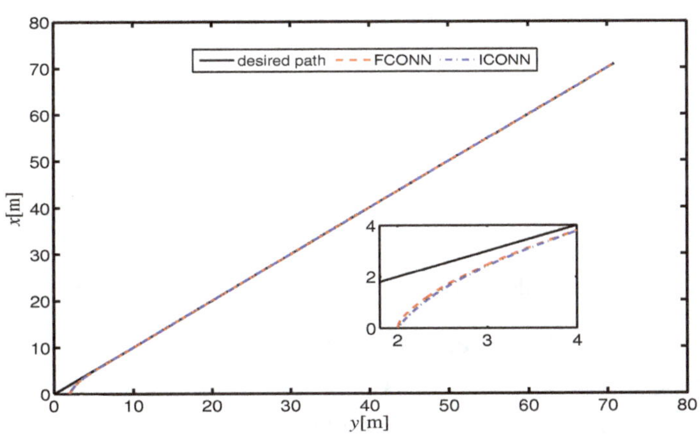

(a) line path following.

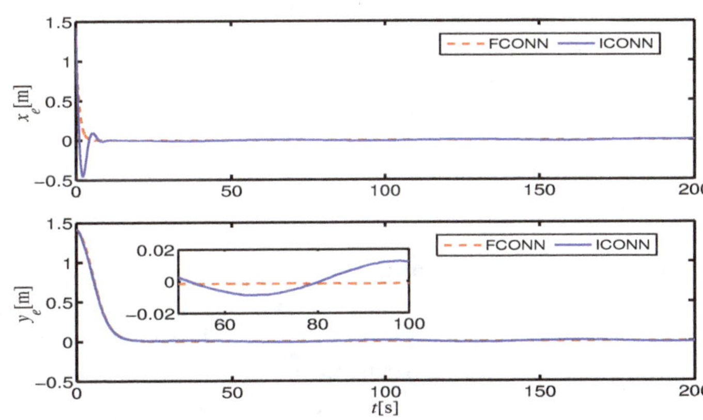

(b) Along- and cross-tracking error in Case 1.

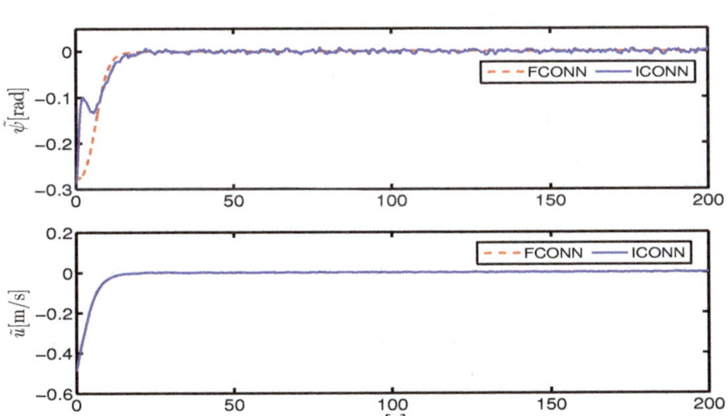

(c) Profile of the heading and surge velocity tracking in Case 1.

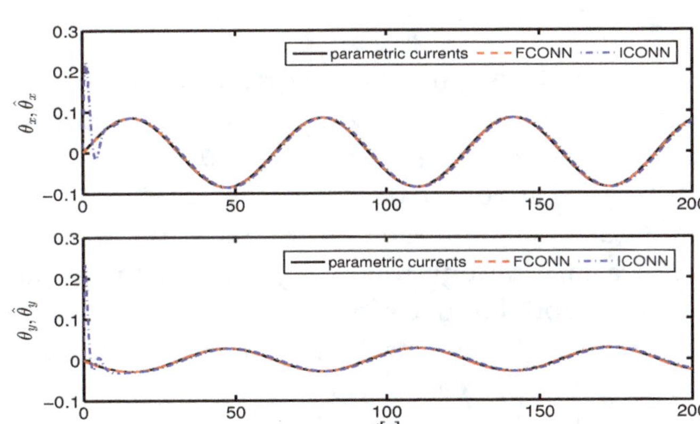

(d) Estimations of parametric currents in Case 1.

Figure 3. *Cont.*

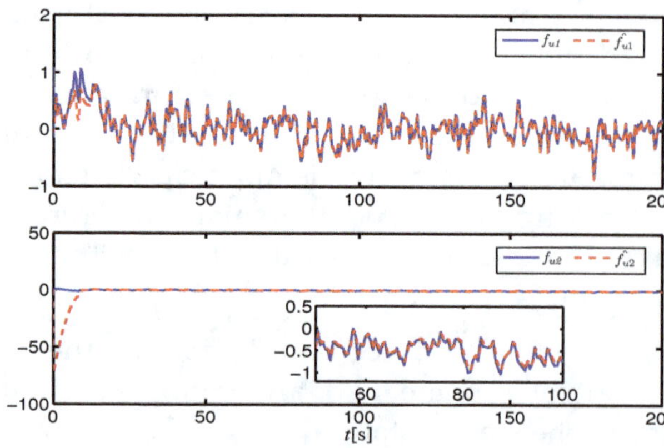

(e) Estimations of dynamical uncertainties using RBFNN in Case 1.

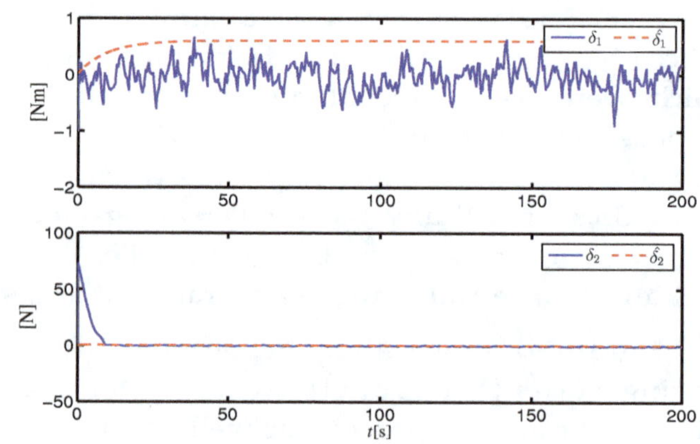

(f) Estimations of compound bound in Case 1.

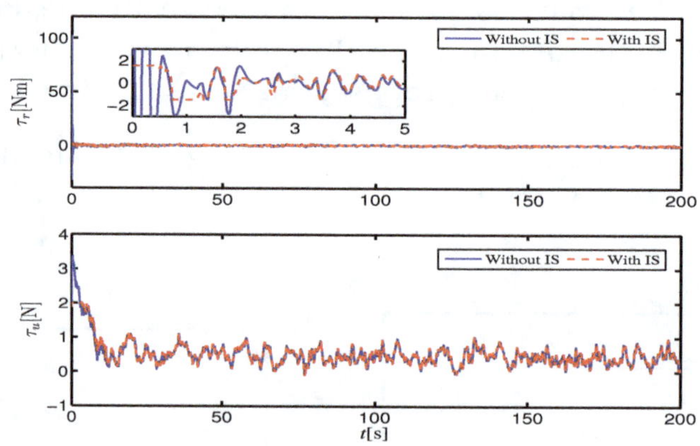

(g) Profile of the control inputs in Case 1.

Figure 3. Line path following results.

Table 2. Performance indices in these two path following scenarios.

Control Law	Line Path Following		Curvilinear Path Following	
	IAE($\cdot10^3$) $\int_0^t \lvert y_e \rvert d\tau$	ITAE($\cdot10^3$) $\int_0^t t \lvert y_e \rvert d\tau$	IAE($\cdot10^4$) $\int_0^t \lvert y_e \rvert d\tau$	ITAE($\cdot10^4$) $\int_0^t t \lvert y_e \rvert d\tau$
FCONN	1.92	5.78	2.58	9.01
IAONN	4.18	13.12	7.32	20.32

Case 2: Similar to Case 1, another comparison is presented to verify the performance in the case of following a curvilinear path with the same design parameters. The desired path and speed assignment were chosen as $\boldsymbol{P}_d = \theta, [10sin(0.1\theta)]^T$ and $u_d = 0.5$ m/s. The initial states were given as $[x(0), y(0), u_r(0), v_r(0), r]^T = [0\text{ m}, 2\text{ m}, 0.01\text{ m/s}, 0\text{ m/s}, 0\text{ rad/s}]^T$. The simulations results are shown in Figure 4a–g and the performance quantification indices are summarized in Table 2. As illustrated in Figure 4a, although following the curvilinear path, the proposed FCONN method behaved almost the same in both control scenarios. As shown in Figure 4b, the along- and cross-tracking errors oscillated to varying degrees for the poor performance of IAONN, whereas the position errors of FCONN could smoothly and steadily converge to a small neighborhood around zero within a short time. Moreover, the smaller IAE and ITAE metrical values verified it. Figure 4c shows the slight oscillation of the heading and surge velocity tracking error of IAONN. Moreover, in Figure 4d, the poor performance of estimating ocean currents of IAONN undoubtedly degraded the tracking performance. Figure 4e,f shows the exceptional performance of the dynamical uncertainties estimation and disturbance rejection

of the FCONN method. In addition, the control inputs were in the specified region for introducing the auxiliary system, as depicted in Figure 4g. Overall, the proposed control method achieved satisfactory performance and robustness in both cases with fast and accurate estimations of the unknown time-varying ocean currents and dynamical uncertainties, and satisfactory rejection of external disturbances.

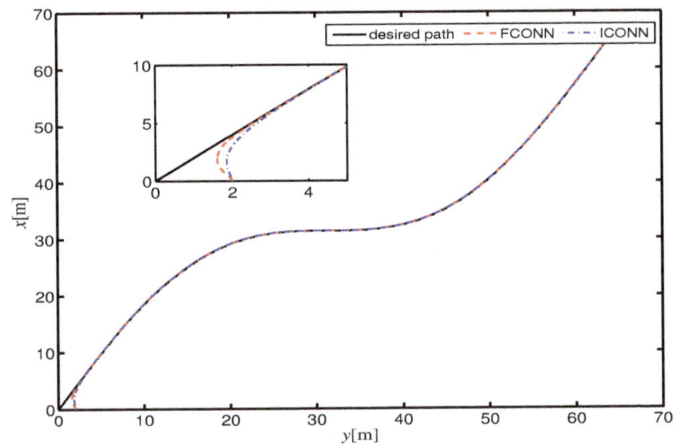

(a) Curvilinear path following.

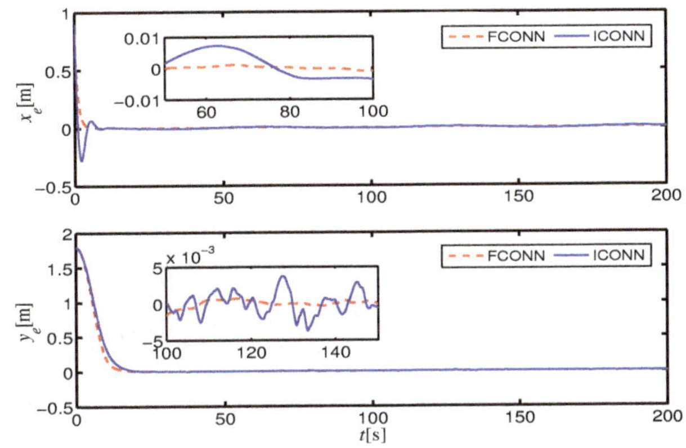

(b) Along- and cross-tracking error in Case 2.

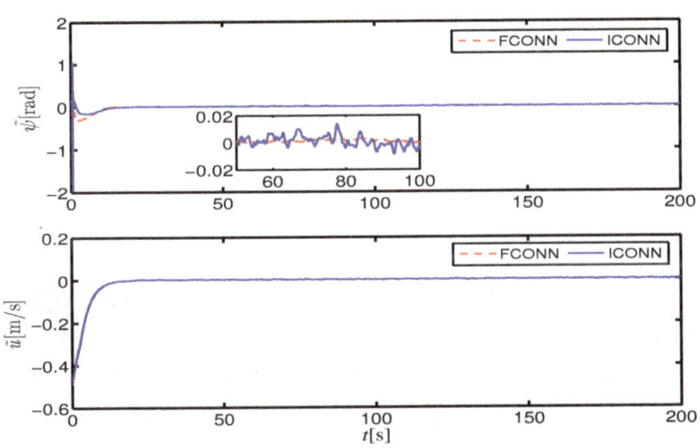

(c) Profile of the heading and surge velocity tracking in Case 2.

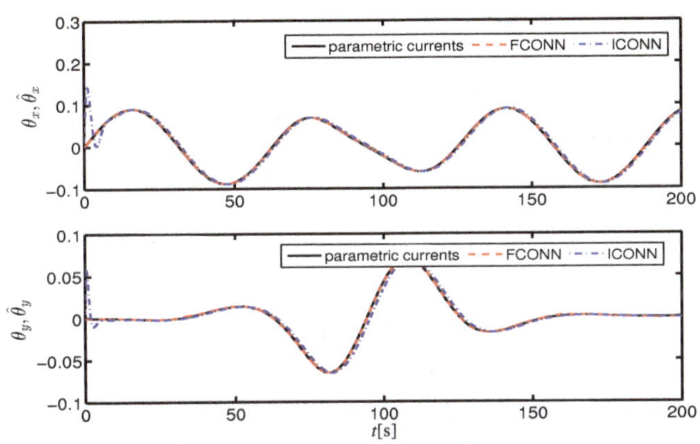

(d) Estimations of parametric currents in Case 2.

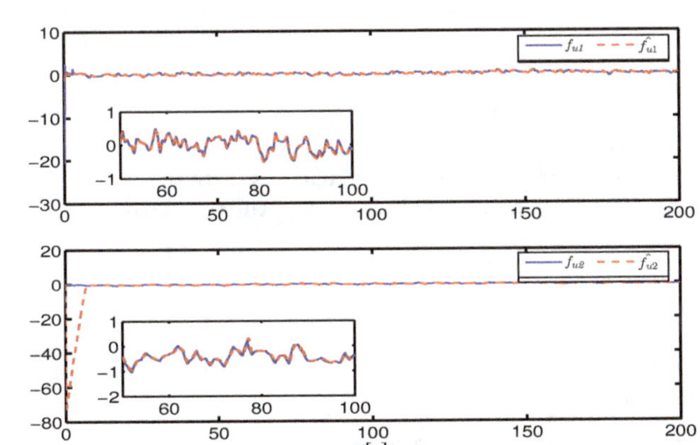

(e) Estimations of dynamical uncertainties using RBFNN in Case 2.

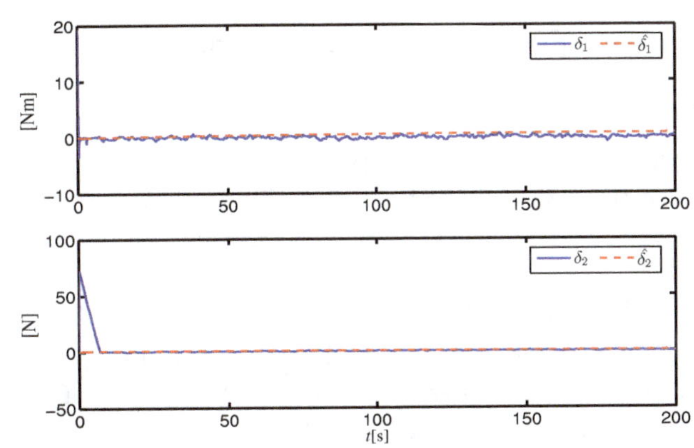

(f) Estimations of compound bound in Case 2.

Figure 4. *Cont.*

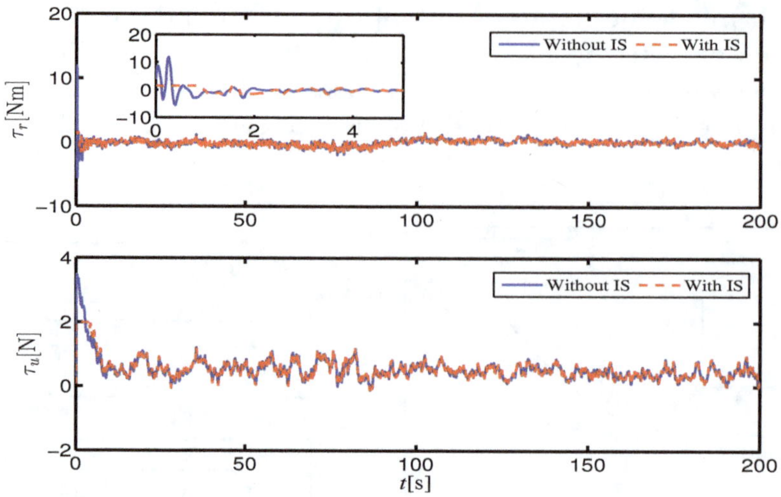

(g) Profile of the control inputs in Case 2.

Figure 4. Curvilinear path following results.

5. Conclusions

In this paper, a path following control scheme for a USV subject to input saturation and uncertainties has been proposed by resorting to the finite-time currents observer based ILOS guidance, the adaptive RBFNN, and the auxiliary dynamic system. The finite-time currents observer based ILOS guidance is applied to obtain the desired yaw angle, where the incorporated finite time currents observer can provide the precise estimations of the unknown time-varying ocean currents. Simultaneously, the RBF neural networks and the adaptive laws with leakages terms can provide the precise estimations of dynamical uncertainties and the compound bounds of the approximation errors and external disturbances, without knowing any prior knowledge of the time-varying disturbance. The auxiliary control system is introduced to handle input saturation of the actuators. It has been proved that all error signals of the closed-loop system are locally UUB. Finally, both linear and curved path following are presented and compared with the preceding control method. Simulations results have verified that the proposed control method can achieve satisfactory performance and robustness. Future work will cover the aspect of the position error constraint to ensure that the USV can work in these situations including the narrow passage and the channel between obstacles, as well as the precise estimation of the sideslip angle.

Author Contributions: Y.F., H.H. and Y.T. conceived the framework and wrote the paper. Y.F. and H.H. designed the controller and H.H. and Y.T. were responsible for the formula analyses. H.H. made the simulations and Y.T. analyzed the data. Y.F., H.H. and Y.T. discussed the results and contributed to the whole manuscript.

References

1. Liu, Z.; Zhang, Y.; Yu, X.; Yuan, C. Unmanned surface vehicles: An overview of developments and challenges. *Annu. Rev. Control.* **2016**, *41*, 71–93. [CrossRef]
2. Ashrafiuon, H.; Muske, K.R.; Mcninch, L.C. Review of nonlinear tracking and setpoint control approaches for autonomous underactuated marine vehicles. In Proceedings of the 2010 American Control Conference, Baltimore, MD, USA, 30 June–2 July 2010; pp. 5203–5211.
3. Do, K.D.; Pan, J. Control of ships and underwater vehicles. In *Advances in Industrial Control*; Springer: London, UK, 2009; pp. 40–42, ISBN 978-1-84882-729-5.
4. Caccia, M.; Bibuli, M.; Bono, R.; Bruzzone, G. Basic navigation, guidance and control of an unmanned surface vehicle. *Auton. Robot.* **2008**, *25*, 349–365. [CrossRef]

5. Breivik, M.; Fossen, T.I. Path following for marine surface vessels. In Proceedings of the Oceans '04 MTS/IEEE Techno-Ocean '04, Kobe, Japan, 9–12 November 2004; pp. 2282–2289.
6. Fossen, T.I.; Breivik, M.; Skjetne, R. Line-of-sight path following of underactuated marine craft. In Proceedings of the IFAC Manoeuvring and control of Marine Craft, Girona, Spain, 17–19 September 2003; pp. 211–216.
7. Fossen, T.I.; Pettersen, K.Y. On uniform semiglobal exponential stability (usges) of proportional line-of-sight guidance laws. *Automatica* **2014**, *50*, 2912–2917. [CrossRef]
8. Borhaug, E.; Pavlov, A.; Pettersen, K.Y. Integral LOS control for path following of underactuated marine surface vessels in the presence of constant ocean currents. In Proceedings of the 47th IEEE conference on Decision and Control, Cancun, Mexico, 9–11 September 2008; pp. 4984–4991.
9. Lekkas, A.M.; Fossen, T.I. Integral los path following for curved paths based on a monotone cubic hermite spline parametrization. *IEEE Trans. Control. Syst. Technol.* **2014**, *22*, 2287–2301. [CrossRef]
10. Fossen, T.I.; Lekkas, A.M. Direct and indirect adaptive integral line-of-sight path-following controllers for marine craft exposed to ocean currents. *Int. J. Adapt. Control. Signal Process.* **2017**, *31*, 445–463. [CrossRef]
11. Fossen, T.I.; Pettersen, K.Y.; Galeazzi, R. Line-of-sight path following for dubins paths with adaptive sideslip compensation of drift forces. *IEEE Trans. Control. Syst. Technol.* **2015**, *23*, 820–827. [CrossRef]
12. Liu, L.; Wang, D.; Peng, Z. Eso-based line-of-sight guidance law for path following of underactuated marine surface vehicles with exact sideslip compensation. *IEEE J. Ocean. Eng.* **2017**, *42*, 477–487. [CrossRef]
13. Wang, N.; Sun, Z.; Yin, J.; Su, S.F.; Sharma, S. Finite-time observer based guidance and control of underactuated surface vehicles with unknown sideslip angles and disturbances. *IEEE Access* **2018**, *6*, 14059–14070. [CrossRef]
14. Wang, W.; Huang, J.; Wen, C.; Fan, H. Distributed adaptive control for consensus tracking with application to formation control of nonholonomic mobile robots. *Automatica* **2014**, *50*, 1254–1263. [CrossRef]
15. Miao, J.; Wang, S.; Tomovic, M.M.; Zhao, Z. Compound line-of-sight nonlinear path following control of underactuated marine vehicles exposed to wind, waves, and ocean currents. *Nonlinear Dyn.* **2017**, *89*, 1–19. [CrossRef]
16. Do, K.D. Global robust adaptive path-tracking control of underactuated ships under stochastic disturbances. *Ocean. Eng.* **2016**, *111*, 267–278. [CrossRef]
17. Liu, S.; Liu, Y.; Wang, N. Nonlinear disturbance observer-based backstepping finite-time sliding mode tracking control of underwater vehicles with system uncertainties and external disturbances. *Nonlinear Dyn.* **2017**, *88*, 1–12. [CrossRef]
18. Du, J.; Hu, X.; Krstić, M.; Sun, Y. Robust dynamic positioning of ships with disturbances under input saturation. *Automatica* **2016**, *73*, 207–214. [CrossRef]
19. Sun, Z.; Zhang, G.; Yi, B.; Zhang, W. Practical proportional integral sliding mode control for underactuated surface ships in the fields of marine practice. *Ocean. Eng.* **2017**, *142*, 217–223. [CrossRef]
20. Roy, S.; Roy, S.B.; Kar, I.N. Adaptive-robust control of euler-lagrange systems with linearly parametrizable uncertainty bound. *IEEE Trans. Control. Syst. Technol.* **2017**, *26*, 1842–1850. [CrossRef]
21. Roy, S.; Roy, S.B.; Kar, I.N. A New Design Methodology of Adaptive Sliding Mode Control for a Class of Nonlinear Systems with State Dependent Uncertainty Bound. In Proceedings of the 15th International Workshop on Variable Structure Systems, Graz, Austria, 9–11 September 2018; pp. 414–419.
22. Wang, N.; Er, M.J.; Sun, J.C.; Liu, Y.C. Adaptive robust online constructive fuzzy control of a complex surface vehicle system. *IEEE Trans. Cybern.* **2017**, *46*, 1511–1523. [CrossRef]
23. Zheng, Z.; Sun, L. Path following control for marine surface vessel with uncertainties and input saturation. *Neurocomputing* **2016**, *177*, 158–167. [CrossRef]
24. Liu, L.; Wang, D.; Peng, Z. Path following of marine surface vehicles with dynamical uncertainty and time-varying ocean disturbances. *Neurocomputing* **2016**, *173*, 799–808. [CrossRef]
25. Zheng, Z.; Feroskhan, M. Path following of a surface vessel with prescribed performance in the presence of input saturation and external disturbances. *IEEE/ASME Trans. Mechatronics* **2017**, *22*, 2564–2575. [CrossRef]
26. Roy, S.; Shome, S.N.; Nandy, S.; Ray, R.; Kumar, V. Trajectory following control of auv: A robust approach. *J. Inst. Eng. India Series C* **2013**, *94*, 253–265. [CrossRef]
27. Shtessel, Y.B.; Shkolnikov, I.A.; Levant, A. Smooth second-order sliding modes: Missile guidance application. *Automatica* **2007**, *43*, 1470–1476. [CrossRef]

28. Chen, M.; Ge, S.S.; Ren, B. Adaptive tracking control of uncertain mimo nonlinear systems with input constraints. *Automatica* **2011**, *47*, 452–465. [CrossRef]
29. Polycarpou, M.M.; Ioannou, P.A. A robust adaptive nonlinear control design. *Automatica* **1996**, *32*, 423–427. [CrossRef]
30. Krstic, M.; Kanellakopoulos, I.; Kokotovic, P.V. Nonlinear and adaptive control. In *Lecture Notes in Control and Information*; Sciences: New York, NY, USA, 1995; pp. 511–514, ISBN 9780471127321.
31. Fredriksen, E.; Pettersen, K.Y. Global k-exponential way-point maneuvering of ships: Theory and experiments. *Automatica* **2009**, *42*, 677–687. [CrossRef]
32. Fossen, T.I. How to incorporate wind, waves and ocean currents in the marine craft equations of motion. In Proceedings of the 9th IFAC Conference on Manoeuvring and Control of Marine Craft, Arenzano, Italy, 19–21 September 2012; pp. 126–131.

8

Cooperative Path Planning for Multiple Mobile Robots via HAFSA and an Expansion Logic Strategy

Yiqing Huang *, Zhikun Li, Yan Jiang and Lu Cheng

College of Electrical Engineering, Anhui Polytechnic University, Wuhu 241000, China; lzk52170@163.com (Z.L.); Jiangyan117@163.com (Y.J.); chenglu073@163.com (L.C.)

* Correspondence: yiqhuang@ahpu.edu.cn

Featured Application: In various unknown environments, cooperative path planning problem of multiple mobile robots is becoming more and more important. The efficiency and reliability can be greatly improved by the cooperation of multiple mobile robots. The novel obstacle avoidance and real-time navigation algorithm presented in this article may be useful for marine exploration, military, aerospace and mining detection. Also, for the developed real-time navigation algorithm, the presented hybrid artificial fish swarm algorithm and expansion logic strategy are helpful not only for accelerating the convergence rate, but also for improving decision-making ability.

Abstract: The cooperative path planning problem of multiple mobile robots in an unknown indoor environment is considered in this article. We presented a novel obstacle avoidance and real-time navigation algorithm. The proposed approach consisted of global path planning and local path planning via HAFSA (hybrid artificial fish swarm algorithm) and an expansion logic strategy. Meanwhile, a kind of scoring function was developed, which shortened the time of local path planning and improved the decision-making ability of the path planning algorithm. Finally, using STDR (simple two dimensional robot simulator) and RVIZ (robot operating system visualizer), a multiple mobile robot simulation platform was designed to verify the presented real-time navigation algorithm. Simulation experiments were performed to validate the effectiveness of the proposed path planning method for multiple mobile robots.

Keywords: path planning; multiple mobile robots; artificial fish swarm algorithm; expansion logic strategy

---

1. Introduction

Mobile robots can be equipped with different sensors and tools to afford a variety of services such as home care, mining detection and object handling [1,2]. One of the fundamental issues with mobile robots performing tasks is ensuring that they can navigate safely in an unknown indoor environment. Therefore, path planning is crucial for the successful application of mobile robots. The goal of mobile robot path planning is to find a motion path from a starting position to a target position in an environment with obstacles [3–5]. For the past two decades, there has been a great deal of research on the problem of mobile robot path planning. For example, a novel motion map was constructed for mobile robots based on the BIE (boundary integral equation) method, and then, a point-to point path planning problem was addressed in a known environment with static obstacles [6]. Furthermore, an improved three-dimensional-like grid map was developed to represent the environment model [7], and then, a simple but efficient path planning algorithm was presented to solve robot navigation problems in a static environment. The authors designed an autonomous multi-goal navigation system for picking up or delivering tasks in mobile robotics and a multi-goal path planning method based on the Lin–Kernighan heuristics (LKH) algorithm for intelligent service mobile robots in Reference [8].

There are also some intelligent methods that can be applied to mobile robot path planning. The cross probability and the mutation probability for GA (genetic algorithm) were improved and the improved algorithm was applied to the path planning problem of mobile robots in Reference [9], whereas the authors proposed an intelligent motion planning and navigation method for omnidirectional mobile robots via a fuzzy logic algorithm in Reference [10]. Moreover, the developed navigation system is especially suitable for real-time path planning applications. A novel optimal hierarchical global path planning method for mobile robots in a cluttered environment was presented in Reference [11]. In this method, a combination of the triangular decomposition approach, constrained multi-objective PSO (particle swarm optimization) and Dijkstra's algorithm is presented in order to obtain an optimal path planning trajectory. In addition, due to good feedback information and better distributed computing, the authors proposed a path planning method for mobile robots via an improved ant colony algorithm in grid maps in Reference [12].

Using multiple mobile robots rather than a single mobile robot can improve working capability and performance. Therefore, recently, research on multiple mobile robots has become a hot topic. In previous studies, only static obstacles in the unknown environment were considered. For holonomic wheeled mobile robots in static environments, an optimal multiple mobile robot path planning method based on adaptive charged system search (CSS) algorithms was addressed [13]. However, path planning methods in an unknown environment with dynamic obstacles are even more acute in multiple mobile robot areas. In Reference [14], a new path planning approach for coordinating multiple mobile robots was presented and the authors developed an online strategy to adjust path planning for avoidance of dynamic obstacles. Furthermore, a biologically inspired neural-network-based intelligent method was proposed for a multiple robot system with moving obstacles [15]. The proposed method could plan the paths of multiple robots to avoid collision with dynamic obstacles.

Although these previously-developed navigation algorithms have shown good performance for solving robot path planning, they have also shown some limitations such as slow convergence and a local optimum. The local optimal problem is the most common problem in solving path planning. Motivated by the aforementioned reasons, we reconstructed an analytical real-time cooperative navigation algorithm to accommodate multiple mobile robot systems. The proposed EAFSA (empirical artificial fish swarm algorithm) was used to avoid falling into local optimal problems and to realize global path planning for a single mobile robot. Then, an expansion logic strategy was introduced to avoid collisions between multiple mobile robots and an environment with obstacles. A multiple mobile robot simulation system was developed using STDR (simple two dimensional robot simulator) and RVIZ (robot operating system visualizer) software. Finally, the presented method was proven to be effective by experiments conducted in a simulated environment.

The main contributions of the article are summarized as follows. (1) EAFSA is presented to solve the global path planning problem for a single mobile robot; (2) an expansion logic strategy and a kind of scoring function are proposed for a multiple mobile robot real-time navigation algorithm. The presented real-time navigation algorithm is helpful not only for accelerating the convergence rate, but also for improving decision-making ability.

The remainder of this paper is organized as follows: Section 2 describes the process of the presented HAFSA. Section 3 presents the developed expansion logic strategy and scoring function. Section 4 shows the results of a simulation to demonstrate the performance of the proposed algorithms. The concluding remarks are given in Section 5.

2. Hybrid Artificial Fish Swarm Algorithm

The artificial fish swarm algorithm (AFSA) is a novel swarm intelligent optimization method inspired by natural fish swarm behavior. It has been successfully used in the field of wireless telemedicine systems [16], fault diagnosis [17], indoor visible light positioning [18], floating wind turbines [19], etc.

The basic idea of AFSA can be described as follows: If the position of each artificial fish is $X = (x_1, x_2, \cdots, x_n)$ and the size of the fish population is *Num*, Y denotes the food concentration of the artificial fish in the current position and $Y = f(X)$ is fitness or the objective function at position X. Each artificial fish tries to find an optimal position to satisfy their food needs using preying behavior, swarming behavior, following behavior and random behavior [20].

(1) Preying Behavior

If the current state of an artificial fish is $X_i(t)$, $X_j(t)$ is the random state of its visual distance, and $X_{i+1}(t)$ is the next position of $X_i(t)$. If food concentration is $Y_i < Y_j$, the artificial fish swims a *step* in the direction of $X_j(t)$. Otherwise, it randomly selects a state again and judges whether it satisfies the aforementioned condition. In other words, preying behavior can be expressed by the following equation:

$$X_{i+1}(t) = \begin{cases} X_i(t) + \frac{X_j(t) - X_i(t)}{\|X_j(t) - X_i(t)\|} \times step \times rand(0,1) & if\ Y_i < Y_j \\ X_i(t) + Visual \times rand(0,1) & if\ Y_i \geq Y \end{cases} . \tag{1}$$

(2) Swarming Behavior

When N_F is the number of artificial fishes in the current position $X_i(t)$, $X_c(t)$ is the center position of the artificial fishes in their current neighborhood. $if\ Y_c/N_F > \delta Y_i$ is satisfied, the artificial fish moves to a center position, according to Equation (2), due to high food concentration and to avoid crowding each other. Otherwise, the artificial fish executes preying behavior.

$$X_i(t) + \frac{X_c(t) - X_i(t)}{\|X_c(t) - X_i(t)\|} \times step \times rand(0,1) \quad if\ Y_c/N_F > \delta Y_i \tag{2}$$

(3) Following Behavior

Let $X_{\max}(t)$ be the local best companion with food concentration $Y_{\max}$ in the current neighborhood of $X_i(t)$. $if\ Y_{\max}/N_F > \delta Y_i$ is satisfied, the artificial fish moves to a position according to Equation (3). Otherwise, the next position of the artificial fish can be obtained by preying behavior.

$$X_i(t) + \frac{X_{\max}(t) - X_i(t)}{\|X_{\max}(t) - X_i(t)\|} \times step \times rand(0,1) \quad if\ Y_{\max}/N_F > \delta Y_i \tag{3}$$

(4) Random Behavior

The artificial fish chooses an arbitrary state or position randomly in its *Visual* field, and then it swims towards the selected state. Random behavior is a default behavior and it can be described as

$$X_i(t) + Visual \times rand(0,1), \tag{4}$$

where $X_i(t)$ is the current state of the artificial fish and $X_{i+1}(t)$ is the next position of $X_i(t)$.

Given the above consideration, an effective hybrid fish swarm algorithm (HFSA) with experiential learning and a detection operator was presented to solve the local optimal problem and realize global path planning for a single mobile robot.

2.1. Experiential Learning

In this section, an experiential learning strategy is presented to improve the performance of AFSA. Experiential learning strategies include adjustment of the step size and food concentration for the artificial fish. The step size of an artificial fish is fixed in traditional AFSA. However, as is known, the step size determines the convergence rate. If the step size is too small, the artificial fish will reach the optimal solution slowly and the global search ability will be decreased. Therefore, the artificial fish is easy to fall into a local optimum. Conversely, if the step size is too large, the convergence speed will be increased and oscillation will occur later in the algorithm iteration. Therefore, it is necessary to select an appropriate step size to ensure the global convergence speed and improve the accuracy of the optimal solution. In this article, a logarithmic function was used to update the step size by Equation (5);

$$b = \log_N p, p = 1,2,3\cdots N, \tag{5}$$

where p indicates the current iteration number, and N is the maximum number of iterations.

Then, the step size in the population update formula could be obtained by

$$step = \log_N p\cdot(X_\varepsilon(t) - X_i(t))p = 1,2,3\cdots N, i = 1,2,3\cdots NUM, \tag{6}$$

where $X_i(t)$ is the current state of the artificial fish, $X_{i+1}(t)$ is the next position of $X_i(t)$ and $X_\varepsilon(t)$ is the state that needs to be searched.

Finally, the position of the updated solution could be expressed as follows:

$$X_{i+1}(t) = X_i(t) + \log_{NC} p\cdot(X_\varepsilon(t) - X_i(t))p = 1,2,3\cdots N, i = 1,2,3\cdots NUM. \tag{7}$$

As shown in Equation (7), with the increase in the number of iterations, the moving step gradually adapts to the change of the iteration numbers. The local search ability is increased by the improved moving factor. As a result, the artificial fish can locate the search direction quickly, move to the target area, maintain the global search ability of the optimal solution and accelerate the convergence speed.

On the other hand, the food concentration of the artificial fish was represented as the ability of the solution to solve an optimization problem. The current position of the artificial fish with the highest food concentration was the optimal solution of the optimization problem. In this article, a weight coefficient function was used to design the food concentration, which could decrease the food concentration of the artificial fish and avoid the problem that the suboptimal solution of the fish swarm algorithm would interfere with the global solution. The food concentration equation was designed as follows:

$$H_{\exp} = w\sqrt{(x_i - x_g)^2 + (y_i - y_g)^2}, \tag{8}$$

where the weight coefficient is $w \in [1, 1.5)$.

2.2. Detection Operator

A detection operator was developed to optimize the resulting path trajectory. If $p(kx, ky)$ is the position coordinate of the optimal fish group solved at the kth time, n is the dimension of the grid-based map and k_t is the number of the solutions that have been optimized. $goal(x, y)$ is the position coordinate of the target point. The detection operator $R(t)$ can be described as follows:

$$R(t) = \begin{cases} \lceil D(k)/(2k_t) \rceil \times \sqrt{2} & 0 < t < count \\ 2\sqrt{2} & t \geq count \end{cases} \tag{9}$$

$$D(k) = \frac{\|p(kx, ky) - goal(x, y)\|}{\sqrt{2}} \tag{10}$$

$$count = \lfloor [0.27n + 0.5] \times 75\% \rfloor, \tag{11}$$

where $D(k) = \frac{\|p(kx,ky) - goal(x,y)\|}{\sqrt{2}}$, $count = \lfloor [0.27n + 0.5] * 75\% \rfloor$, $\lfloor \cdot \rfloor$ and $\lceil \cdot \rceil$ round toward negative or positive infinity.

The outline of the presented algorithm is described in the following steps:

Step 1. Initialize the population size *NUM*, the parameters *step* and *visual*, and the maximum number of iterations *N*.
Step 2. Update the step size and position of the artificial fish using Equation (7).
Step 3. Calculate the food concentration for each artificial fish using Equation (8) and record the optimal value in the bulletin board.
Step 4. Perform preying behavior, swarming behavior, following behavior and random behavior.
Step 5. Check the termination condition. If the stopping condition is satisfied, terminate the iteration process and output optimal solution. Otherwise, return to Step 2.

3. Local Path Planning Based on an Expansion Logic Strategy

In this section, an expansion logic strategy is presented to avoid collisions between multiple mobile robots and an environment with obstacles. It plays a decisive role in real-time navigation of mobile robots. Figure 1 shows the obstacle information in an unknown environment. For any polygonal obstacle, the minimum circumscribed circle (MCCI) method and wire envelopes method can be used to perform obstacle expansion operations. In this article, as we sought a rapid expansion method, we adopted the endpoint connection method to generate a circular equation instead of the MCCI method. The expansion logic strategy is described as follows:

If the vertices of n-sided polygonal obstacles are denoted as $p_i(x_i, y_i)$, the Euclidean distance d_{ij} between two vertices $p_i(x_i, y_i)$ and $p_j(x_j, y_j)$ can be expressed by the following equation:

$$d_{ij} = \sqrt{(x_i - x_j)^2 + (y_i - y_j)^2}. \tag{12}$$

Therefore, the diameter of a range circle is given by

$$d_{\max} = \max\{d_{ij}\}, i, j = 1, 2, \cdots, n \tag{13}$$

and the circle equation is given by Equation (10).

$$(x - \frac{x_i + x_j}{2})^2 + (y - \frac{y_i + y_j}{2})^2 = (\frac{d_{\max}}{2})^2 \tag{14}$$

In this paper, an environmental map with n-sided polygonal obstacles is shown in Figure 1a. Following the circle Equation (14), two concentric circles with diameters $1.2d_{\max}$ and $1.4d_{\max}$ could be obtained (Figure 1b). As the figure shows, the collision probability was 0.85 for the complementary set area of the intersection between the first circle and the polygonal obstacles. Furthermore, the collision probabilities were 0.65 and 0.45 when the robots moved in the other two grid areas.

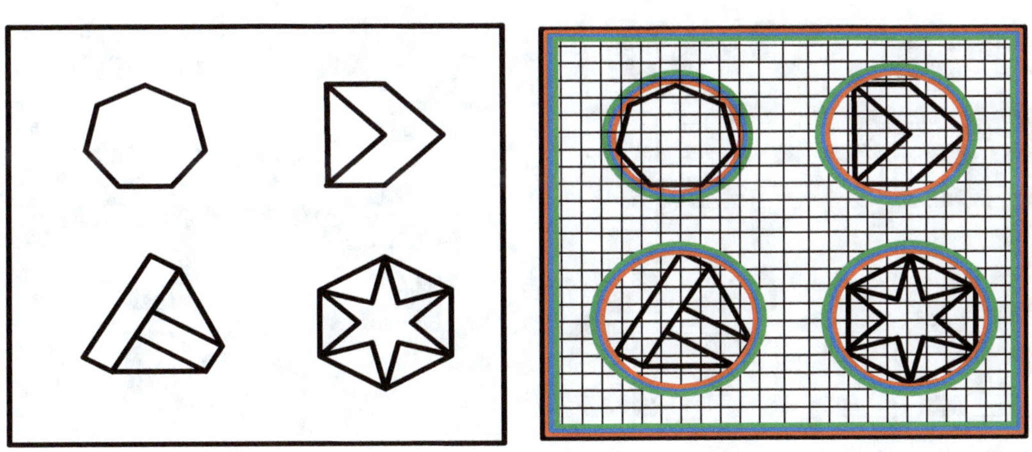

(a) (b)

Figure 1. (**a**) Environmental map information; (**b**) expansion logic operation for obstacles.

To evaluate the grid-based environment map, a kind of scoring function is given by Equation (15), which shortens the time of local path planning and improves decision-making ability.

$$score = 100 - \lceil \frac{dist}{l} \rceil, \tag{15}$$

where *dist* denotes the Euclidean distance between the starting point and the target point, and l represents the Euclidean distance between the starting point and the current position of the robot.

From the vertical line of the motion direction, we scored the surrounding grids by Equation (15) and the obtained grid scores are shown in Figure 2a. Then, in accordance with the current position information and Equation (12), the mobile robot selected the grid that was the closest to the target point as the position of the next moment. If the distance between the grids was equal, preference was given to the high score grid. Finally, the local path planning trajectory, which is indicated by a dotted line, could be obtained using the expansion logic strategy (Figure 2b).

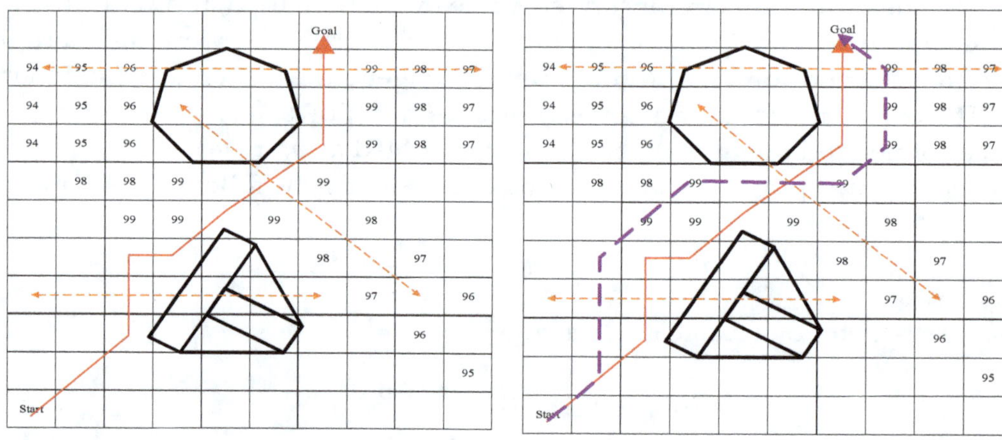

(**a**) (**b**)

Figure 2. (**a**) Local path planning and grid scores; (**b**) local path planning using the expansion logic strategy.

4. Simulation Experiments

In this section, to verify the superiority of the presented algorithm for a single mobile robot and 20×20 grid-based environment maps, the path planning results under the presented hybrid artificial fish swarm algorithm and the traditional fish swarm algorithm are shown in Figure 3.

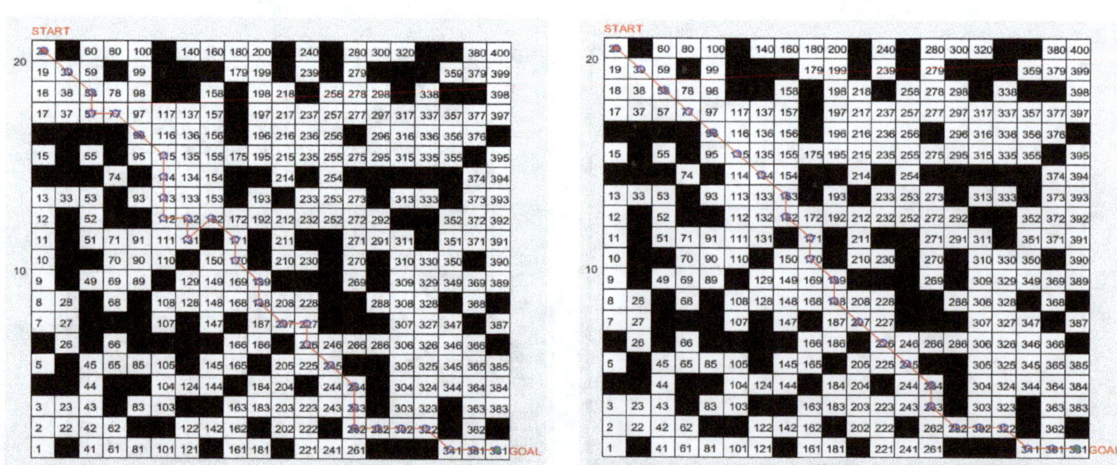

(**a**) (**b**)

Figure 3. (**a**) Artificial fish swarm algorithm; (**b**) the presented hybrid artificial fish swarm algorithm.

For 20 × 20 and 40 × 40 grid-based environment maps, the path planning results based on the adaptive artificial fish swarm algorithm (AAFA) [21], fuzzy logic (FL) [22], the improved genetic algorithm (IGA) [23] and our method are shown in Table 1. As can be seen in Figure 3 and Table 1, optimization performance and iteration time can be improved by using the presented method.

Table 1. Performance comparison of four algorithms.

Environment Map	Algorithms	The Longest Path Length	The Optimal Path Length	The Average Path Length	Iteration Time/s
20 × 20 grids	AAFA	35.1283	30.0348	33.6231	16.5182
	FL	34.3848	29.7990	32.0919	12.2304
	IGA	32.3254	29.6325	30.8652	16.1826
	Our method	30.3848	29.2132	29.7990	9.3102
40 × 40 grids	AAFA	80.2372	73.7103	79.2293	98.5621
	FL	75.1838	69.4975	72.3407	90.4073
	IGA	66.4723	62.9002	65.7213	97.7652
	Our method	64.0833	61.4264	62.7549	74.5801

Furthermore, simulation results were performed on a group of mobile robots. A multiple mobile robot navigation system, which included three mobile robots labeled as Robot 0, Robot 1 and Robot 2, was designed using STDR (simple two dimensional robot simulator) and RVIZ (robot operating system visualizer) software. Each mobile robot was equipped with four ultrasonic sensors and a radar detector with laser detection capabilities. The environment map and obstacle information of the simulation experiments are shown in Figure 4a. If the size of the empirical fish swarm was $N = 50$, and the visual distance of an artificial fish was $v = 10$, the crowd factor was $\delta = 0.618$.

As shown in Figure 4, the initial poses of the three mobile robots were (1, 1, 0), (1, 5, 0) and (1, 9, 0). The goal positions were (10, 14), (17, 1) and (18, 13).

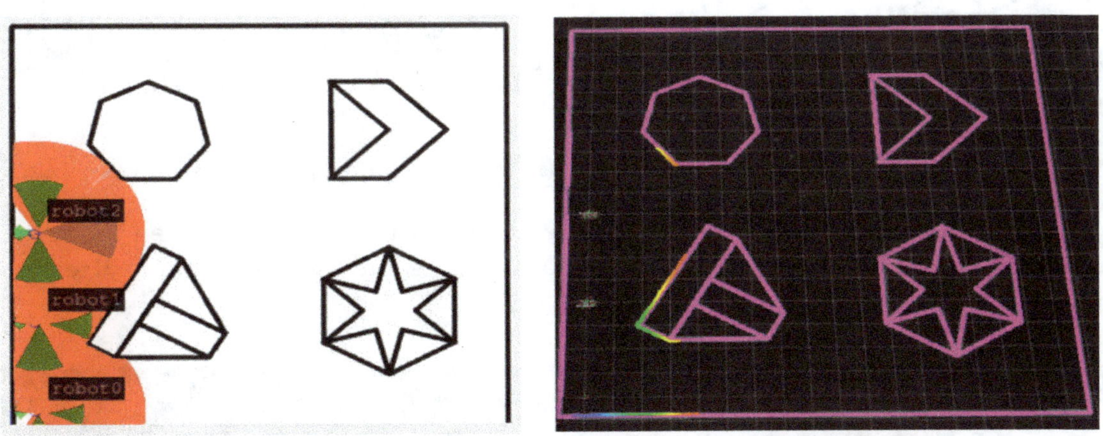

(**a**) (**b**)

Figure 4. (**a**) The initial poses of the three mobile robots; (**b**) simulation visual interface of multiple mobile robots.

The global path planning trajectories of the three mobile robots are shown in Figure 5. As shown in Figure 5a, Robot 0 (initial pose was (1, 1, 0)), Robot 1 (initial pose was (1, 5, 0)) and Robot 2 (initial pose was (1, 9, 0)) started to move, and then they updated their motion paths using the expansion logic strategy to avoid mutual collision. Meanwhile, many feasible paths could be obtained by the presented EFSA. Therefore, we can see that the global path planning results for mobile robots were not unique (Figure 5b).

Figure 6a illustrates the position information (the poses of the three mobile robots were (3, 10, 0), (9, 1, 0) and (10, 10, 0)) at a specific time in the simulation experiment. Local and global paths of multiple mobile robots are shown in Figure 6b.

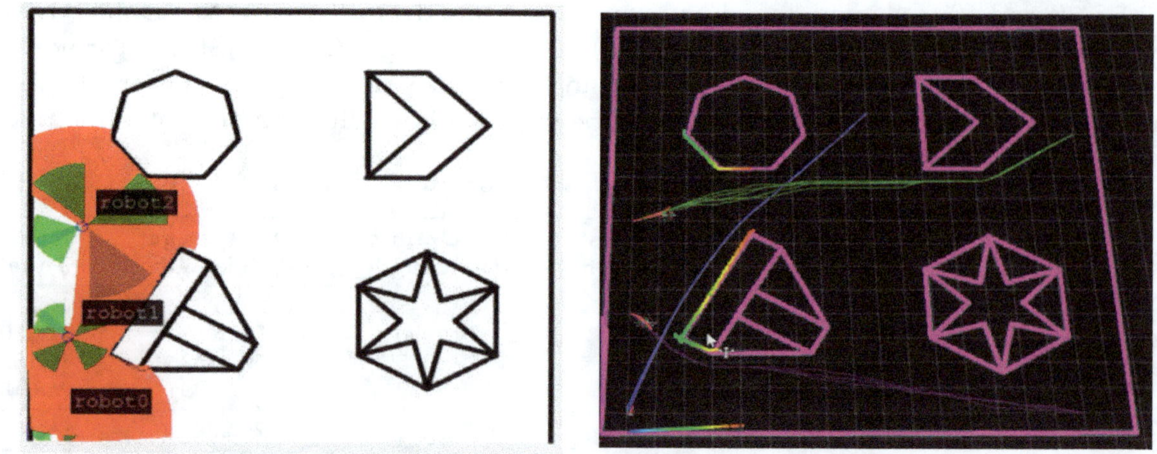

(a) (b)

Figure 5. (**a**) The initial poses of the three mobile robots; (**b**) global path planning trajectories.

(a) (b)

Figure 6. (**a**) Position information at a specific time; (**b**) local and global paths of multiple mobile robots.

The positions of Robot 0, Robot 1 and Robot 2 are shown in Figure 7a when the robots reached the target points. Figure 7b illustrates the final paths of the three mobile robots after reaching the target points.

(a) (b)

Figure 7. (**a**) The robots reached the target points; (**b**) the final paths of the three mobile robots.

5. Conclusions

This article focused on the cooperative path planning problem of multiple mobile robots in an unknown environment with obstacles. HAFSA (hybrid artificial fish swarm algorithm) was proposed to solve the local optimal problem and realize cooperative path planning for multiple mobile robots. An experiential learning strategy was presented to improve the performance of AFSA and a detection operator was developed to optimize the resulting path trajectory. In particular, an expansion logic strategy was used to avoid collision between multiple mobile robots and an environment with obstacles. In order to evaluate a grid-based environment map, a kind of scoring function was designed. Finally, a multiple mobile robot simulation system was developed by utilizing STDR and RVIZ software; simulation experimental results validated the effectiveness of the proposed real-time navigation algorithm.

Author Contributions: Validation, Z.L. and Y.J.; formal analysis, L.C.; writing—original draft preparation, Y.H.; supervision, Y.H.; project administration, Y.H.

References

1. Cheng, H.; Chen, H.; Liu, Y. Topological Indoor Localization and Navigation for Autonomous Mobile Robot. *IEEE Trans. Autom. Sci. Eng.* **2015**, *12*, 729–738. [CrossRef]
2. Jhang, J.; Lin, C.; Lin, C.; Young, K. Navigation Control of Mobile Robots Using an Interval Type-2 Fuzzy Controller Based on Dynamic-group Particle Swarm Optimization. *Int. J. Control Autom. Syst.* **2018**, *16*, 2446–2457. [CrossRef]
3. Raja, P.; Pugazhenthi, S. Optimal path planning of mobile robots: A review. *Int. J. Phys. Sci.* **2012**, *7*, 1314–1320. [CrossRef]
4. Han, J.; Seo, Y. Mobile robot path planning with surrounding point set and path improvement. *Appl. Soft Comput.* **2017**, *57*, 35–47. [CrossRef]
5. Cherni, F.; Boujelben, M. Autonomous mobile robot navigation based on an integrated environment representation designed in dynamic environments. *Int. J. Autom. Control* **2017**, *11*, 35–53. [CrossRef]
6. Mantegh, I.; Jenkin, M.R.M.; Goldenberg, A.A. Path Planning for Autonomous Mobile Robots Using the Boundary Integral Equation Method. *J. Intell. Robot. Syst.* **2010**, *59*, 191–220. [CrossRef]
7. Wang, Y.; Cao, W. A global path planning method for mobile robot based on a three-dimensional-like map. *Robotica* **2014**, *32*, 611–624. [CrossRef]
8. Hernández, K.; Bacca, B.; Posso, B. Multi-goal Path Planning Autonomous System for Picking up and Delivery Tasks in Mobile Robotics. *IEEE Lat. Am. Trans.* **2017**, *15*, 232–238. [CrossRef]
9. Wu, M.; Chen, E.; Shi, Q.; Zhou, L.; Chen, Z.; Li, M. Path planning of mobile robot based on improved genetic algorithm. In Proceedings of the Chinese Automation Congress (CAC), Jinan, China, 20–22 October 2017; pp. 6696–6700.
10. Zavlangas, P.G.; Tzafestas, S.G. Motion control for mobile robot obstacle avoidance and navigation: A fuzzy logic-based approach. *Syst. Anal. Model. Simul.* **2003**, *43*, 1625–1637. [CrossRef]
11. Mac, T.T.; Copot, C.; Tran, D.T.; de Keyser, R. A hierarchical global path planning approach for mobile robots based on multi-objective particle swarm optimization. *Appl. Soft Comput.* **2017**, *59*, 68–76. [CrossRef]
12. Akka, K.; Khaber, F. Mobile robot path planning using an improved ant colony optimization. *Int. J. Adv. Robot. Syst.* **2018**, *15*. [CrossRef]
13. Precup, R.E.; Petriu, E.M.; Radae, M.B.; Voisan, E.I.; Dragan, F. Adaptive Charged System Search Approach to Path Planning for Multiple Mobile Robots. In Proceedings of the 2nd IFAC Conference on Embedded Systems, Computer Intelligence and Telematics, Maribor, Slovenia, 22–24 June 2015; Volume 48, pp. 294–299.
14. Liu, S.; Sun, D.; Zhu, C. Coordinated Motion Planning for Multiple Mobile Robots along Designed Paths with Formation Requirement. *IEEE/ASME Trans. Mechatron.* **2011**, *16*, 1021–1031. [CrossRef]
15. Li, H.; Yang, S.X.; Seto, M.L. Neural-network-based path planning for a multi-robot system with moving obstacles. *IEEE Trans. Syst. Man Cybern. Part C Appl. Rev.* **2009**, *39*, 410–419. [CrossRef]
16. Umarani, P.; Thangaraj, P. Improving Group Communication in Wireless Telemedicine System Using Fish Swarm Algorithm. *J. Med. Imaging Health Inform.* **2016**, *6*, 1576–1580. [CrossRef]

17. Zhu, J.; Wang, C.; Hu, Z.; Kong, F.; Liu, X. Adaptive variational mode decomposition based on artificial fish swarm algorithm for fault diagnosis of rolling bearings. *Proc. Inst. Mech. Eng. Part C J. Mech. Eng. Sci.* **2016**, *231*, 635–654. [CrossRef]
18. Wen, S.; Cai, X.; Guan, W.; Jiang, J.; Chen, B.; Huang, M. High-precision indoor three-dimensional positioning system based on visible light communication using modified artificial fish swarm algorithm. *Opt. Eng.* **2018**, *57*, 106102. [CrossRef]
19. Jin, X.; Xie, S.; He, J.; Lin, Y.; Wang, Y.; Wang, N. Optimization of tuned mass damper parameters for floating wind turbines by using the artificial fish swarm algorithm. *Ocean Eng.* **2018**, *167*, 130–141. [CrossRef]
20. He, Q.; Hu, X.; Ren, H.; Zhang, H. A novelartificial fish swarm algorithm for solving large-scale reliability-redundancy application problem. *ISA Trans.* **2015**, *59*, 105–113. [CrossRef] [PubMed]
21. Yiqing, H.; Panpan, W.; Mengru, Y. Path planning of mobile robots based on logarithmic function adaptive artificial fish swarm algorithm. In Proceedings of the 36th Chinese Control Conference, Dalian, China, 26–28 July 2017; pp. 4819–4823.
22. Singh, N.H.; Thongam, K. Mobile Robot Navigation Using Fuzzy Logic in Static Environments. *Procedia Comput. Sci.* **2018**, *125*, 11–17. [CrossRef]
23. Tuncer, A.; Yildirim, M. Dynamic path planning of mobile robots with improved genetic algorithm. *Comput. Electr. Eng.* **2012**, *38*, 1564–1572. [CrossRef]

9

MIM_SLAM: A Multi-Level ICP Matching Method for Mobile Robot in Large-Scale and Sparse Scenes

Jingchuan Wang [1,2,*], Ming Zhao [1,2] and Weidong Chen [1,2]

1 Department of Automation, Shanghai Jiao Tong University, Shanghai 200240, China; mzhao1993@sjtu.edu.cn (M.Z.); wdchen@sjtu.edu.cn (W.C.)

2 Laboratory of System Control and Information Processing, Ministry of Education of China, Shanghai 200240, China

* Correspondence: jchwang@sjtu.edu.cn.

Abstract: In large-scale and sparse scenes, such as farmland, orchards, mines, and substations, 3D simultaneous localization and mapping are challenging matters that need to address issues such as maintaining reliable data association for scarce environmental information and reducing the computational complexity of global optimization for large-scale scenes. To solve these problems, a real-time incremental simultaneous localization and mapping algorithm called MIM_SLAM is proposed in this paper. This algorithm is applied in mobile robots to build a map on a non-flat road with a 3D LiDAR sensor. MIM_SLAM's main contribution is that multi-level ICP (Iterative Closest Point) matching is used to solve the data association problem, a Fisher information matrix is used to describe the uncertainty of the estimated pose, and these poses are optimized by the incremental optimization method, which can greatly reduce the computational cost. Then, a map with a high consistency will be established. The proposed algorithm has been evaluated in the real indoor and outdoor scenes as well as two substations and benchmarking dataset from KITTI with the characteristics of sparse and large-scale. Results show that the proposed algorithm has a high mapping accuracy and meets the real-time requirements.

Keywords: data association; 3D-SLAM; localization; mapping

---

1. Introduction

In recent years, the simultaneous localization and mapping (SLAM) based on a 3D LIDAR sensor have become an important topic in robotics due to the rise of autonomous driving technology. Meanwhile, the building of a map has become the basis for an autonomous mobile robot to complete tasks such as inspection and autonomous navigation [1] in some harsh environments.

The research on SLAM can be traced back to Smith et al. [2] of Stanford University in the 1980s. They published a seminal paper on SLAM, and later generations have done considerable work on it. Among them, the algorithms such as Extended Kalman Filters [3], Extended Information Filters [4], and Rao-Blackwellized Particle Filters [5] based on a filtering idea all adopt the Bayesian state estimation theory to estimate the posterior probability of the system's state. However, the essence of this kind of algorithm is only optimizing the local state of the system. As the distance traveled by the robot increases, it is difficult to establish a high consistency map. Another kind of algorithm is the graph optimization that was first proposed by Lu and Milio [6]. The essence of this algorithm is maintaining all the observation and space constraints between observations, then using the maximum likelihood method to estimate the pose of the robot. Better mapping results can be obtained in large-scale scenes using the global optimization method.

Although the SLAM problem has been thoroughly researched, there are still many problems to be solved, including the fact that computational complexity will increase from two-dimensions to three-dimensions, and how to ensure a consistent map with the increase of the size of mapping and reliable data association with less information in sparse environments. Combining these issues, Zlot [7] utilizes a rotating 2D LIDAR sensor to achieve high-precision map construction in a wide mine, but it cannot be applied online.

Olufs [8] proposes a data association method in sparse environments, but it is only suitable for solving the localization problem under known maps. Huang [9] proposes a sparse local submap joining filter (SLSJF) for map-building in large-scale environments. However, this depends on environmental features.

Moosmann [10] and Nüchter [11] propose a real-time SLAM framework, and the experiments in some scenarios have achieved good results, but the results of the mapping are not tested for large-scale and sparse environments. Zhang [12] estimates the pose of the robot by extracting the feature points in the scene, but it is difficult to establish a closed-loop map as the map size expands. Wang [13] proposes a multi-layer matching SLAM in large-scale and sparse environments, but it only solves the data association in a 2D scene. The algorithm will be ineffective when the robot drives on a non-flat road. Therefore, this paper is dedicated to solving the following two problems:

- In sparse environments, due to the scarce environmental information perceived by a 3D LIDAR sensor and the non-flat road, data association is easy to fall into a local minimum. Liu [14] proposes a LiDAR SLAM method in natural terrains which fuses multiple sensors including two 3D LiDAR. Therefore, the robustness of data associated with a lower accumulated error should be strengthened.

- In large-scale environments, it is difficult to obtain a reasonable pose estimation when the robot returns to a region it has previously explored and then generates an inconsistent map. Liang [15] addresses the laser-based loop closure problem by fusing visual information. Hess [16] can achieve real-time mapping in indoor scenes, and it may fail in the closing loop due to the incremental computation in large-scale outdoor scenes. How to reduce the computational complexity of graph optimization is also a problem that needs to be improved.

To solve these problems, we propose a real-time incremental SLAM algorithm called the MIM_SLAM based on a multi-level ICP matching method. This paper is organized as follows. We continue in the next section with an overview of the algorithm. Section 3 details the process of the MIM_SLAM algorithm. Then, we follow up with the experimental results and analysis in Section 4 and finally provide a summary in Section 5.

2. Algorithm Overview

Figure 1 shows the overall MIM_SLAM algorithm framework. The MIM_SLAM algorithm simplifies the SLAM problem into the data association and the incremental pose graph optimization. Firstly, the data association problem is solved by the multi-level ICP matching method which includes matching the time-neighbor frame, matching the current frame with the map, and matching the current frame with an area-neighbor keyframe.

Then the uncertainty of the estimated pose is described via the Fisher information matrix. After the above steps, this algorithm can obtain the transformation matrix T_{ij} and covariance matrix Σ_{ij} between pose x_i and pose x_j, which can be saved in the pose graph. Finally, the incremental QR decomposition method [14] is used to optimize the pose graph. In the next section, the MIM_SLAM algorithm will be described in detail.

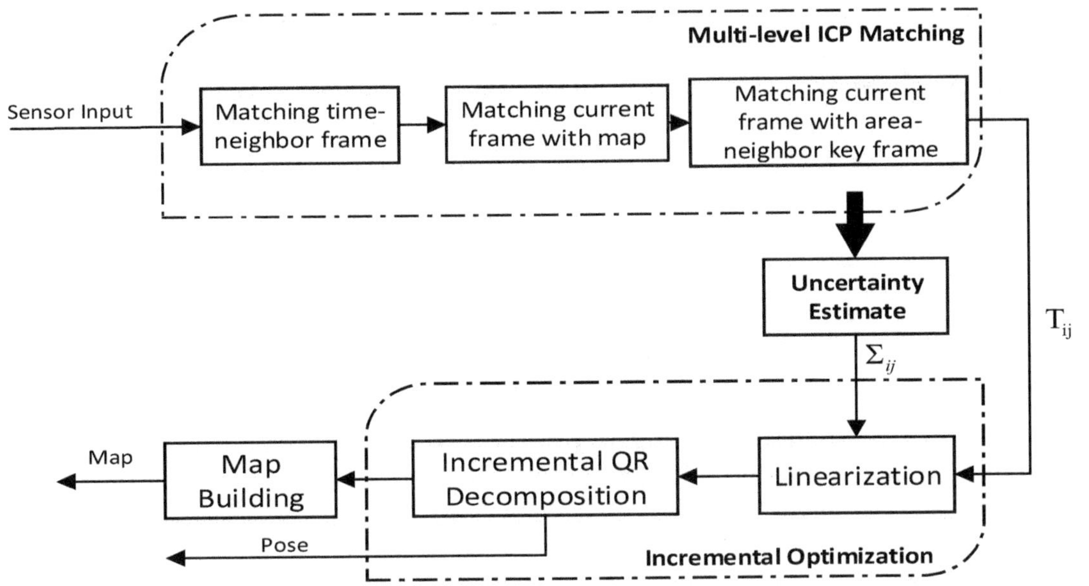

Figure 1. The proposed MIM_SLAM algorithm framework.

3. Algorithm Description

3.1. SLAM as an Incremental Optimization Problem

This part analyzes the SLAM problem from the probabilistic model and transforms the problem into a least squares problem based on the maximum a posteriori estimate. Then, the incremental QR decomposition method [17] is used to solve the least squares problem.

We denote robot pose by $X = \{x_i\}$ with $i \in 0 \ldots M$, the landmark by $L = \{l_j\}$ with $j \in 1 \ldots N$, control input given through wheeled odometry and IMU by $U = \{u_i\}$ with $i \in 1 \ldots M$, landmark measurement by $Z = \{z_i\}$ with $i = 1 \ldots M$. The SLAM problem is equivalent to estimating the posterior probability distribution:

$$\begin{aligned} &P(X,L|Z,U) \\ &= P(L|Z,X,U)P(X|Z,U) \\ &= P(L|Z,X)P(X|Z,U) \end{aligned} \tag{1}$$

From Equation (1), it can be seen that the landmark L depends on the observations and robot pose sequences, so the SLAM problem is further simplified to the pose estimation problem utilizing Baye's rule $P(X|Z,U) = P(Z|X,U)P(X|U)/P(Z|U)$. In this paper, we just consider the online SLAM problem [4], which only involves estimating the posterior over the momentary pose, this is $P(x_M|z_{1:M}, u_{1:M})$.

$$\begin{aligned} &P(x_M|z_{1:M}, u_{1:M}) \\ &= \tfrac{P(z_{1:M}|x_M, u_{1:M})P(x_M|u_{1:M})}{P(z_{1:M}|u_{1:M})} \\ &= \tfrac{P(z_M|x_M, z_{1:M-1}, u_{1:M})P(x_M|z_{1:M-1}, u_{1:M})}{P(z_M|z_{1:M-1}, u_{1:M})} \\ &= \eta P(z_M|x_M)P(x_M|z_{1:M-1}, u_{1:M}) \\ &= \eta P(z_M|x_M)\int P(x_M|x_{M-1}, z_{1:M-1}, u_{1:M})P(x_{M-1}|z_{1:M-1}, u_{1:M})dx_{M-1} \\ &= \eta P(z_M|x_M)\int P(x_M|x_{M-1}, u_M)P(x_{M-1}|z_{1:M-1}, u_{1:M-1})dx_{M-1} \\ &= \eta P(x_0)\prod_{i=1}^{M} P(x_i|x_{i-1}, u_i)P(z_i|x_i) \end{aligned} \tag{2}$$

In Equation (2) line 4 to 5, it is derived using the theorem of total probability. $P(x_0)$ is a prior on the initial state, $P(x_i|x_{i-1}, u_i)$ is the motion model, and $P(z_i|x_i)$ is the measurement model.

To estimate the pose of the robot, we transform the SLAM problem into the least-squares problem based on the maximum a posteriori estimate. The maximum a posteriori estimate X^* and L^* for the trajectory and map are obtained by minimizing the negative log of the joint probability from Equation (2):

$$\begin{aligned} X^* = -\underset{X}{\operatorname{argmin}} \eta \Big\{ \sum_{i=1}^{M} \log P(x_i|x_{i-1}, u_i) \\ + \sum_{i=1}^{M} \log P(z_i|x_i) \Big\} \end{aligned} \tag{3}$$

$$L^* = \{(x_1^*, z_1), \ldots, (x_i^*, z_i)\} \tag{4}$$

Both of the motion model and measurement model are generally assumed such that they meet the Gaussian distribution. Then, both of them can be converted to the following form:

$$x_j \sim N(x_i \oplus T_{ij}, \Sigma_{ij}) \tag{5}$$

where the operator $\oplus$ denotes the coordinate transformation. T_{ij} and Σ_{ij} represent respectively the transformation matrix and covariance matrix between pose x_i and pose x_j, which will be obtained through the multi-level ICP matching method and uncertainty estimation method in the next two paragraphs. Additionally, it can be written as follows:

$$\begin{cases} x_i = f(x_{i-1}, u_i) + m_i \\ z_i = g(x_i) + n_i \end{cases} \tag{6}$$

where m_i is the motion noise, and n_i is the measurement noise. This leads to the following nonlinear least squares problem:

$$\begin{aligned} X^* = \underset{X}{\operatorname{argmin}} \Big\{ \sum_{i=1}^{M} ||f(x_{i-1}, u_i) - x_i||_{m_i}^2 \\ + \sum_{i=1}^{M} ||g(x_i) - z_i||_{n_i}^2 \Big\} \end{aligned} \tag{7}$$

Since the above model has non-linear functions that are not easy to solve, they must be linearized, as Gauss-Newton [18] and Levenberg-Marquardt [19] has done. It can eventually be transformed into the general least-squares problem as follows, and the specific derivation can refer to the literature [14].

$$\theta^* = \underset{\theta}{\operatorname{argmin}} ||A\theta - b||^2 \tag{8}$$

where the vector θ contains all pose variables, and the matrix A is a large and sparse Jacobian matrix.

The problem in Equation (8) is apparently solved by the Cholesky decomposition method, but the essential problem is the number of calculations involved in solving the information matrix $A^T A$. Thus, applying standard QR decomposition to matrix A:

$$A = Q \begin{bmatrix} R \\ 0 \end{bmatrix} \tag{9}$$

Apply Equation (9) to the least squares problem in Equation (8). As Q is an orthonormal matrix, $Q^TQ = I$ (I is identity matrix). Thus, an additional term Q^T is added to the second line in (10), which doesn't change the length of the vector $A\theta - b$.

$$\begin{aligned} \|A\theta - b\|^2 &= \|Q\begin{bmatrix} R \\ 0 \end{bmatrix}\theta - b\|^2 \\ &= \|Q^TQ\begin{bmatrix} R \\ 0 \end{bmatrix}\theta - Q^Tb\|^2 \\ &= \|\begin{bmatrix} R \\ 0 \end{bmatrix}\theta - \begin{bmatrix} d \\ e \end{bmatrix}\|^2 \\ &= \|R\theta - d\|^2 + \|e\|^2 \end{aligned} \tag{10}$$

where $[d, e]^T = Q^Tb$ is defined, and (10) obtain the minimum $\|e\|^2$ if and only if $R\theta = d$. Now, the SLAM problem has been solved by the above analysis. However, as a new observation arrives, the Jacobian matrix A does not significantly change. Thus the results of the QR decomposition at previous times can be used as the iterative initial value and the process is incremental, which can significantly reduce computational complexity.

3.2. The Multi-Level ICP Matching Method

Before detailing this method, it is useful to understand the ICP algorithm. The standard ICP [20] (Iterative Closest Point) algorithm firstly establishes the point-to-point correspondence between the two point clouds by the nearest neighbor principle and then establishes the matching error function.

Finally, it computes a transformation matrix so that the error function is minimized. In the past 30 years, many ICP variant algorithms [21] have been proposed. In this paper, an ICP variant algorithm that calls point-to-plane ICP [22] is used. Compared with the standard ICP, the robustness and accuracy are better. This algorithm is listed as Algorithm 1.

Algorithm 1: Point-to-plane ICP.

Input: Two point cloud: $A = \{a_i\}$, $B = \{b_i\}$; An initial transformation: T_0
Output: Transformation T_{AB} which aligns A and B; Fitness Score: $Score_{fitness}$

$T \leftarrow T_0$
while not converged **do**
    for $i \leftarrow 1$ *to* N **do**
        $m_i \leftarrow \text{FindClosestPointInA}(T \cdot b_i)$
        if $\|m_i - T \cdot b_i\| \leq d_{\max}$ **then** $w_i \leftarrow 1$
        else $w_i \leftarrow 0$
        end
    end
    $T \leftarrow \text{argmin}\left\{\sum_i w_i \|\eta_i \cdot (T \cdot b_i - m_i)\|^2\right\}$
end
$T_{AB} \leftarrow T$
$Score_{fitness} = \sum_i w_i \|\eta_i \cdot (T \cdot b_i - m_i)\|^2$

The observation of the 3D LiDAR sensor is denoted as z_k at time k. There are data associations between all previous observations $z_{1:k-1}$ and the current observation included in z_k. A traditional scan matching method associates z_k with z_{k-1} at last time, or associates z_k with all previous observations $z_{1:k-1}$. These two methods either have a substantial accumulated error, or have large amounts of calculations which cannot meet online applications. In this paper, the data association is divided into two categories: continuous in time and continuous in area.

For observations that are continuous in the time, the method firstly utilizes the ICP algorithm to associate z_k and z_{k-1}, and a rough estimation of the current robot's pose can be obtained. The initial value of ICP is obtained by using a Kalman filter to merge the wheeled odometry with IMU. Because the inter-frame matching cannot obtain the accurate pose estimation, the next step will match the current frame with the map to further eliminate the accumulated error.

As shown in Figure 2, M_i denotes the global map at time i and T_i^W denotes the pose of the robot in the world coordinate system at time i. T_{i+1}^L denotes the transformation matrix from time $i+1$ to time i (that is, the output of inter-frame matching). z_{i+1} denotes the observation at time $i+1$. The process for matching the current frame with the map is as follows:

- Finding the nearest neighbor in M_i for each point in z_{i+1}, and saving it as m_{i+1}. We use the combination of octree and approximate nearest neighbor algorithm in PCL [23] for speeding up.
- Taking z_{i+1} and m_{i+1} as the input of Algorithm 1, the transformation matrix T_{opt} can be obtained after registering the two point clouds.
- The pose of the robot at time $k+1$ is $T_{i+1}^W = T_i^W \cdot T_{i+1}^L \cdot T_{opt}$.

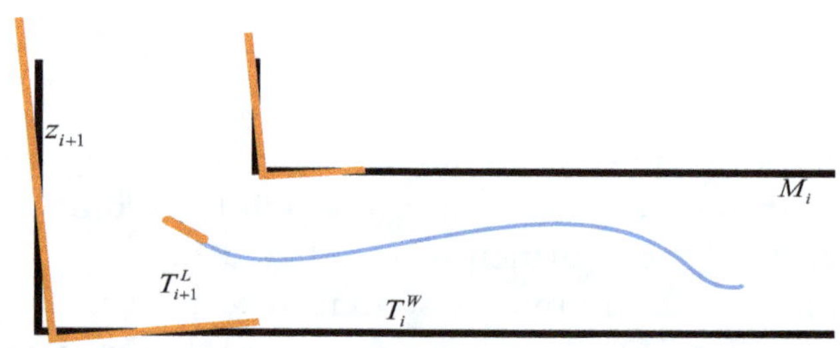

Figure 2. The matching process of the current frame and map.

The effectiveness of the above scan matching method is verified as shown in Figure 3. The white point clouds denote the map that SLAM has established. The blue point clouds denote the point clouds after inter-frame matching. The redpoint clouds denote the point clouds after matching the current frame with the map. It can be seen that the accumulated error in the process of inter-frame matching is obviously eliminated.

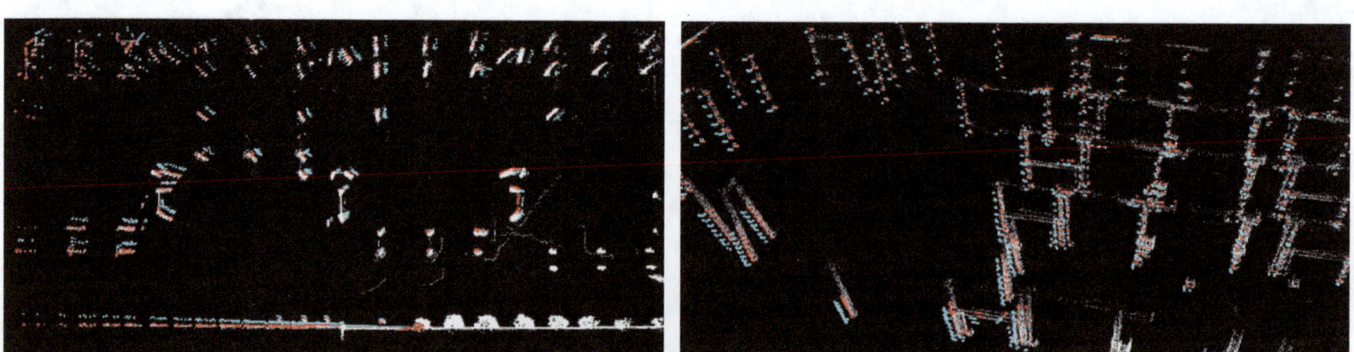

(**a**) Scene 1 (**b**) Scene 2

Figure 3. An example of the matching process.

For observations that are continuous in the area, the robot returns to a region that it has previously explored. In particular, in large-scale and sparse environments, the accumulative error cannot be completely eliminated when the robot travels a long distance if only the above two matching processes are considered. So the following method matches the current frame z_i with the area-neighbor keyframe

$z_\mathcal{T}$ based on a priori of pose estimation. If the matching score is less than a certain threshold, the pose constraint between x_i and $x_\mathcal{T}$ will be established. The process is listed as Algorithm 2. The initial parameter is: $L = \{(0, x_0, z_0)\}, d = 0$.

Algorithm 2: Matching the current frame with an area-neighbor keyframe.

Input: Current pose and observation: x_i, z_i
Output: Transformation $T_{i\mathcal{T}}$ which aligns z_i and $z_\mathcal{T}$

calculate $d = d + \|x_i - x_{i-1}\|$
if $d < d_{thre}$ **then**
  return null
end
foreach $(k_\mathcal{T}, x_\mathcal{T}, z_\mathcal{T})$ **in** L **do**
  if $k - k_\mathcal{T} > k_{thre}$ **then**
    if $\|x_t - x_\mathcal{T}\| < x_{thre}$ **then**
      put z_t and $z_\mathcal{T}$ into Algorithm 1, get T and $Score_{fitness}$
      if $Score_{fitness} < S_{thre}$ **then**
        $T_{i\mathcal{T}} \leftarrow T$
        return $T_{i\mathcal{T}}$
      end
    end
  end
end
$L = L \cup (k_i, x_i, z_i), d = 0, k = k + 1$
return null

3.3. Uncertainty Estimation

After multi-level ICP matching, the transformation matrix T_{ij} between pose x_i and pose x_j can be obtained, but we still need to know the exact uncertainty estimation, namely the covariance matrix Σ_{ij}. However, an incorrect covariance matrix may damage the established map.

In this paper, the inversion of the Fisher information matrix is used as the covariance. The Fisher information matrix is defined as the function of the expected measurement and the surface slope scanned by the laser sensor [24]. Liu [24] and Wang [25] give the derivation of the Fisher information matrix based on a two-dimensional probability grid. It is now extended to three-dimensions. The Fisher information matrix is discretized:

$$\hat{L}(p) = \sum_{i=1}^{N} \frac{1}{\sigma_i^2} \left(\frac{\Delta r_{iE}}{\Delta p}\right)^T \left(\frac{\Delta r_{iE}}{\Delta p}\right) \tag{11}$$

$$\frac{\Delta r_{iE}}{\Delta p} = \left[\frac{\Delta r_{iE}}{\Delta x}, \frac{\Delta r_{iE}}{\Delta y}, \frac{\Delta r_{iE}}{\Delta z}, \frac{\Delta r_{iE}}{\Delta \varphi}, \frac{\Delta r_{iE}}{\Delta \psi}, \frac{\Delta r_{iE}}{\Delta \theta}\right] \tag{12}$$

where $p = [x, y, z, \varphi, \psi, \theta]$ is the pose of the robot, and σ_i^2 is the noise variance of the i-th 3D laser range finder scan ray. N is the total number of scan rays. r_{iE} is the expected distance from the robot to the nearest obstacle along the i-th scan ray. To calculate Equation (12) efficiently, the point cloud map is converted into a three-dimensional grid map. Thus r_{iE} can be computed as follows:

$$r_{iE} = \frac{\sum_{j=1}^{s} r_{ij}\mu_{ij}}{\sum_{j=1}^{s} \mu_{ij}} \tag{13}$$

where r_{ij} is the distance between the robot and the j-th voxel along the direction of an i-th scan ray. μ_{ij} is the occupancy value of the corresponding voxel. s is the sequence number of the ending voxel. By combining Equation (11) and Equation (12), we obtain:

$$\hat{L}(p) = \sum_{i=1}^{N} \frac{1}{\sigma_i^2} \begin{bmatrix} \frac{\Delta r_{iE}^2}{\Delta x^2} & \frac{\Delta r_{iE}^2}{\Delta x \Delta y} & \frac{\Delta r_{iE}^2}{\Delta x \Delta z} & \frac{\Delta r_{iE}^2}{\Delta x \Delta \varphi} & \frac{\Delta r_{iE}^2}{\Delta x \Delta \psi} & \frac{\Delta r_{iE}^2}{\Delta x \Delta \theta} \\ \frac{\Delta r_{iE}^2}{\Delta x \Delta y} & \frac{\Delta r_{iE}^2}{\Delta y^2} & \frac{\Delta r_{iE}^2}{\Delta y \Delta z} & \frac{\Delta r_{iE}^2}{\Delta y \Delta \varphi} & \frac{\Delta r_{iE}^2}{\Delta y \Delta \psi} & \frac{\Delta r_{iE}^2}{\Delta y \Delta \theta} \\ \frac{\Delta r_{iE}^2}{\Delta x \Delta z} & \frac{\Delta r_{iE}^2}{\Delta y \Delta z} & \frac{\Delta r_{iE}^2}{\Delta z^2} & \frac{\Delta r_{iE}^2}{\Delta z \Delta \varphi} & \frac{\Delta r_{iE}^2}{\Delta z \Delta \psi} & \frac{\Delta r_{iE}^2}{\Delta z \Delta \theta} \\ \frac{\Delta r_{iE}^2}{\Delta x \Delta \varphi} & \frac{\Delta r_{iE}^2}{\Delta y \Delta \varphi} & \frac{\Delta r_{iE}^2}{\Delta z \Delta \varphi} & \frac{\Delta r_{iE}^2}{\Delta \varphi^2} & \frac{\Delta r_{iE}^2}{\Delta \varphi \Delta \psi} & \frac{\Delta r_{iE}^2}{\Delta \varphi \Delta \theta} \\ \frac{\Delta r_{iE}^2}{\Delta x \Delta \psi} & \frac{\Delta r_{iE}^2}{\Delta y \Delta \psi} & \frac{\Delta r_{iE}^2}{\Delta z \Delta \psi} & \frac{\Delta r_{iE}^2}{\Delta \varphi \Delta \psi} & \frac{\Delta r_{iE}^2}{\Delta \psi^2} & \frac{\Delta r_{iE}^2}{\Delta \psi \Delta \theta} \\ \frac{\Delta r_{iE}^2}{\Delta x \Delta \theta} & \frac{\Delta r_{iE}^2}{\Delta y \Delta \theta} & \frac{\Delta r_{iE}^2}{\Delta z \Delta \theta} & \frac{\Delta r_{iE}^2}{\Delta \varphi \Delta \theta} & \frac{\Delta r_{iE}^2}{\Delta \psi \Delta \theta} & \frac{\Delta r_{iE}^2}{\Delta \theta^2} \end{bmatrix} \quad (14)$$

According to the Cramér–Rao Bound theory [26], the lower bound of covariance can be determined by the inversion of the Fisher information matrix. Finally, the uncertainty of the pose estimation can be obtained:

$$\text{cov}(p) = \hat{L}^{-1}(p) \quad (15)$$

4. Experimental Results and Analysis

As shown in Figure 4, the mobile robot called SmartGuard [27] is used to verify the proposed algorithm. It is a completely autonomous robotic system that can inspect substation equipment. SmartGuard is equipped with a wheeled odometry, an IMU, and a 3D LIDAR (RS-Lidar-16). The RS-Lidar-16 can measure out to 150 m with a high precision, ±2 cm, and it has a +15 to −15-degree vertical field of view. It continuously scans the 360-degree surrounding environment at a 10 Hz frame rate and at 300,000 points/sec. The capability of the computer that runs the SLAM algorithm is as follows: Intel Core i5-6300HQ CPU 2.3 Hz and 8 G DDR3 RAM.

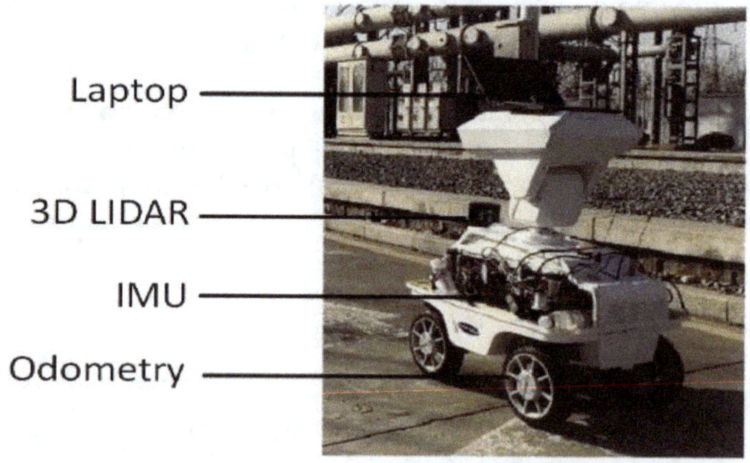

Figure 4. The mobile robot (SmartGuard) for experiments.

4.1. *Indoor and Outdoor Mapping Test*

To verify the mapping performance, the proposed algorithms have been tested in several different indoor and outdoor environments (Figure 5). Figure 5a shows that the proposed MIM_SLAM algorithm can build a consistency map with lower accumulated errors after a 180 m long loop. Figure 5b shows that the consistency map can also be generated from a narrow and long corridor. As shown in Figure 5c,d, in the wide range of the parking lot, the robot traveled approximately 1600 m at a speed of 0.6 m/s; the trajectory is shown in Figure 5c. The map shown in Figure 5c is still accurate and clear.

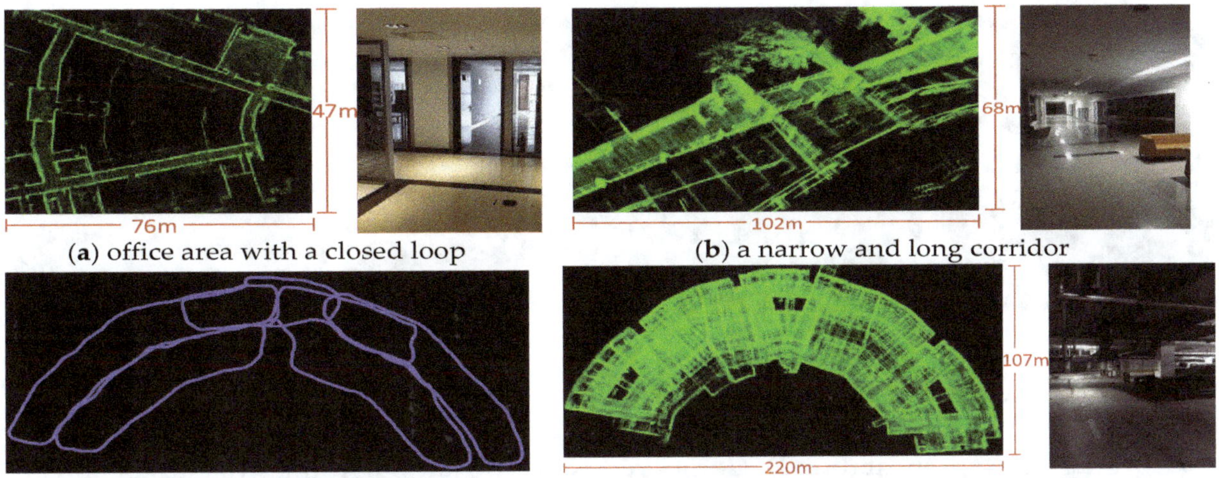

(**a**) office area with a closed loop (**b**) a narrow and long corridor

(**c**) mapping trajectory in a parking lot (**d**) a wide range of parking lot

Figure 5. The maps generated in the indoor and outdoor environments.

4.2. Large-Scale and Sparse Scenes Test

Two substations are selected to verify the validity of the MIM_SLAM algorithm in the larger-scale and sparse environments. A general substation comprises thousands of square meters, and its electrical equipment needs to include a sufficient distance for electrical safety. We first tested in substation A shown in Figure 6. During the experiment, the robot traveled approximately 600 m at a speed of 0.6 m/s and the 3D LIDAR sensor collected a total of 10,080 frames. The plan and the environment of substation A are shown in Figure 6. The red line denotes the robot's trajectory ABCDA. The point cloud maps established by the MIM_SALM and the LOAM [12] which has been considered state-of-the-art in LiDAR SLAM are shown in Figure 7.

Intuitively, the MIM_SLAM algorithm has a high consistency map compared with the LOAM. Due to the accumulated error, the estimated pose has a great deviation when the robot travels from area A along the red track shown in Figure 6, and then returns back to area A. The LOAM cannot effectively handle the data association at this moment, which leads to the wrong association and damage to the established map. In this paper, the multi-level ICP matching method takes into account this situation and generates a reliable constraint between the two poses. The incremental optimization method is used to optimize the global pose. Thus the final map has a high consistency.

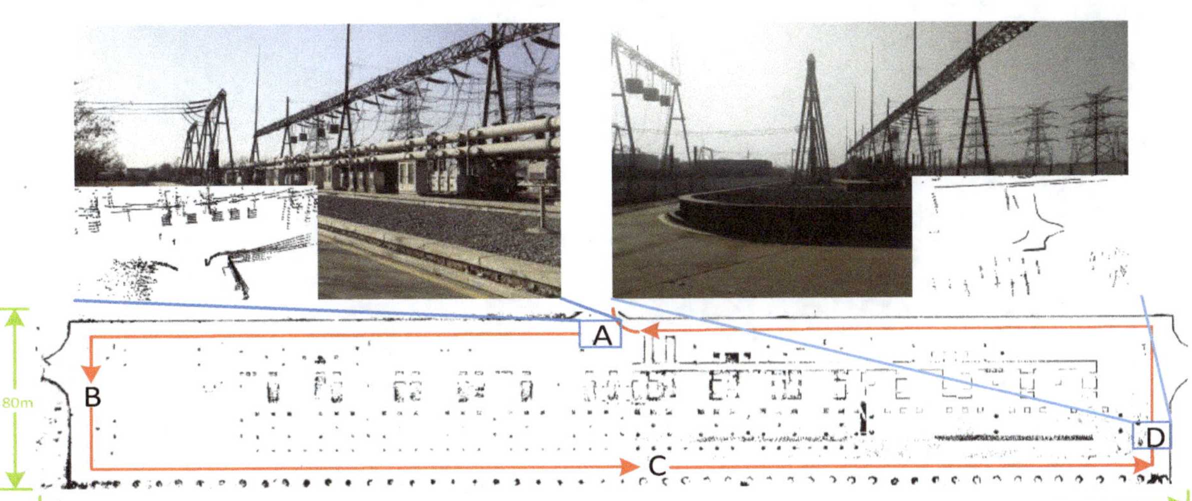

Figure 6. The plan and the environment of substation A.

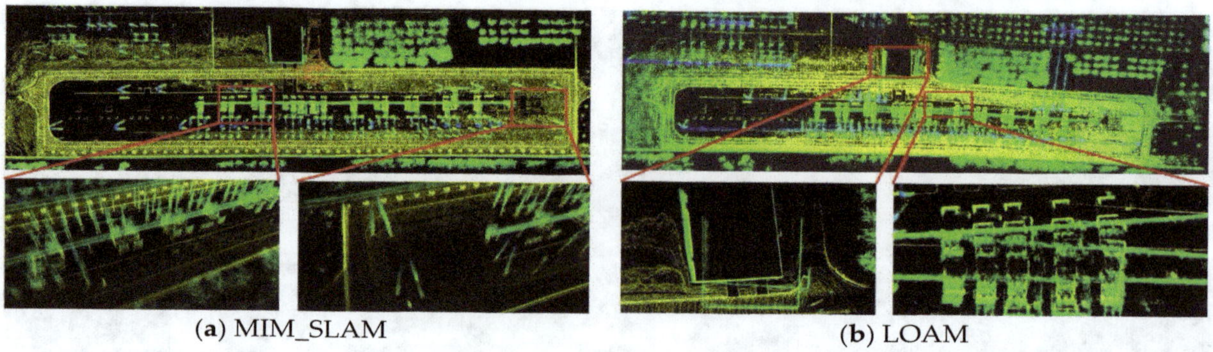

Figure 7. The point cloud maps established by the MIM_SLAM and the LOAM [12] algorithms.

Next, we tested it in substation B. Compared with substation A, this was larger. Meanwhile, the electrical equipment is placed more sparsely, and less information is gathered by the 3D LIDAR sensor. During the experiments, the robot traveled approximately 1450 m at a speed of 0.6 m/s, and the LIDAR sensor collected a total of 24,220 frames. The plan and the environment of substation B are shown in Figure 8. The red line denotes the robot's trajectory ABCDEABFGDC. The point cloud maps established by the MIM_SLAM algorithm and the LOAM are shown in Figure 9.

It can be seen from the observation of the two scenes given in Figure 8 that the electrical equipment is distributed more sparsely, and more than 70% of the laser rays reach the maximum measurement range. As shown in Figure 9a, both contour lines of the map established by the MIM_SLAM algorithm are at right angles at areas F and G. Additionally, when the robot returned back to the area C, the details of the point cloud map at area C showed that the telegraph poles and the road edge are clearly visible, and there is no incorrect point cloud accumulation. Because the LOAM cannot effectively process the data association of the closed-loop area, the performance is very poor in this situation, which has more closed-loop areas in large-scale environments, as shown in Figure 9b.

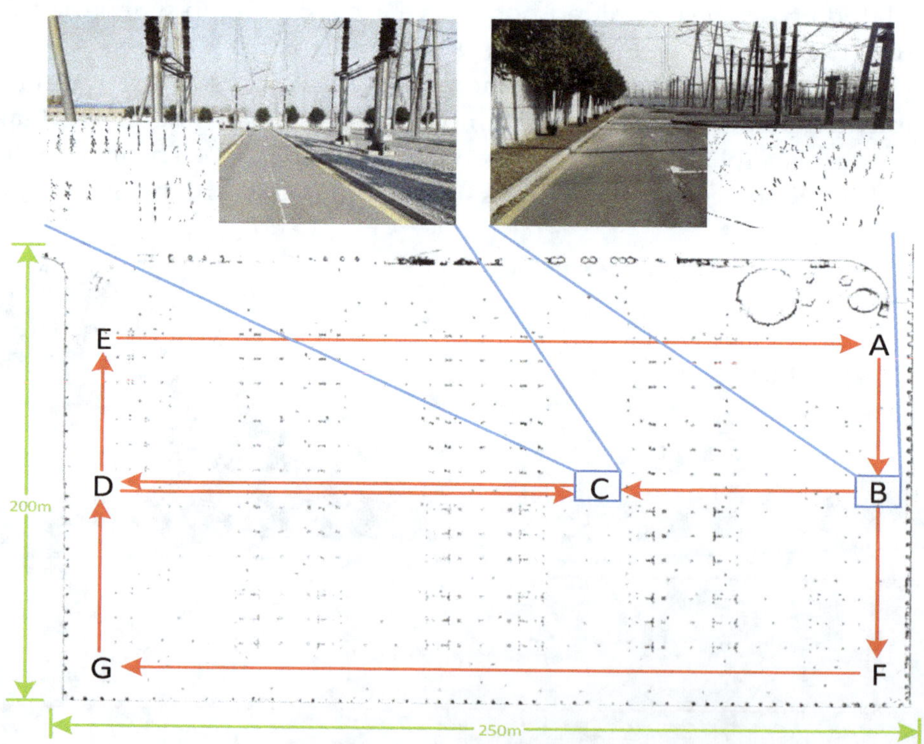

Figure 8. The plan and the environment of substation B.

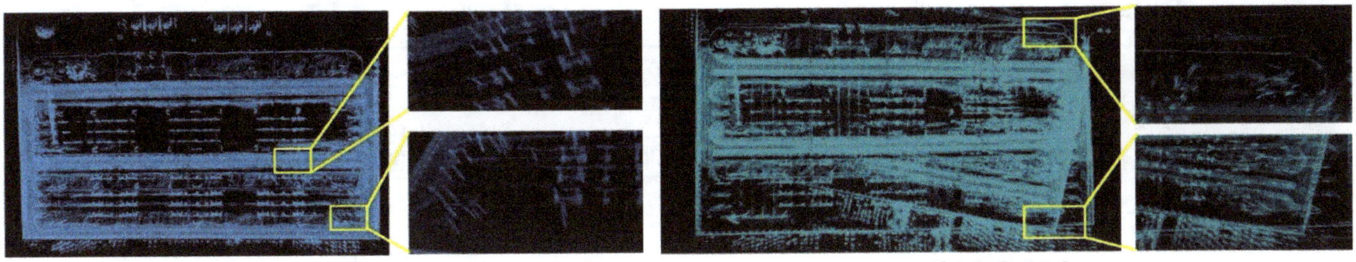

(**a**) MIM_SLAM (**b**) LOAM

Figure 9. The point cloud maps established by the MIM_SLAM and the LOAM algorithms.

4.3. Mapping Accuracy Test

In order to analyze the mapping accuracy quantitatively, we selected two substations with large-scale and sparse characteristics which are shown in Figures 6 and 8, also the standard ICP [20] and the LOAM [12] algorithm was selected to compare with the MIM_SLAM algorithm. In the experiments, the robot is controlled running at a speed of 0.6 m/s on a flat road along a straight line with a length of 10 m. The accuracy of mapping is defined as the robot's localization error ε [28], expressed as follows:

$$\begin{aligned} \varepsilon &= \frac{1}{N}\sum_{i=0}^{N-1}\|e_i\| \\ e_i &= (x_0 \oplus T_i) - (x_0^* \oplus T_i^*) \end{aligned} \tag{16}$$

where x_0 and x_0^* denote the initial pose. $\oplus$ is the standard motion composition operator. T_i denotes the transformation matrix of the estimated pose x_i relative to the initial pose x_0. T_i^* denotes the transformation matrix of the true pose x_i^* relative to the initial pose x_0^*. $\|\cdot\|$ is the 2-Norm used in this paper. For the robot to navigate in three-dimensional environments, e_i is represented as $(x_e, y_e, z_e, \varphi_e, \psi_e, \theta_e)$. Then ε can be divided into two parts: translation error and rotation error.

$$\varepsilon = \frac{1}{N}\sum_{i=0}^{N-1}\|trans(e_i)\| + \frac{1}{N}\sum_{i=0}^{N-1}\|rot(e_i)\| \tag{17}$$

Since it is difficult to obtain the true pose of the robot, 20 points are selected from the robot's trajectory. The first point is taken as the initial pose. For the next 19 points, the difference value of the localization results relative to the initial pose is calculated. The difference value of the true pose can be measured by the tape. In Table 1, the localization error is tested with the standard ICP, the LOAM and the MIM_SLAM algorithms under substation A and B. It can be seen that the MIM_SLAM algorithm is more precise. If there is a closed-loop area on the robot's path, the MIM_SLAM algorithm will be better than the LOAM. The processing time per frame using different approaches is calculated and shown in Table 2. MIM_SLAM has a similar time consuming compared to the LOAM and meets the real-time requirements. In Section 4.4, this will be further analyzed in the benchmark dataset.

Table 1. The localization error results of different approaches/scenes.

	Standard ICP	**LOAM**	**MIM_SALM**
		Trans. Error (unit: m)	
A	0.213 ± 0.148	0.064 ± 0.057	0.052 ± 0.043
B	0.282 ± 0.177	0.075 ± 0.066	0.066 ± 0.062
		Rot. Error (unit: deg)	
A	2.5 ± 1.9	1.8 ± 1.2	1.6 ± 1.4
B	3.1 ± 2.2	2.5 ± 2.1	2.2 ± 1.8

Table 2. The average time consumption per frame for different approaches/scenes.

	Standard ICP	LOAM	MIM_SALM
		Processing Time (unit: s)	
A	0.0405	0.1036	0.0886
B	0.0416	0.1067	0.0932

4.4. Benchmarking Datasets Test

Then, datasets from the KITTI odometry benchmark [29] are used to evaluated MIM_SLAM. The datasets take advantage of the autonomous driving platform Annieway to develop novel challenging real-world computer vision benchmarks. The autonomous driving platform is equipped with a 360° Velodyne HDL-64E laser scanner and two high-resolution color and grayscale video cameras. Accurate ground truth is provided by a GPS localization system. We selected three typical types of environments: "urban" with building around (sequence 07), "country" on small roads with vegetation in this scene (sequence 03), and "highway" where roads are wide and lower dynamic (sequence 06).

(1) Sequence 03: This dataset is designed to verify that MIM_SLAM can achieve a low drift pose estimation in sparse vegetation environment. The mapping result is shown in Figure 10a and the trajectory and the ground truth are shown in Figure 11a. Due to the scarce stable features in this scene, there is a bit of position deviation after 420 m of traveling compared with the ground truth in Figure 11a. To evaluate the pose estimation accuracy, we use the evaluation method in the KITTI odometry benchmark which calculated the translational and rotational errors for all possible subsequences of length (100, 200, . . . , 800) meters. As is shown in Figure 12a, both LOAM and MIM_SLAM achieve lower drift values compared with the standard ICP method, and MIM_SLAM is slightly worse than LOAM.

(2) Sequence 06: We use this dataset to verify that MIM_SLAM can effectively process the data association of the closed-loop area shown in Figures 10 and 11. From Figure 12b, it can be seen that the translational and rotational errors are lower compared to the standard ICP, and slightly better than LOAM.

(3) Sequence 07: This urban road scene is highly dynamic and large-scale. The vehicle travels approximately 660 m at a speed of 6.2 m/s. The mapping result is shown in Figure 10c and the trajectory and the ground truth are shown in Figure 11c. Intuitively, the details of the point cloud map are clear. The estimated pose and the ground truth are almost overlapping. The pose estimation error is shown in Figure 12c quantitatively.

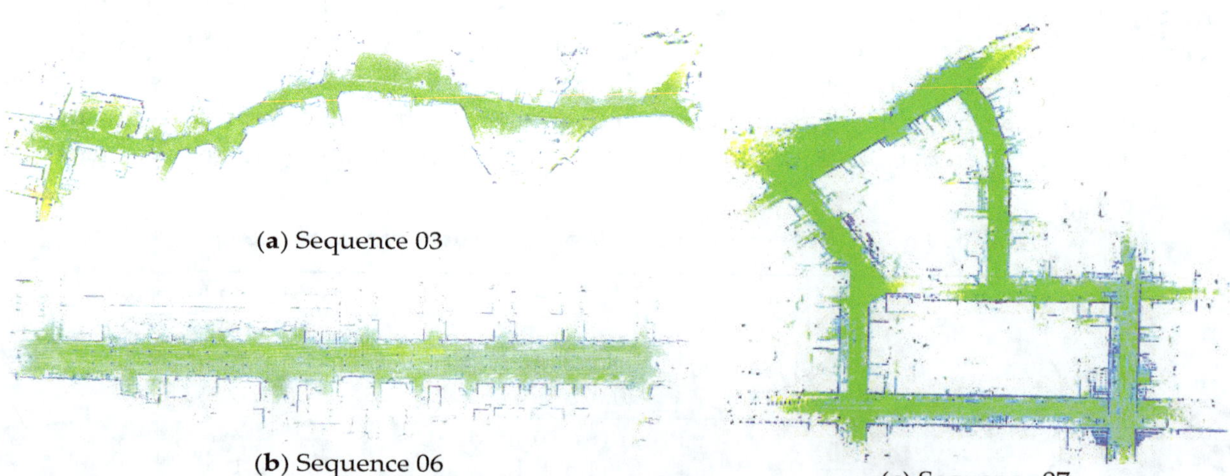

(**a**) Sequence 03

(**b**) Sequence 06

(**c**) Sequence 07

Figure 10. The mapping results in the three sequences.

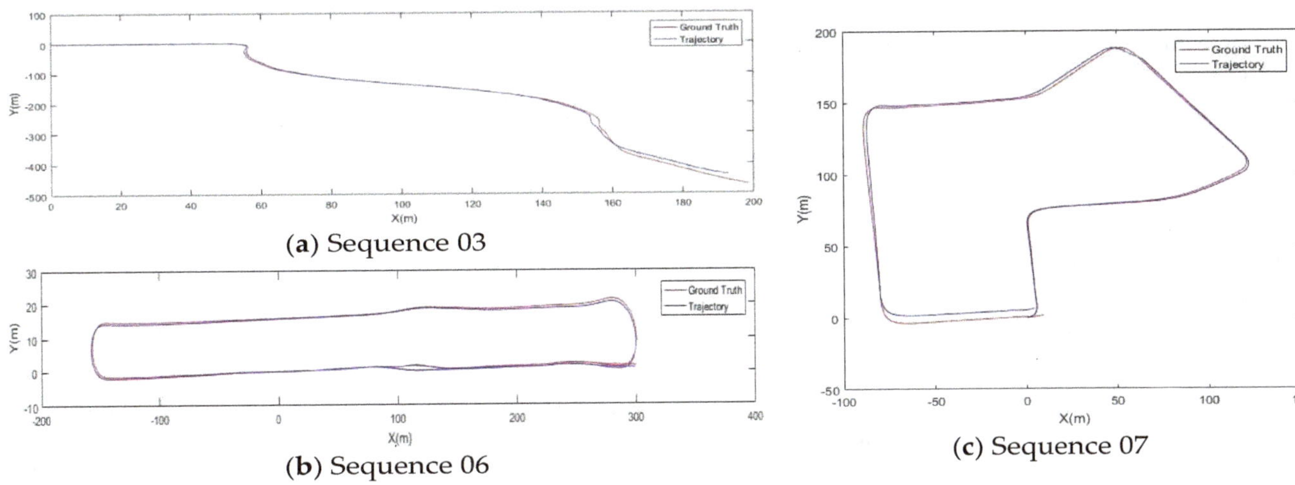

(a) Sequence 03

(b) Sequence 06

(c) Sequence 07

Figure 11. The trajectory produced by MIM_SLAM in the three sequences.

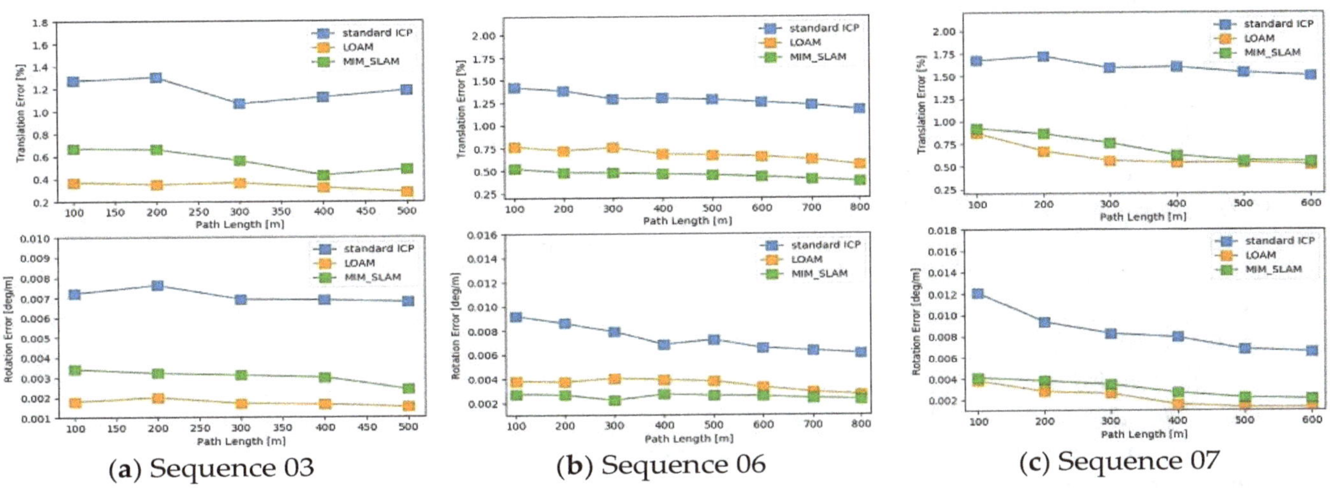

(a) Sequence 03

(b) Sequence 06

(c) Sequence 07

Figure 12. The pose estimation error in the three sequences.

5. Conclusions

For large-scale and sparse environments, we propose a novel MIM_SLAM algorithm in this paper, which simplifies the SLAM problem to the least-squares optimization problem. The contribution of this paper is that the multi-level ICP matching method is proposed to solve the data association problem and the uncertainty estimation is handled by the Fisher information matrix. It considers the accumulated errors in two aspects: matching between time-neighbor frames (building a low-drift LiDAR odometry) and matching between area-neighbor frames (dealing with the data association at a revisited area). Moreover, multi-level matching and incremental optimization reduce the computational complexity while ensuring mapping accuracy, we can achieve accurate mapping and real-time applications. Additionally, its application is not only limited to mobile robots, and it can potentially be extended to other vehicles, e.g., UAVs. Experimental results show that it can effectively build a high consistency map with a smaller amount of environmental information and the large-scale scenes, and it also achieves similar or better accuracy compared with the standard ICP and the state-of-the-art LOAM algorithm in the KITTI dataset. Additionally, there are some limitations in high dynamic scenes. Our future work will focus on the construction of a more robust data association method, we will integrate the output of our multi-level ICP matching with an IMU in a Kalman filter to further reduce the accumulated error.

Author Contributions: This study was completed by the co-authors. J.W. conceived and led the research. The major experiments and analyses were undertaken by J.W. and M.Z. W.C. supervised and guided this study. J.W., M.Z. and W.C. wrote the paper. All the authors have read and approved the final manuscript.

References

1. Martínez, J.L.; Morán, M.; Morales, J.; Reina, A.J.; Zafra, M. Field Navigation Using Fuzzy Elevation Maps Built with Local 3D Laser Scans. *Appl. Sci.* **2018**, *8*, 397. [CrossRef]
2. Smith, R.; Cheeseman, P. On the Estimation and Representations of Spatial Uncertainty. *Int. J. Robot. Res.* **1987**, *5*, 56–68. [CrossRef]
3. Shinsuke, M.M.D.; Noboru, I.M.D.; Yoshito, I.M.D. Estimating Uncertain Spatial Relationships in Robotics. *Mach. Intell. Pattern Recognit.* **2013**, *5*, 435–461.
4. Thrun, S.; Liu, Y.; Koller, D.; Ng, A.Y.; Ghahramani, Z.; Durrant-Whyte, H. Simultaneous localization and mapping with sparse extended information filters. *Int. J. Robot. Res.* **2004**, *23*, 693–716. [CrossRef]
5. Montemerlo, M.S. Fastslam: A factored solution to the simultaneous localization and mapping problem with unknown data association. *Arch. Environ. Contam. Toxicol.* **2015**, *50*, 240–248.
6. Gutmann, J.S.; Konolige, K. Incremental mapping of large cyclic environments. In Proceedings of the 1999 IEEE International Symposium on Computational Intelligence in Robotics and Automation (CIRA '99), Monterey, CA, USA, 8–9 November 2002; pp. 318–325.
7. Zlot, R.; Bosse, M. Efficient large-scale 3D mobile mapping and surface reconstruction of an underground mine. In *Field and Service Robotics*; Springer: Berlin/Heidelberg, Germany, 2014; pp. 479–493.
8. Olufs, S.; Vincze, M. An efficient area-based observation model for monte-carlo robot localization. In Proceedings of the 2009 IEEE/RSJ International Conference on Intelligent Robots and Systems (IROS 2009), St. Louis, MO, USA, 10–15 October 2009; pp. 13–20.
9. Huang, S.; Wang, Z.; Dissanayake, G. Sparse local submap joining filter for building large-scale maps. *IEEE Trans. Robot.* **2008**, *24*, 1121–1130. [CrossRef]
10. Moosmann, F.; Stiller, C. Velodyne slam. In Proceedings of the 2011 IEEE Intelligent Vehicles Symposium (IV), Baden, Germany, 5–9 June 2011; pp. 393–398.
11. Nüchter, A.; Lingemann, K.; Hertzberg, J.; Surmann, H. 6D SLAM-3D mapping outdoor environments. *J. Field Robot.* **2007**, *24*, 699–722. [CrossRef]
12. Zhang, J.; Singh, S. LOAM: Lidar Odometry and Mapping in Real-time. In Proceedings of the 2014 Robotics: Science and Systems, Berkeley, CA, USA, 12–16 July2014; Volume 2.
13. Wang, J.; Li, L.; Zhe, L.; Weidong, C. Multilayer matching SLAM for large-scale and spacious environments. *Int. J. Adv. Robot. Syst.* **2015**, *12*, 124. [CrossRef]
14. Liu, Z.; Chen, H.; Di, H.; Tao, Y.; Gong, J.; Xiong, G.; Qi, J. Real-Time 6D Lidar SLAM in Large Scale Natural Terrains for UGV. In Proceedings of the 2018 IEEE Intelligent Vehicles Symposium (IV), Changshu, China, 26–30 June 2018; pp. 662–667.
15. Liang, X.; Chen, H.; Li, Y.; Liu, Y. Visual laser-SLAM in large-scale indoor environments. In Proceedings of the IEEE International Conference on Robotics and Biomimetics, Qingdao, China, 3–7 December 2017; pp. 19–24.
16. Hess, W.; Kohler, D.; Rapp, H.; Andor, D. Real-time loop closure in 2D LIDAR SLAM. In Proceedings of the IEEE International Conference on Robotics and Automation, Stockholm, Sweden, 16–21 May 2016; pp. 1271–1278.
17. Kaess, M.; Ranganathan, A.; Dellaert, F. iSAM: Incremental smoothing and mapping. *IEEE Trans. Robot.* **2008**, *24*, 1365–1378. [CrossRef]
18. Hartley, H.O. The modified Gauss-Newton method for the fitting of non-linear regression functions by least squares. *Technometrics* **1961**, *3*, 269–280. [CrossRef]
19. Moré, J.J. The Levenberg-Marquardt algorithm: Implementation and theory. In *Numerical Analysis*; Springer: Berlin/Heidelberg, Germany, 1978; pp. 105–116.
20. Besl, P.J.; Mckay, N.D. Method for registration of 3-D shapes. *IEEE Trans. Pattern Anal. Mach. Intell.* **2002**, *14*, 239–256. [CrossRef]
21. Li, L.; Liu, J.; Zuo, X.; Zhu, H. An Improved MbICP Algorithm for Mobile Robot Pose Estimation. *Appl. Sci.* **2018**, *8*, 272. [CrossRef]
22. Chen, Y.; Medioni, G. Object modelling by registration of multiple range images. *Image Vis. Comput.* **1992**, *10*, 145–155. [CrossRef]

23. Rusu, R.B.; Cousins, S. 3D is here: Point Cloud Library (PCL). In Proceedings of the 2011 IEEE International Conference on Robotics and Automation, Shanghai, China, 9–13 May 2011; pp. 1–4.
24. Liu, Z.; Chen, W.; Wang, Y.; Wang, J. Localizability estimation for mobile robots based on probabilistic grid map and its applications to localization. In Proceedings of the 2012 IEEE Conference on Multisensor Fusion and Integration for Intelligent Systems (MFI), Hamburg, Germany, 13–15 September 2015; pp. 46–51.
25. Wang, Y.; Chen, W.; Wang, J.; Wang, H. Active global localization based on localizability for mobile robots. *Robotica* **2015**, *33*, 1609–1627. [CrossRef]
26. Bobrovsky, B.; Zakai, M. A lower bound on the estimation error for certain diffusion processes. *IEEE Trans. Inf. Theory* **1976**, *22*, 45–52. [CrossRef]
27. Wang, B.; Guo, R.; Li, B.; Han, L.; Sun, Y.; Wang, M. SmartGuard: An autonomous robotic system for inspecting substation equipment. *J. Field Robot.* **2012**, *29*, 123–137. [CrossRef]
28. Burgard, W.; Stachniss, C.; Grisetti, G.; Steder, B.; Kümmerle, R.; Dornhege, C.; Ruhnke, M.; Kleiner, A.; Tardös, J.D. A comparison of SLAM algorithms based on a graph of relations. In Proceedings of the 2009 IEEE/RSJ International Conference on Intelligent Robots and Systems (IROS 2009), St. Louis, MO, USA, 10–15 October 2009; pp. 2089–2095.
29. Geiger, A. Are we ready for autonomous driving? The KITTI vision benchmark suite. In Proceedings of the 2012 Computer Vision and Pattern Recognition, Providence, RI, USA, 16–21 June 2012; pp. 3354–3361.

10

Design and Experiment of a Variable Spray System for Unmanned Aerial Vehicles based on PID and PWM Control

Sheng Wen [1,2], Quanyong Zhang [2,3], Jizhong Deng [2,3,*], Yubin Lan [2,3], Xuanchun Yin [2,3] and Jian Shan [2,3]

1 Engineering Foundation Teaching and Training Center, South China Agricultural University, Guangzhou 510642, China; vincen@scau.edu.cn

2 National Center for International Collaboration Research on Precision Agriculture Aviation Pesticides Spraying Technology, Guangzhou 510642, China; qy_zhang@stu.scau.edu.cn (Q.Z.); ylan@scau.edu.cn (Y.L.); xc_yin@scau.edu.cn (X.Y.); shanj@stu.scau.edu.cn (J.S.)

3 Engineering College, South China Agricultural University, Guangzhou 510642, China

* Correspondence: jz-deng@scau.edu.cn

Abstract: Unmanned aerial vehicle (UAV) variable-rate spraying technology, as the development direction of aviation for plant protection in the future, has been developed rapidly in recent years. In the actual agricultural production, the severity of plant diseases and insect pests varies in different locations. In order to reduce the waste of pesticides, pesticides should be applied according to the severity of pests, insects and weeds. On the basis of explaining the plant diseases and insect pests map in the target area, a pulse width modulation variable spray system is designed. Moreover, the STMicroelectronics-32 (STM32) chip is invoked as the core of the control system. The system combines with sensor technology to get the prescription value through real-time interpretation of prescription diagram in operation. Then, a pulse square wave with variable duty cycles is generated to adjust the flow rate. A closed-loop Proportional-Integral-Derivative (PID) control algorithm is used to shorten the time of system reaching steady state. The results indicate that the deviation between volume and target traffic is stable, which is within 2.16%. When the duty cycle of the square wave is within the range of 40% to 100%, the flow range of the single nozzle varies from 0.16 L/min to 0.54 L/min. Variable spray operation under different spray requirements is achieved. The outdoor tests of variable spray system show that the variable spray system can adjust the flow rapidly according to the prescription value set in the prescription map. The proportion of actual droplet deposition and deposition density in the operation unit is consistent with the prescription value, which proves the effectiveness of the designed variable spray system.

Keywords: UAV; variable spray; prescription map translation; PID algorithm

---

1. Introduction

Pests and diseases of crops are a major factor affecting the yield and quality of crops, and chemical pesticides are the main means for their prevention and control. The pesticide application method currently used in China is mainly based on uniform spraying. Therefore, the utilization rate of pesticides and herbicides is still low [1]. According to statistics, as of 2017, the pesticide utilization rate in china was only 36.6%, which was lower than the level of 50% in developed countries [2–5]. The extensive use of pesticides directly endangers the ecological environment and human health. Therefore, pesticide reduction and efficiency have been realized worldwide. In the field of plant protection, the variable spray technology can be applied on demand. It has definite prospects and potentialities in improving the utilization rate of pesticides and reducing pesticide residues [6]. Chen et al. [7] developed a field information processing system based on a Beidou positioning embedded vehicle variable sprayer. The system was able to complete the spray operation according to the generated job prescription map. However, the performance of field information processing system was not completely verified by the field testing. Perez-Ruiz et al. [8] used a geospatial prescription map prepared for olive trees, along with Real-Time Kinematic-Global Positioning System (RTK-GPS)-based position information to control the spray rate. This system only implements variable spray based on tree shape, and does not combine the degree of disease and pest with tree shape. Qiu et al. [9] used a variable-pressure spray system that controls the flow rate with an electric control valve. The step responses of the five target flows were measured experimentally. The results showed that the rise time, peak time and overshoot of the nonlinear system have amplitude correlation. However, the way of regulating flow by an electronic control valve is only applicable to ground machinery. Compared with the high-speed flight of plant protection unmanned aerial vehicles (UAV), the response time of the electronic control valve has a greater impact on the system. Gonzalez et al. [10] designed a nonlinear variable spray system based on pressure regulation. By establishing the transfer function of the open-loop system, the nonlinear control of the variable spray system was realized. However, for the low volume spray of plant protection UAV, the droplet size and the effect of spraying operation will be affected by the pressure. Shahemabadi and Moayed [11] proposed an algorithm to improve the Pulse width modulation (PWM) algorithm. By controlling the rising or falling state of the valve opening corresponding to the high and low pulse levels, an adjustment range of the flow rate from 0% to 100% can be realized according to the adjustment precision of 2.5%. The response time by using pulse to adjust valve opening cannot meet the requirements of UAV high-speed flight.

At present, the precision variable spraying technology for ground plant protection machinery in China is developing more rapidly. However, the technology of agricultural aviation plant protection is still in its infancy, mainly because the plant protection UAV is faster than the ground plant protection machinery. UAV has the characteristics of high control precision and fast response speed to the variable spray system. In this study, a plant protection UAV variable spray system is designed based on the interpretation of the work prescription map. The system obtains prescription information for real-time location through graphic interpretation of plant protection UAV, and PWM-Proportional–Integral–Derivative (PID) control is used to adjust the spray volume quickly and accurately. The stability and feasibility of the system are verified by experiments, which provide a theoretical reference for the research of plant protection UAV variable spray system and extend to the field of modern agricultural aviation for plant protection.

The paper is organized as follows: Section 2 introduces the working principle of the plant protection UAV variable spray system based on prescription diagram interpretation. Section 3 introduces the components, design and simulation of PID algorithm in detail. Section 4 presents the results of the experiments and discussion. Finally, Section 5 provides the concluding remarks.

2. System Composition and Working Principle

The plant protection drone variable spray system designed by the research team is presented in Figure 1. It comprises a prescription graphic translation subsystem and a variable spraying subsystem. At first, the prescription value of the medication is interpreted by the prescription graphic translation subsystem from the prescription map of the work area. Subsequently, the prescription value is transmitted to the variable spray system. At last, the closed-loop PID control algorithm is utilized to adjust the duty cycle according to the prescription value received. By assigning a value to the single-chip timer counter, a square wave signal with an adjustable duty cycle is generated to adjust the rotation speed of the micro-diaphragm pump. Therefore, the way of variable spraying is realized. In the spray system pipeline, the flow information in the spray pipeline is fed back to the STM32 controller by the Hall flow sensor. Meanwhile, the instantaneous flow rate, real-time position, flight parameters and prescription values are displayed through the liquid crystal display (LCD).

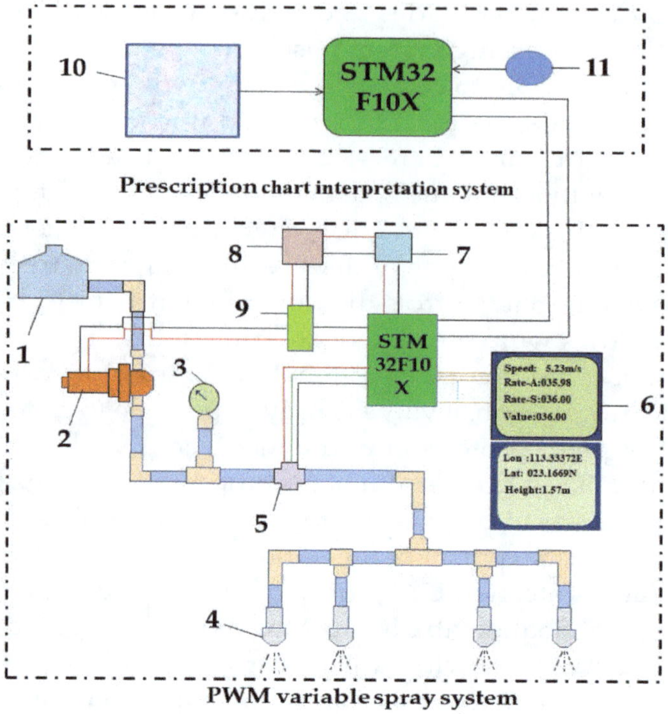

Figure 1. Schematic diagram of the structure of a prescription graphical interpretation of a variable spray system. PWM: pulse width modulation. **Note:** 1. Medicine box; 2. micro-diaphragm pump; 3. digital pressure gague; 4. pressure nozzle; 5. hall flow sensor; 6. liquid crystal display (LCD); 7. buck module; 8. 12 V direct-current (DC) power; 9. drive amplification module; 10. prescription figure; 11. GPS.

3. System Design

3.1. Prescription Map Generation and Interpretation

In order to get the prescription map, the ArcMap software (Environment System Research Institute, ESRI) was used to generate a prescription map with different prescription values, so that the variable spray system can be guided by the prescription map. At the same time, in order to verify the effectiveness of the system, a 40 m × 60 m field experimental field was selected in the Zengcheng Experimental Teaching Base of South China Agricultural University in Guangzhou, China (113°38′15″ E, 23°14′37″N). The target area was divided into small slices of 10 m × 10 m. The UAV flew over the center of each operation unit. The flight path and unit division are shown in Figure 2.

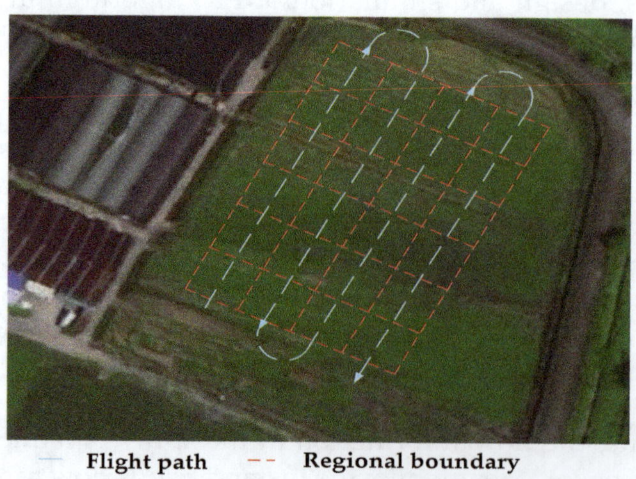

Figure 2. Sketch map of target plot.

Each 10 m × 10 m area is an operational unit, and a spraying amount is presented in each operational unit. According to the performance of each component of the variable spray system, five different levels of dosage (7.5 L/hm^2, 15 L/hm^2, 22.5 L/hm^2, 30 L/hm^2, and 37.5 L/hm^2) were set up. The prescription value of each operation unit was selected randomly from the above 5 gradient prescription values, which are shown in Figure 3.

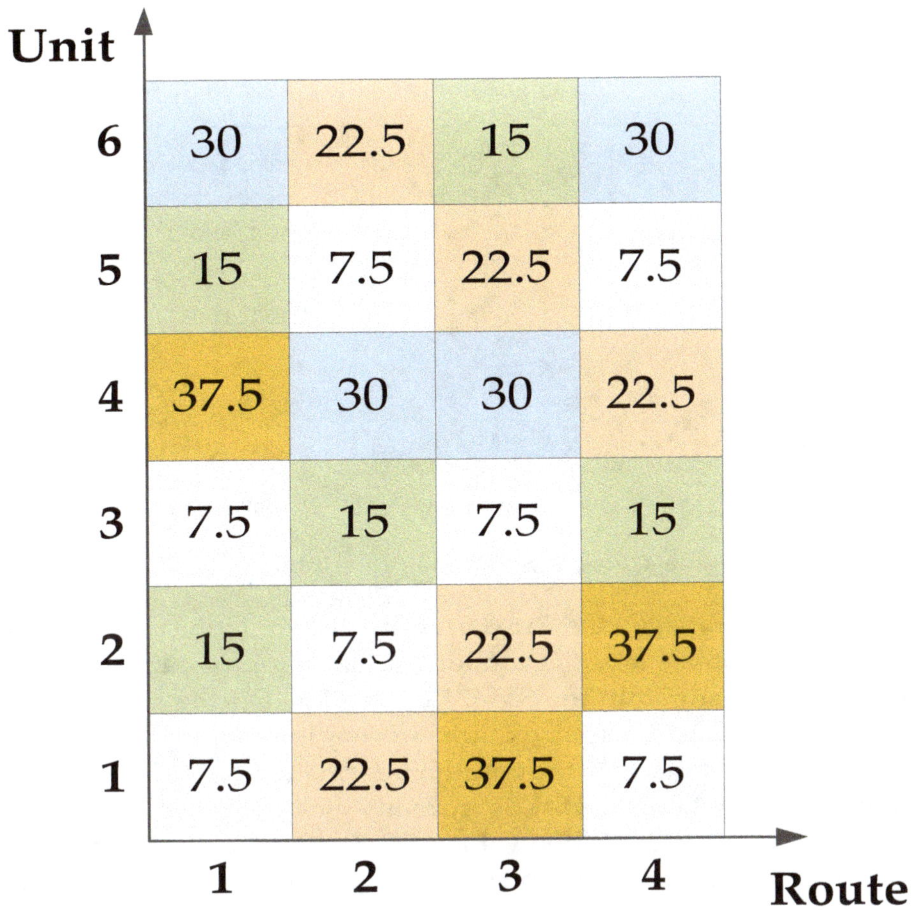

Figure 3. Prescription value setting in each operation unit.

Before the experiment, the working area was located by LocaSpace Viewer software (Beijing 3D Vision Technology Co., Ltd.). Then, the geographic position information of the working field was obtained, and the longitude and latitude were input into Microsoft Excel (Microsoft Corporation, USA) file. As a result, the ArcMap software was imported to generate the line map layer grid [12–14]. Eventually, the prescription map was created. By adding element classes to generate fishing nets, the grid can be edited and read. Finally, the longitude and latitude information of each unit and the dosage information were inserted by the linear difference method to generate the prescription map of the work area.

The prescription diagram mainly contains three layers of information, as shown in Figure 4. The first layer is raster information, which is a rectangular raster of equal size according to the effective spray amplitude and flight speed of UAV. The second layer is prescription value information layer, which is based on the grid dosage obtained by the expert system of pesticide application, and different colors represent different prescription values. The third level is the geographic information layer, which is mainly the latitude and longitude information of raster rows and columns.

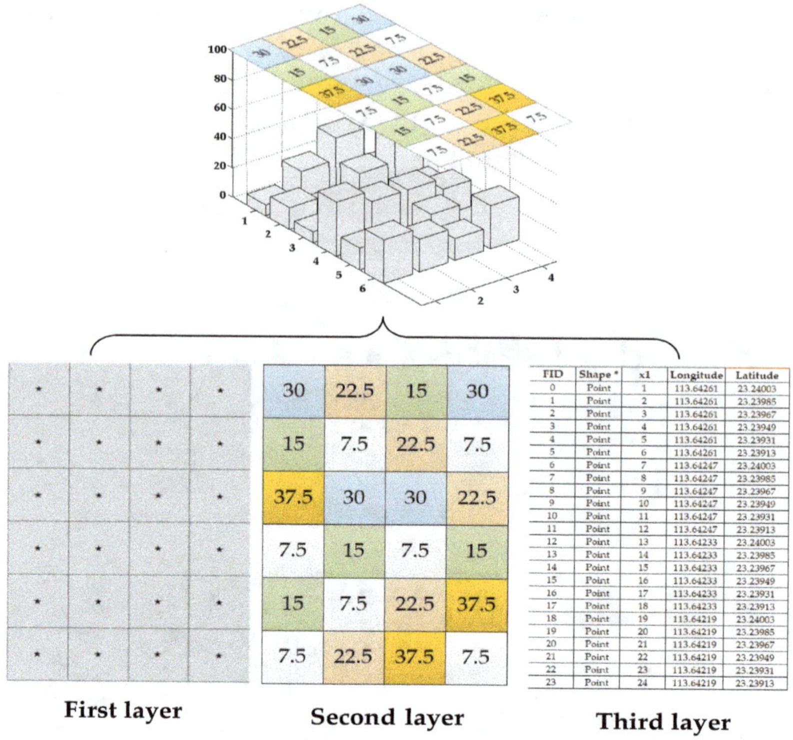

FID	Shape *	x1	Longitude	Latitude
0	Point	1	113.64261	23.24003
1	Point	2	113.64261	23.23985
2	Point	3	113.64261	23.23967
3	Point	4	113.64261	23.23949
4	Point	5	113.64261	23.23931
5	Point	6	113.64261	23.23913
6	Point	7	113.64247	23.24003
7	Point	8	113.64247	23.23985
8	Point	9	113.64247	23.23967
9	Point	10	113.64247	23.23949
10	Point	11	113.64247	23.23931
11	Point	12	113.64247	23.23913
12	Point	13	113.64233	23.24003
13	Point	14	113.64233	23.23985
14	Point	15	113.64233	23.23967
15	Point	16	113.64233	23.23949
16	Point	17	113.64233	23.23931
17	Point	18	113.64233	23.23913
18	Point	19	113.64219	23.24003
19	Point	20	113.64219	23.23985
20	Point	21	113.64219	23.23967
21	Point	22	113.64219	23.23949
22	Point	23	113.64219	23.23931
23	Point	24	113.64219	23.23913

Figure 4. Structure chart of prescription map.

The prescription map generated by the ArcMap software simulation mainly includes information such as the latitude and longitude and the amount of spraying of each unit, and programs the information of each unit to be stored in the STM32 controller. During the spraying operation, the on-board high-precision GPS transmits the position information of the plant protection drone to the STM32 controller in real time, and matches the position information contained in the work prescription map to determine the unit area where the UAV is currently located. When the position information of UAV is successfully matched, the prescription value of the current location is obtained. The obtained spray quantity information is transmitted to the variable spray controller to perform spray operation. The schematic diagram of hardware representing the prescription graphic translation system is shown in Figure 5.

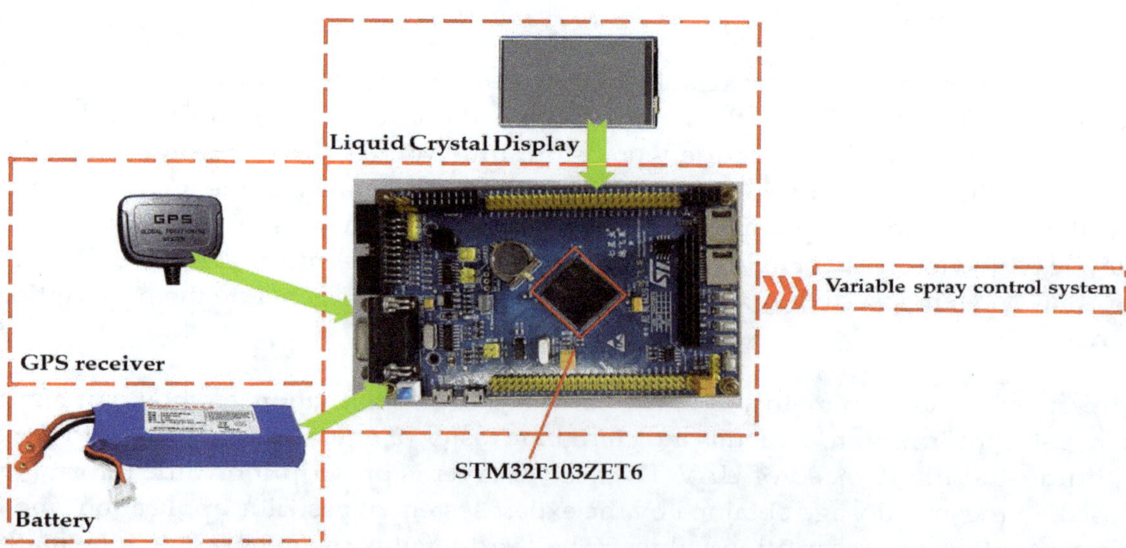

Figure 5. Schematic diagram of hardware representing a prescription graphic translation system.

3.2. Variable Spray System Design

The overall structure of the plant protection drone variable spray system is illustrated in Figure 6. The self-developed subsystems of prescription graphic translation and spray controller were installed in the M23 UAV (Shenzhen Hi-tech New Agriculture Technologies Co, Ltd., Shenzhen, China). Along with the medicine box, a miniature diaphragm pump (PLD-2201, Shijiazhuang Prandi Co, Ltd., Shijiazhuang, China) and a pressure nozzle (110-015 types, LECHLER Company, Germany) were used. Among these, the STM32 chip (STM32F103ZET6, Langyi Electronic Technology Co, Ltd., Shanghai, China) was the spray controller core. The Hall flow sensor (MJ-HZ06K, Mocho Technology Co, Ltd., Shenzhen, China) was used to measure the flow of the system, and the measuring range of the hall flow sensor was 0.1–1 L/min.

Figure 6. Physical diagram of the variable spraying system. **Note:** 1. Prescription map interpretation system and spray controller; 2. medicine case; 3. Hall flow sensor; 4. miniature diaphragm pump; 5. pressure nozzle.

The micro-diaphragm pump was controlled by the signal sent by the interpretation system. The liquid in the medicine box was transported to the nozzles and fractured into tiny droplets under pressure. The Hall flow sensor was used to measure the flow inside the system, and the flow rate information was fed back to the spray controller. Subsequently, the PWM was adjusted according to the deviation between the actual flow and the target flow. The schematic diagram of hardware representing the variable spray controller is shown in Figure 7.

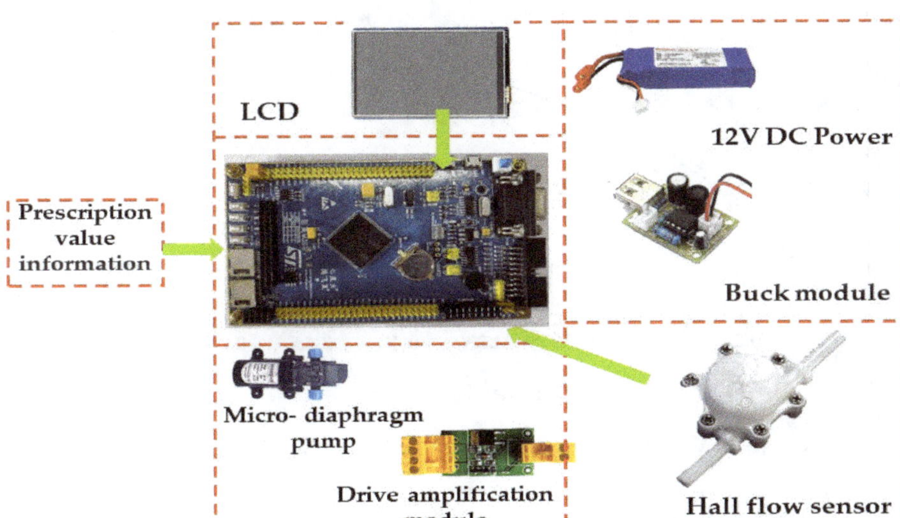

Figure 7. Schematic of hardware representing the variable spray controller.

3.3. Micro-Diaphragm Pump Drive

The micro-diaphragm pump was controlled by the PWM technology. When the spray controller received the prescription value information, the high-level time was used as the percentage of the whole cycle of change, thus obtaining the PWM square wave signals with different duty cycles. Since the PWM signal generated by the STM32 series microcomputer (MCU) was 5 V and the rated voltage of the micro-diaphragm pump was 12 V, the PWM square wave output from the I/O port of the MCU cannot directly drive the micro-diaphragm pump. As a result, the driving amplification effect of the metal–oxide–semiconductor (MOS) transistor was used. Moreover, the rotation of the micro-diaphragm pump was drive by the PWM square wave signal.

When the UAV was in flight operation, the variable spray system adjusted the rotational speed of the diaphragm pump by changing the duty ratio of the PWM square wave signal according to the prescription value. As a result, the flow rate of the system was adjusted. The period of the PWM square wave signal was set to 200 ms. When the duty ratio was lower than 40%, the micro-diaphragm pump could not be started. To ensure normal operation of micro-diaphragm pump, the duty ratio changed within the range of 40% to 100% during the test. In order to measure the change of flow clearly, the duty cycle of the PWM square wave signal was increased by 5% each time.

The Hall flow sensor was used to measure the single nozzle flow rate of the micro-diaphragm pump under different duty ratios. Additionally, the relationship between the flow rate of the micro-diaphragm pump and the duty ratio of PWM square wave signal was obtained. The result is shown in Figure 8. By using the cubic polynomial to fit the actual flow rate and duty cycle curve [15,16], the relationship between the flow rate of the variable system and the duty cycle of the PWM square-wave signal is:

$$duty = (20.776v^3 - 21.452v^2 + 8.242v - 0.431) \times 100\% \quad (1)$$

where v is the flow rate of a nozzle in L/min, and *duty* is the duty cycle of the PWM square wave signal in %.

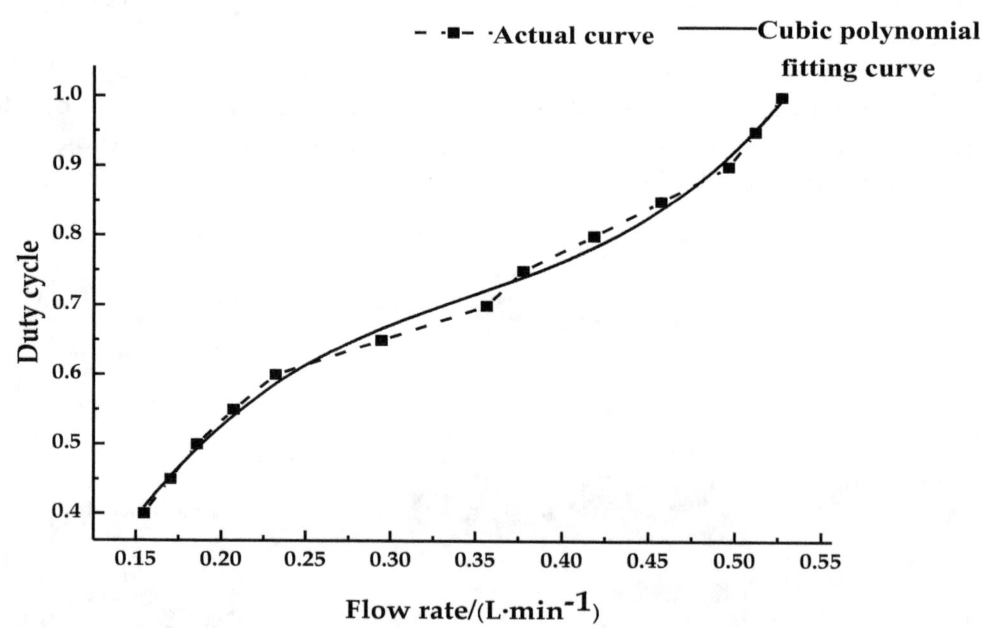

Figure 8. Velocity curve of the miniature diaphragm pump.

3.4. Nozzle Installation of Variable Spray System

The spray amplitude of the plant protection UAV is related to the installation distance of the nozzle. In order to get the theoretical spray size of the variable spray system designed by the project group, four standard angle sector nozzles (110-015 types, LECHLER Company, Germany) were installed side by side. The spray was gradually sharpened at the edge, and the theoretical spray angle of the nozzle was 110 degrees. The plant protection UAV spray system was dismantled and carried out by spraying with self-designed sprays. According to the basic requirement of the spray nozzle, the installation space of the adjacent sprinkler must be more than 50 cm, and thus a better spraying effect can be obtained when spraying [17,18]. Therefore, the installation space of the nozzle of the project group was 50 cm. A schematic diagram of the nozzle installation is shown in Figure 9.

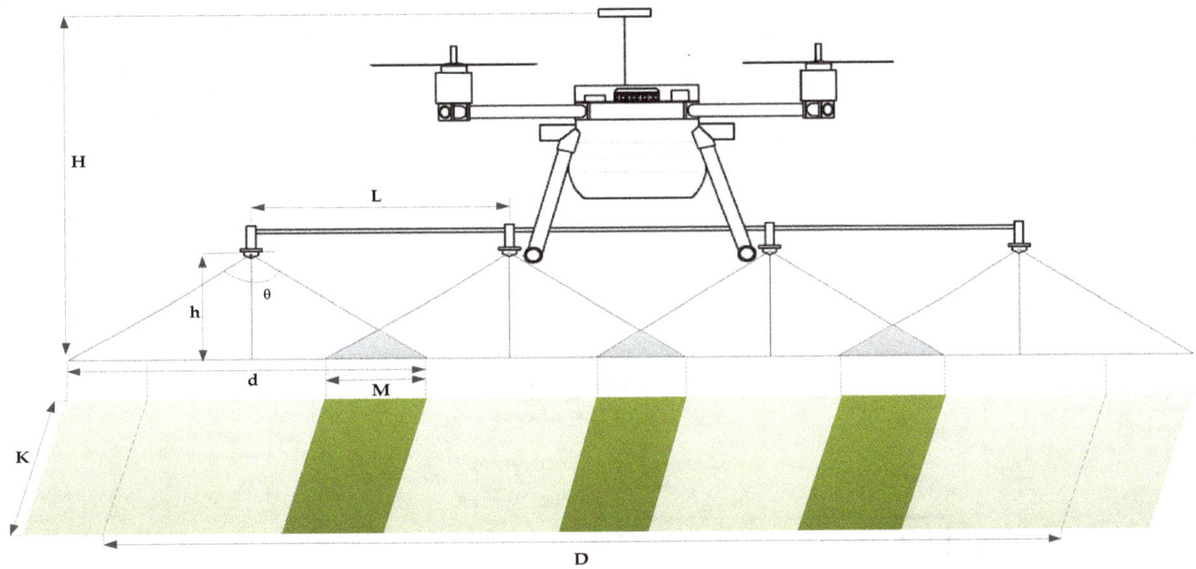

Figure 9. Schematic diagram of nozzle installation. **Note:** H is the height of the GPS from the top of the crop in me; h is the height of the nozzle from the top of the crop in m; L is the installation distance of the adjacent nozzle in m; d is the spray nozzle of the corresponding height in m; M is the width of the overlap area of the adjacent nozzles in m; K is a certain time, with the distance of the drone flight in m; D is the effective spray width of the drone in m.

It can be seen from Figure 9 that the equation to calculate the spray width of a single nozzle is:

$$d = 2h \tan \frac{\theta}{2} \tag{2}$$

where d is the nozzle spray of the corresponding height in m; h is the height of the nozzle at the tip of the crop in m; and θ is the number of spray angles in °.

In order to achieve a uniform spraying effect, the complementary spraying range of the nozzle should be 25–30% of the single spraying range [19,20]. As shown in Figure 9, when the adjacent two nozzles are located in the complementary region, the effective spraying range of the four nozzles is 25% of the single spraying range. The effective pray width of the four nozzles can be expressed as:

$$D = \frac{11}{2} h \tan \theta \tag{3}$$

The height of the plant protection UAV is the height of the GPS from the ground. The height of rice, wheat and other crops is about 50 cm. The spray bar designed by the research team was installed on the frame of the drone. The height of the nozzle is about 70 cm away from the GPS. Therefore, when the flying operation height of the drone is about 2 m, regardless of the influence of other external environmental factors, it can be discerned from Equation (3) that the effective spray width of the four nozzles is about 6 m.

3.5. Control Program Design of Variable Spray System

The program was utilized to control the variable spray system and process various signals. It was written in Keil Software (ARM Germany GmbH, USA), which was mainly composed of the PWM duty cycle adjustment module, serial communication module, LCD module and PID control module. The program control flow chart is shown in Figure 10. When the system works, the GPS communication protocol (NMEA0183) is used to analyze the information acquired by the GPS in real time [21], so the position information of the current plant protection drone is obtained. Then, the prescription value is extracted through the prescription graphic translation system which processes the information. The prescription values are sent to the variable spray system. When the variable spray system receives the prescription value information, the timing and counter are turned on. The number of pulses of the flow sensor is fed back every 50 ms, and the current system instantaneous flow is obtained through calculation. The deviation between the instantaneous flow rate and the target flow of the system is used as the input of the PID controller. After the PID control algorithm is operated, a duty value is output, and the variable spray controller generates a PWM square wave with the corresponding duty ratio.

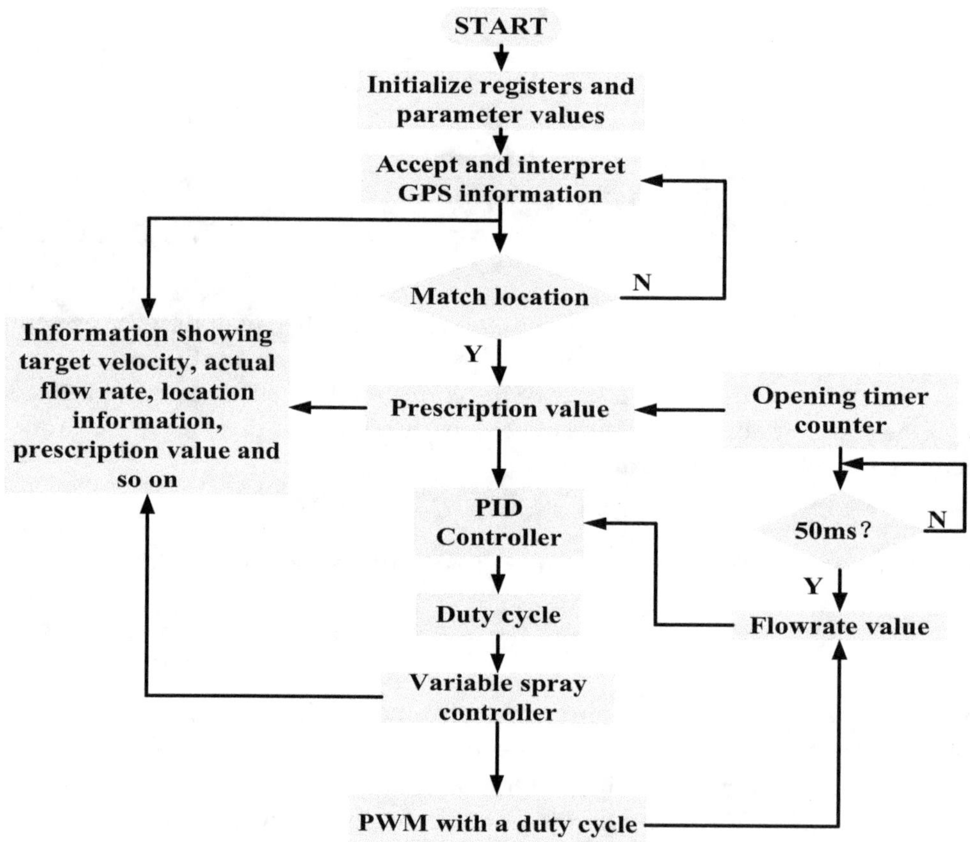

Figure 10. Program flow chart.

3.6. PID Control Algorithm

Due to the fact that the actual spray operation is affected by many factors, the system has a certain delay from receiving the flow change signal to adjusting the flow to the target value. Therefore, the actual flow will fluctuate around the target flow. The PID control algorithm is based on a control law of the system error. The algorithm is an optimal control adjusted by proportion (P), integration (I), and differentiation (D). It has the characteristics of a simple principle, high control precision, easy implementation and strong practicability [22]. The PID control algorithm has obvious effects on the process of the dynamic system calibration of continuous systems. In order to accurately control the flow of the system and ensure the accuracy and stability of the variable spray operation, the PID control algorithm is used to achieve the closed-loop control of the system [23]. The equation of PID control is as follows:

$$\begin{aligned} u(k) &= K_P^{e(k)} + \frac{TK_P}{T_I}\sum_{i=0}^{k} e(i) + K_P T_D \frac{e(k)-e(k-1)}{T} \\ &= K_P^{e(k)} + K_I \sum_{i=0}^{k} e(i) + K_D^{[e(k)-e(k-1)]} \end{aligned} \tag{4}$$

where K_P is the proportional gain; $K_I = TK_P/T_I$ is the integral time constant; $K_D = K_P T_D$ is the differential time constant; $u(k)$ is the output of the control system at sampling time k; $e(k)$ is the system output deviation from the input quantity at sampling time k, and is described as $e(k) = y(k) - r(k)$, $y(k)$ is the output feedback value, and $r(k)$ is the reference input value.

The flow control process of the miniature diaphragm pump has the characteristics of large inertia lag and time variance. The process of adjusting the variable spray of the micro-diaphragm pump can be described as a second-order pure lag system. The transfer function is as follows:

$$G(s) = \frac{C(s)}{R(s)} = \frac{Ke^{-\tau s}}{(T_1 s + 1)(T_2 s + 1)} \tag{5}$$

where K is the amplification factor; τ is the pure lag time in second; T_1 and T_2 are time coefficients.

In the process of adjusting the flow rate of the diaphragm pump, K, τ, T_1 and T_2 are closely related to the performance of the system components and environmental factors [24]. The numerical value of the parameters was obtained through simulation. Therefore, the value of the amplification factor K is 1, the value of the pure lag time τ is 0.25, and the values of the time constants T_1 and T_2 are 0.05 and 1.849, respectively. The variable spray system PID control model was created by the simulink component of MATLAB software (MathWorks, USA), as shown in Figure 11. According to the simulation model, when the given signal is a unit step signal, the PID simulation coefficient is debugged by the optimized parameter design method to obtain the optimal control effect. In actual use, the algorithm needs to be run through the STM32 chip. After repeated simulation and experiments, the values of K_P, K_I, and K_D are set to 6, 0.8, and 0.5, respectively. When the input is the unit "1", the response curve obtained from the step response output of the system is shown in Figure 12.

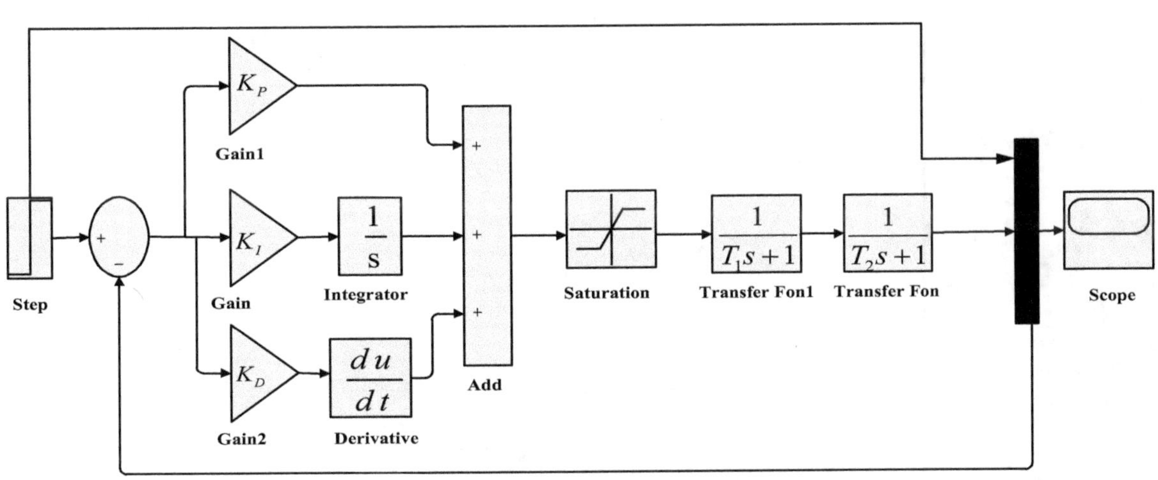

Figure 11. PID regulation simulation model.

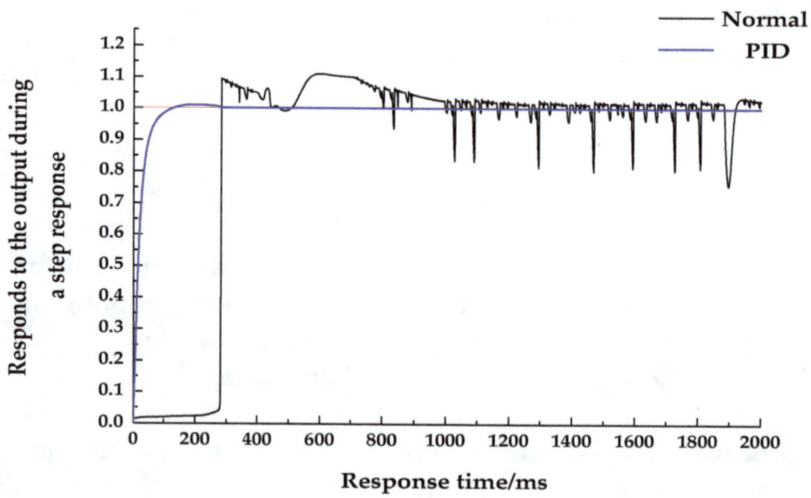

Figure 12. PID control and unregulated step response curve.

According to the step response curve of the system in Figure 12, the target flow is set to "1". When the PID algorithm is not added, the system receives the change signal of the flow and has a certain delay in response. About 300 ms of time is needed to change the flow to the target value; moreover, after reaching the target value, the actual flow value fluctuates around the target value, which has a certain deviation from the target flow. However, after the PID control, the response time is shortened to about 15 ms. At the same time, the actual flow reaches the target flow value and then stabilizes in the vicinity of the target value. To a certain extent, the system response speed and stability are improved.

4. Experiment

4.1. Effect of Duty Ratio on Droplet Size and Spray Angle

The spray droplet volume and spray angle are important parameters for evaluating atomization effect [25]. Since the particle size and spray angle of the pressure nozzle change with the change of the fluid pressure, a laser particle size analyzer (DP-02, Zhuhai Omega Instrument Co., Ltd.) was used to measure droplet volume. Additionally, a single-lens reflex camera (ILCE-5100, Sony Corporation of Japan) was used to photograph the spray angle. The average particle size was tested at different duty cycles, and the results are shown in Figure 13.

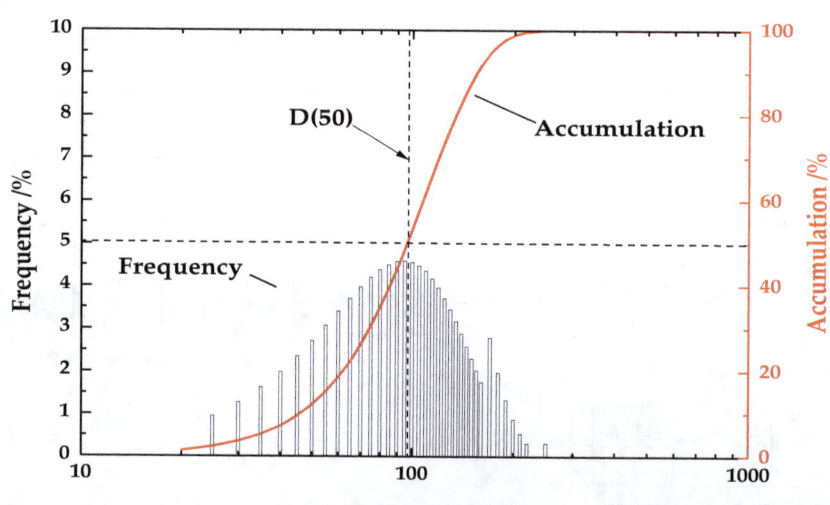

(a)

Figure 13. *Cont.*

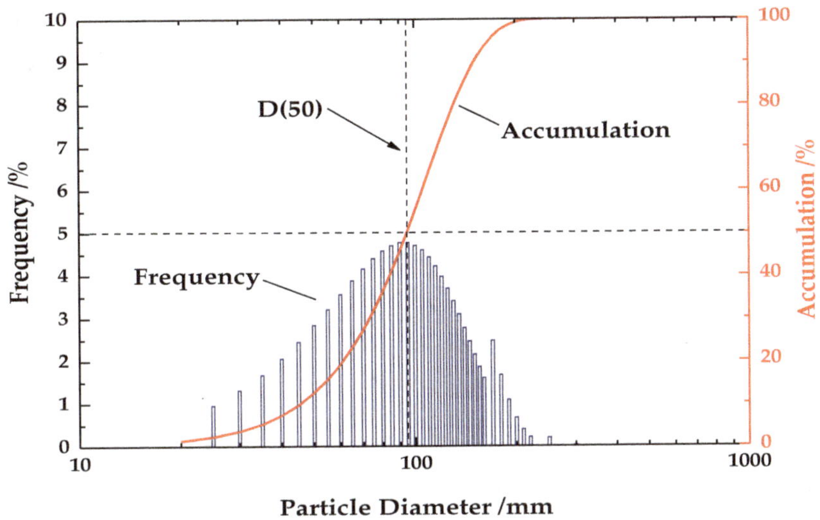

(b)

Figure 13. Droplet spectra of different PWM square wave duty ratios: (**a**) droplet spectrum with a duty ratio of 40%; (**b**) droplet spectrum with a duty ratio of 100%.

In Figure 13, frequency denotes the percentage of droplets with a certain size in the whole droplet group, and accumulation denotes the percentage of droplets arranged from small to large, to a certain size of droplets accumulating in the whole droplet group. The droplet size at 50% position is denoted as droplet volume diameter D (50). It can be seen from Figure 13 that the droplets produced by different nozzles are more uniform and the droplet spectrum is normal. The droplet size percentage is accumulated according to the order of droplet size from small to large. When the PWM duty ratio is 40% and 100%, the droplet volume diameter values are 99.81 and 96.26 μm, respectively.

The spray angles of PWM duty ratios between 40% and 100% are shown in Figure 14, and the particle size and spray angle of the other duty ratios are shown in Table 1. The experimental data in Table 1 demonstrate that the volume of the spray droplets varies between 94 and 100 μm, when the duty ratio is varied from 40% to 100%. Moreover, the volume of the volume gradually decreases as the duty ratio increases. The coefficient of variation is 5.52%, the spray angle varies between 97° and 105°, and the coefficient of variation is 2.77%. As the duty cycle increases, the flow rate of the system increases, and the pressure at the nozzle also increases, increasing atomization energy and promoting droplet breakage. The experimental results of droplet size show that different flow rates can be achieved by changing the duty cycle of PWM square wave signal to regulate the flow rate.

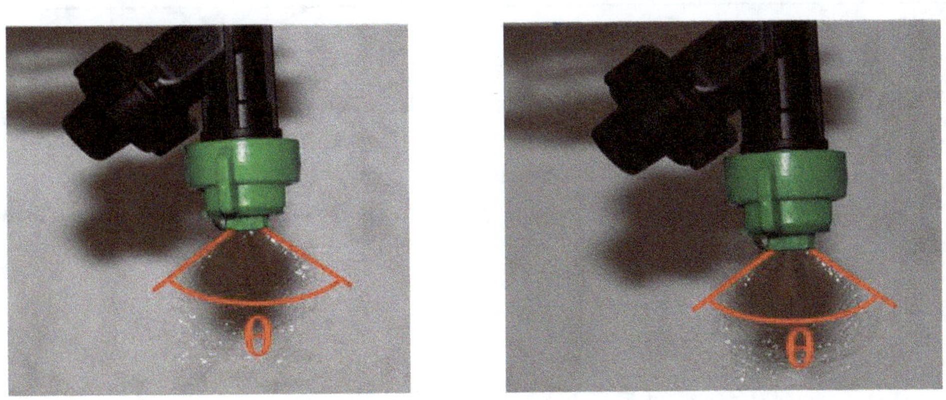

(a) (b)

Figure 14. Spray angle of different PWM square wave duty ratios with a duty cycle of 40% (**a**) and a duty cycle of 100% (**b**). **Note:** θ is the spray angle.

Table 1. Diameter of droplet volume and spray angle at different duty ratios of PWM square waves.

Duty Cycle (%)	Volume Diameter (μm)	Spray Angle (°)
40	99.81	98
45	98.72	99
50	97.85	97
55	97.46	99
60	97.32	98
65	97.07	99
70	96.73	100
75	96.77	98
80	96.49	101
85	96.26	102
90	95.79	104
95	95.65	105
100	94.59	105

4.2. *Analysis of Actual Flow and Theoretical Flow Error*

A laboratory experiment was used to observe the real-time flow of the system conveniently. The flow information measured by the Hall flow sensor was displayed through the LCD screen in real time. The target flow rates under different duty cycles can be obtained by using the relationship between the duty cycle of the PWM square wave signal and nozzle flow rate. Table 2 is a comparison of the actual flow and the target flow of a single nozzle under different duty cycles.

Table 2. System flow deviation.

Duty Cycle (%)	Target Flow Rate ($L \cdot min^{-1}$)	Actual Flow Rate ($L \cdot min^{-1}$)	Deviation (%)
40	0.1550	0.1508	2.71
45	0.1705	0.1682	1.35
50	0.1860	0.1805	1.34
55	0.2072	0.1950	5.89
60	0.2324	0.2290	1.46
65	0.2945	0.2894	1.73
70	0.3564	0.3487	2.16
75	0.3774	0.3683	2.41
80	0.4184	0.4097	2.08
85	0.4572	0.4493	1.73
90	0.4959	0.4850	2.20
95	0.5116	0.5064	1.02
100	0.5268	0.5159	2.07

The actual flow measurement data measured by the Hall flow sensor were fed back to the controller chip. The PID algorithm adjusts the system flow according to the error between the target flow and the actual flow. It can be seen from Table 2 that the system flow is regulated by the duty cycle of PWM square wave signal. The actual flow changes with the change of duty cycle and is stable near the target flow value. Additionally, the average deviation of flow regulation is 2.16%, which indicates that the system can adjust the flow well.

4.3. *Experiments Outdoors*

4.3.1. Experimental Scheme

The stability and sensitivity of variable spray system were tested by outdoor spray deposition experiment, the experiment was carried out in a paddy field of the Zengcheng Research and Teaching Base (113°38′15″ E, 23°14′37″ N) of South China Agricultural University, Guangzhou. The experimental site is the same as the prescription map. The outdoor experiment site photo of the plant protection drone variable spray system is shown in Figure 15.

Figure 15. Spray test site.

According to the prescription diagram, the prescription values of adjacent units are different. In order to explore the uniformity of droplet deposition in each unit and the droplet deposition at the boundary of adjacent units, the sampling bands such as S1, S2 were shown in Figure 16, in which the operational units each were set to 10 m × 10 m. Figure 16 shows only the layout of the sampling points in the adjacent operational units, and the other units were arranged in the same way. In UAV flying one sortie, the total number of sampling bands was 124. A sampling band was set up in the center of each operation unit, such as S4 and S10. In order to study the change of spray volume at the boundary of the operation unit, the sensitivity of the variable spray system was evaluated. The sampling bands were set on the boundary of the adjacent operation unit and to be 1 m and 2 m away from the boundary line on its both sides, as shown in Figure 16. S1, S7 and S13 were the sampling bands on the boundary line; S5, S6, S8 and S9 were the sampling bands on the left and right sides of the dividing line, respectively. The amount of droplet deposition, the sediment density and the sedimentation uniformity of aviation plant protection operations are important parameters reflecting the quality of spraying [26]. In order to study the law of droplet deposition between various regions, a Rhodamine B (soluble fluorescent tracer) solution with a concentration of 5 g/L was used instead of the pesticide solution, and a Mylar card with a size of 50 mm × 80 mm and water-sensitive paper with a size of 28 mm × 75 mm were used to collect the sprayed droplets to analyze the deposition density and deposition uniformity. Nine sampling points were evenly arranged at an interval of 1 m on each collection belt, and both the Mylar card and water-sensitive paper were fixed at a height of about 50 cm from each sampling point. The layout of Mylar card and water-sensitive paper is shown in Figure 17.

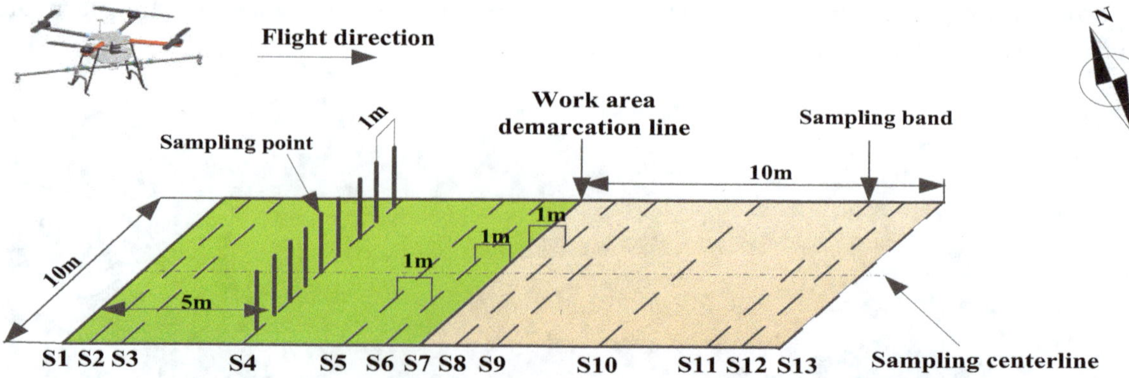

Figure 16. Inner sampling bands of adjacent operation units. **Note:** S1, S2, S3, S4, S5, S6, S7, S8, S9, S10, S11, S12, and S13 are the numbers of the collection bands.

Figure 17. Layout of the Mylar card and water-sensitive paper.

The experiment was conducted on 5 November 2018, and the mature rice was selected as the tested crop. A total of four sorties were tested, the flight speed of the plant protection drone was stable at 5 m/s and the flight altitude was 2 m. At the same time, a portable ultrasonic micro-automatic weather station (Hberw6-3, Shenzhen Hongyuan Technology Co, Ltd., China) was used to measure the environmental information, and the height of the weather station was set at 2 m above the ground. The test was carried out four times in total. The environment parameters are shown in Table 3.

Table 3. Test environment parameter list.

Sorties	Temperature (°C)	Humidity (%)	Wind Speed and Direction ($m \cdot s^{-1}$)
1	19.5	54.3	0.54/SW
2	20.1	54.2	0.78/SW
3	21.3	53.9	0.84/SW
4	22.6	53.7	0.47/SW

4.3.2. Test Data Processing

The amount of droplet deposition is a significant parameter reflecting the quality of droplet deposition per unit area [27]. The Mylar card that is a resin card collected by the test is eluted with 20 mL of distilled water, and six Rhodamine B solutions with different concentrations were set for calibration in the range of absorbance of the fluorescence spectrophotometer (F-380, Tianjin Gangdong Technology Development Co., Ltd., China). Concentrations of Rhodamine B solutions were 0.002 μg/mL, 0.005 μg/mL, 0.01 μg/mL, 0.02 μg/mL, 0.05 μg/mL, and 0.1 μg/mL. The specific method was as follows: three parts of each standard solution were prepared, and each repeated measurement was performed twice to monitor the influence of the cuvette on the measurement result. Therefore, the actual number of repeated measurements for each standard solution amount to six times, and the Rhodamine B solution concentration absorbance curve is shown in Figure 18.

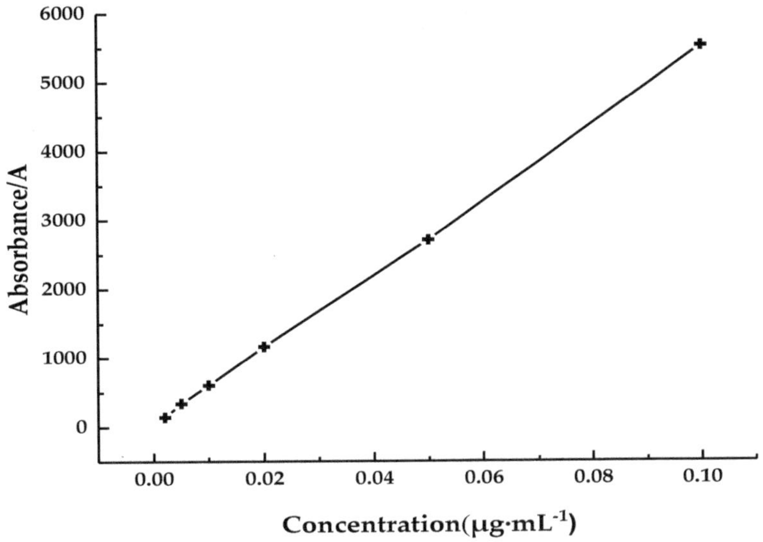

Figure 18. Rhodamine B solution standard concentration absorbance curve.

By linear regression fitting, the coefficient of determination R^2 is 0.9997. The standard curve gives the concentration of the sample solution, thereby calculating the deposition amount per unit area [28]. The calculation equation is as follows:

$$\beta_{dep} = \frac{(\rho_{sampl} - \rho_{blk})F_{cal}V_{dil}}{\rho_{spray}A_{col}} \tag{6}$$

where β_{dep} is the amount of droplet deposition in g·cm^{-2}; ρ_{sampl} is the reading of the sample solution fluorescence meter; ρ_{blk} is the reading of the fluorescence meter of the eluent (distilled water for this test); F_{cal} is the calibration coefficient in g·L^{-1}; V_{dil} is the volume of the solution used to elute the collected sample in L; ρ_{spray} is the concentration of the fluorescent tracer in the spray solution in %; A_{col} is the area of the collected card in cm^2.

The water-sensitive paper collected by the test was scanned and analyzed by DepositScan software (USDA-ARS Application Technology Research Unit, Wooster, OH, USA), and the droplet deposition rate and deposition density under different prescription values were obtained [29]. The coefficient of variation is usually used to represent the uniformity of droplet deposition between different collection points for the same collection [30]. The calculation equation is:

$$CV = \frac{S}{\overline{\overline{X}}} \times 100\% \tag{7}$$

$$S = \sqrt{\sum_{i=1}^{n} (X_i - \overline{X})^2 / (n-1)} \tag{8}$$

where S is the standard deviation of the sampled specimens in the same collection zone; X_i is the deposition amount of each collection point in μL·cm^{-2}; $\overline{X}$ is the average value of the deposition amount of the sampling points of the same collection zone in μL·cm^{-2}; and n is the group sampling with the number of collection points.

4.4. Analysis of Experiments Results

4.4.1. Droplet Deposition Density Analysis

During the experiment, the number of droplets per square centimeter of the water-sensitive paper in each sampling point was collected. The distribution uniformity of the droplet coverage tester was calculated by Equation (7). The grayscale image the water-sensitive paper after scanning is shown in Figure 19. The central sampling band of 6 working orders in the next route of the one or two sorties was selected, as shown in Table 4. When the water droplet coverage density is less than 15 per cm^2, it is regarded as an invalid sampling point.

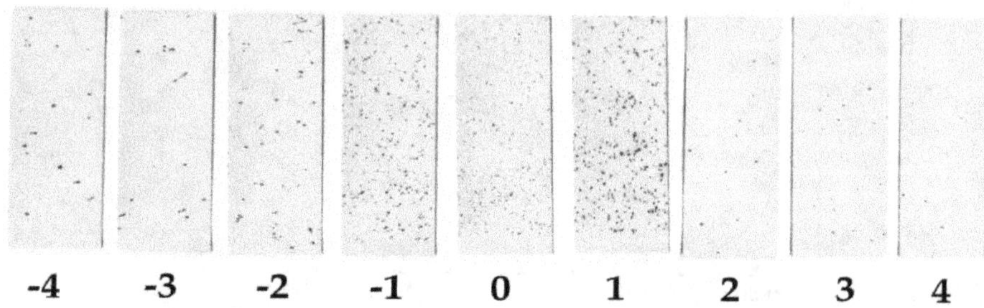

Figure 19. Grayscale image of water-sensitive paper.

Table 4. Statistics of droplet coverage density.

Sortie	Unit	Sampling Point									Average	Coefficient of Variation
		−4	−3	−2	−1	0	1	2	3	4		
First	1	6	16	19	29	44	31	23	14	9	25.14	
	2	2	17	22	37	81	54	22	10	8	34.71	38.07%
	3	0	15	19	26	47	34	17	6	0	23.43	32.5%
	4	13	26	57	123	169	115	41	23	12	79.14	237.77%
	5	1	9	16	39	73	27	18	7	0	27	56.88%
	6	0	14	23	45	137	67	24	16	3	46.57	72.48%
Second	1	0	12	19	24	41	31	20	16	4	23.29	59.47%
	2	9	16	22	37	93	49	29	14	0	37.14	28.08%
	3	0	15	23	36	55	29	16	13	1	26.71	73.83%
	4	4	21	33	59	143	37	20	12	0	46.43	
	5	1	6	15	23	69	31	18	7	0	24.15	47.9%
	6	3	14	29	67	115	89	32	15	3	51.57	113.54%

According to the data in Table 4, when the prescription values are different, the distribution of droplet deposition density is more intense in the middle and less intense on both sides. Because of the small natural wind speed during the experiment, the droplet deposition does not have obvious migration, and the peak value of deposition concentration is gathered near the sampling center line. The prescription values of operational units 1–6 are 7.5 L/hm^2, 15 L/hm^2, 7.5 L/hm^2, 37.5 L/hm^2, 15 L/hm^2, and 30 L/hm^2, and the normalized ratio of prescription value is 1:2:1:5:2:4.

In the first sortie, the normalized ratio of effective droplet deposition density of these six operation units is 1:1.38:0.93:3.15:1.07:1.85. In the second sortie, the normalized ratio of their effective droplet deposition density is 1:1.59:1.15:1.99:1.04:2.21. The actual normalized ratio of fog droplet deposition density of these six operation units is smaller than its theoretical ratio. The main reason is that the droplet does not fall due to the drift of droplets in the process of operation. The data in the table show that the effective sampling point number is from −3 to 3. According to the Civil Aviation Industry Standards of the People's Republic of China (MH/T1002.1-2016) with regard to the technical specifications for the quality of agricultural aviation ultra-low-capacity pesticide-spraying operations [31], it can be estimated that the actual spraying swath is about 5 m.

4.4.2. Droplet Deposition Analysis

A total of four sorties were tested, with four routes under each sortie. The spraying prescription values of each unit were the same as those of the prescription map. The droplet depositions of four sorties were collected by the Mylar card. The average droplet deposition in various operational units was calculated by elution analysis of the Maylar card. The droplet deposition in the sampling zone at the center of six operational units on the third and fourth sorties was analyzed. The collection of droplet deposition data is shown in Figure 20.

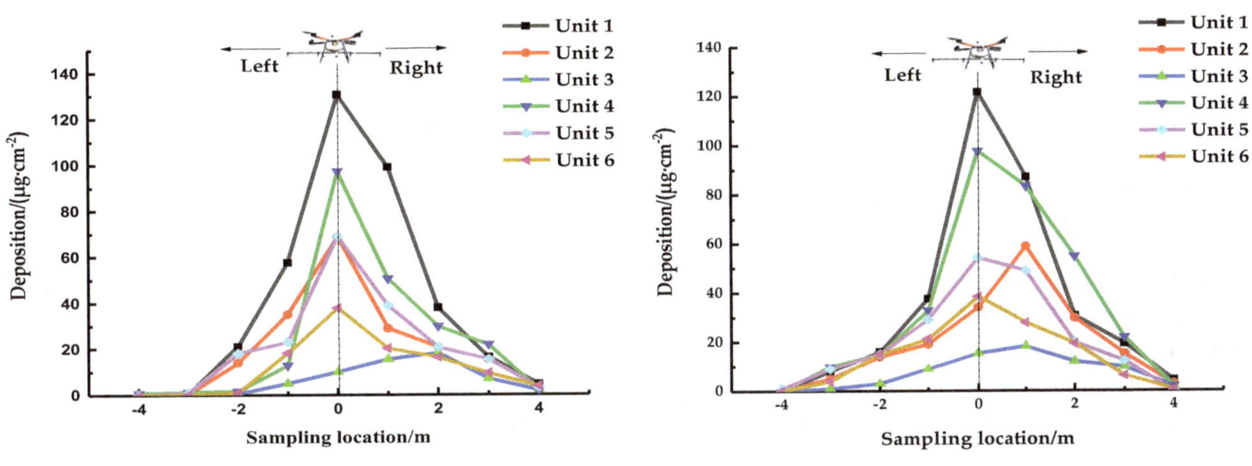

(a) (b)

Figure 20. Droplet deposition in center line of operational unit: (**a**) third sortie; (**b**) fourth sortie.

The prescription values of the operational units 1–6 were set by the prescription map to be 37.5 L/hm^2, 225 L/hm^2, 7.5 L/hm^2, 30 L/hm^2, 22.5 L/hm^2, and 15 L/hm^2, respectively. It can be observed in Figure 20 that the distribution of droplet deposition in each area is basically the same. The amount of droplet deposition is related to the prescription value of each area. The peak value of spray deposition appears below the fuselage. The main reason is the drift of droplets on both sides of the fuselage due to the influence of the rotor wind field. The droplet distribution appears to be near the sampling center line. The droplets on each collection belt are mainly distributed at the sampling points of −2#, −1#, 0#, 1#, 2#, 3#. It can be estimated that the actual spraying swath is about 5 m. This is similar to the result of Table 4.

In order to visualize the droplet deposition changes in different operational units under four sorties, the average droplet deposition in each operation unit was calculated, which is shown in Figure 21.

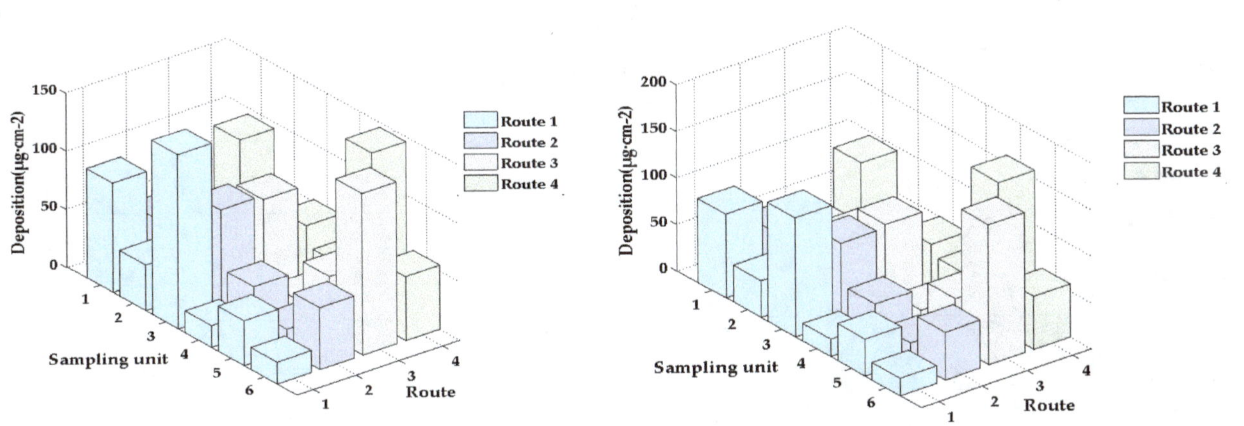

(a) (b)

Figure 21. *Cont.*

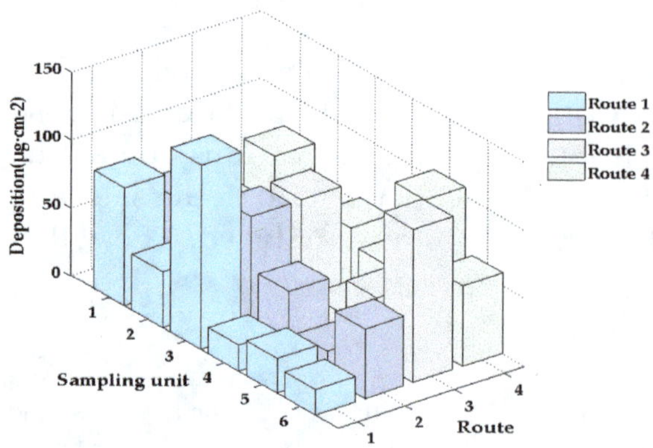

(c)

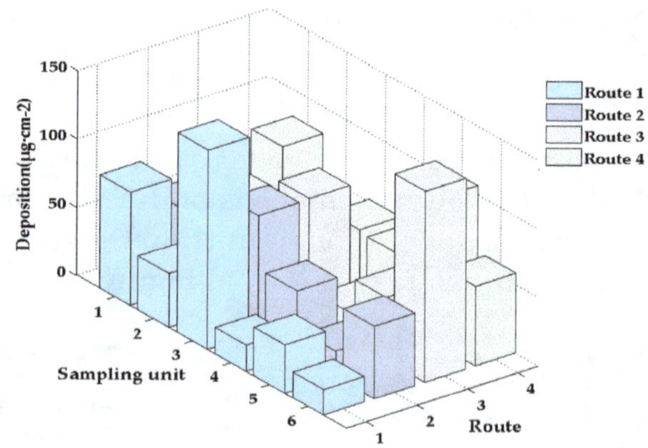

(d)

Figure 21. Droplet depositions in various sorties: (**a**) first sortie; (**b**) second sortie; (**c**) third sortie; (**d**) fourth sortie.

The average deposition amounts of fog droplets in each operational unit of four sorties are basically the same. The variation trends of the deposition amount of fog droplets between different operational units on each route are the same, and the average deposition amounts of fog droplets in each operation unit are different when placed in different places. It can be observed that the variable spray system designed by the research group can accomplish variable spray operation according to the prescription value set in the prescription map.

4.4.3. Droplet Deposition at the Boundary of Operation Units

In order to verify the sensitivity of the designed variable spray system, the sampling bands are set up to collect droplet deposition. The sampling bands are located 1 m and 2 m away from the operation unit boundary line. The droplet deposition amount of the units 1–4 of the first sortie was selected for analysis. And the test data are shown in Table 5. In Table 5, S4, S10 and S16 are the acquisition bands at the demarcation line of the operation unit. S3, S5, S9, S11, S15 and S17 are the numbers of acquisition band numbers which are 1 m away from the demarcation line on both sides of the demarcation line. S2, S6, S8, S12, S14 and S18 are the numbers of acquisition bands which are about 2 m away from the demarcation line on both sides of the demarcation line. S7, S13 and S19 are the numbers of acquisition bands located in the central position of the operation unit respectively. The layout of the sampling zone from S2 to S19 is in accordance with the method shown in Figure 16.

Table 5. Droplet deposition at the boundary of operational unit.

Unit	Prescription Value ($L \cdot hm^{-2}$)	Sampling Band	Deposition ($\mu g \cdot cm^{-2}$)
1	7.5	S2	22.58
		S3	21.89
Boundary		S4	20.63
2	15	S5	35.68
		S6	39.43
		S7	39.21
		S8	40.28
		S9	38.64
Boundary		S10	39.46
3	7.5	S11	27.64
		S12	23.64
		S13	24.06
		S14	21.96
		S15	22.68

Table 5. *Cont.*

Unit	Prescription Value (L·hm^{-2})	Sampling Band	Deposition (μg·cm^{-2})
Boundary		S16	23.07
4	37.5	S17	69.76
		S18	129.87
		S19	134.26

The diagram of droplet deposition in operation units 1–4 is shown in Figure 22.

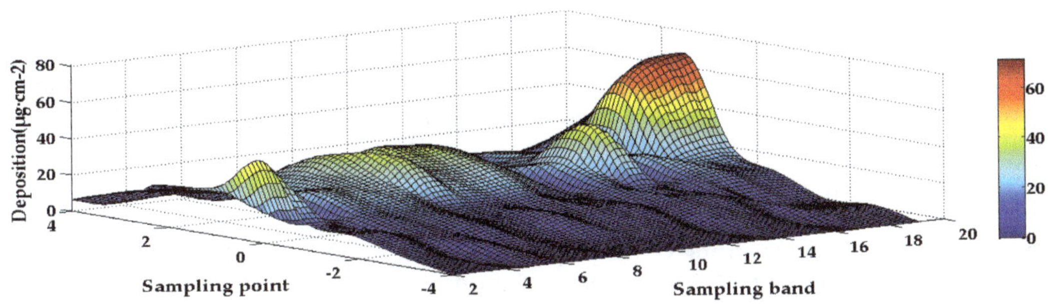

Figure 22. Droplet depositions in operation units.

It is known from Table 5 and Figure 22, when the UAV flies over the boundary band, the prescription interpretation system is interpreted to get the prescription value of the next operation unit and sent to the variable spray control system. The variable spray control system regulates the flow rate of the spray system according to the spraying prescription value. From Table 5, the droplet deposition at the boundary is similar to that of the working unit. The droplet deposition at the 1 m position behind the boundary is between the values at the center of the two working units. For example, the variation rates of the droplet deposition at S5, S11 and S17 are 63%, 29.95% and 217.67% respectively. On the other hand, the deposition of droplets on the acquisition belt, which are 2 m away from the demarcation line, is similar to that on the central line of the operation unit. Since the plant protection drone flies at 4 m/s during the test, the new prescription value is received. When the target value of the spray volume change value is about 2 m, the operation time of the system is 0.4s from receiving to square value to arriving at the target value of flow. Which reflects the sensitivity of the system. From Table 5 and Figure 22, it can be seen that the droplet depositions in the four operation units 1–4 vary with the prescription value. The normalized ratio of the droplet deposition in the center line of operation units 1–4 is 1:1.79:1.07:5.93, and the ratio of prescription value for these four operation units is 1:5:1:7. Due to the drift and adherence of droplets to fuselage, the actual deposition value is smaller than the prescription value, but the ratio is consistent, which reflects the effectiveness of variable spray system. The above analysis shows that the PWM-PID variable spray system designed based on the prescription can quickly adjust the flow according to the prescription map. The system has certain sensitivity and stability.

5. Conclusions

The variable spray system based on PWM-PID control can achieve rapid and accurate change of flow according to prescription information, effectively reducing herbicide use and enhance chemical effect.

(1) Using serial communication technology to receive the prescription value information after the prescription translation, the PWM technology was used to adjust the rotation speed of the micro-diaphragm pump to realize the variable spray, and the spray effect of the spray system was tested. The results show that the variable spray system designed by the research group ensures that the atomization effect is stable under the duty cycle of different PWM square wave signals,

and the coefficient of variation of the system flow rate with the duty cycle of the PWM square wave signal is 39.21%, which can satisfy various kinds of different spray requirements;

(2) The PID algorithm was used to control the flow adjustment process to reduce the steady-state time of the system, so that the deviation between the actual flow and the target flow is stable at 2.16%, indicating that the system can adjust the flow well;

(3) The outdoor sedimentation test shows that the variable spray system can quickly change the spray flow according to the prescription value of the working plot. Variable pulse spraying can be realized by PWM technology.

(4) Based on the data of experimental deposition and deposition density, the variable spray system can be stabilized within 0.4 s from receiving the prescription value to adjusting the flow rate to a predetermined value, and the effective injection rate of actual operation is about 5 m.

Author Contributions: Conceptualization, S.W., Q.Z., J.D. and Y.L.; methodology, W.S., Q.Z. and J.S.; software, S.W., Q.Y., and J.S.; validation, Q.Z., X.Y. and J.S.; formal analysis, S.W. and Y.L.; investigation, W.S.; resources, J.Z. and Y.L.; data curation, Q.Z. and J.S.; writing of the original draft preparation, Q.Y.; writing of review and editing, S.W.; visualization, S.W. and J.Z.; supervision, S.W.; project administration, S.W.; funding acquisition, S.W. and J.Z.

Acknowledgments: Thanks to Yilong Zhan, et al. from the National Center for International Collaboration Research on Precision Agricultural Aviation Pesticides Spraying Technology for helping the authors to complete the outdoor experiment.

References

1. Guo, Y.W.; Yuan, H.Z.; He, X.K.; Shao, Z.R. Analysis on the development and prospect of agricultural aviation protection in China. *Chin. J. Plant Prot.* **2014**, *10*, 78–82. (In Chinese)
2. Zhou, Z.Y.; Ming, R.; Zang, Y.; He, X.G.; Luo, X.W.; Lan, Y.B. Development status and countermeasures of agricultural aviation in China. *Trans. Chin. Soc. Agric. Eng.* **2017**, *33*, 1–13. (In Chinese)
3. Song, Y.; Sun, H.; Li, M.; Zhang, Q. Technology Application of Smart Spray in Agriculture: A Review. *Intell. Autom. Soft Comput.* **2015**, *21*, 319–333. [CrossRef]
4. Mogili, U.R.; Deepak, B.B.V.L. Review on application of drone systems in precision agriculture. *Procedia Comput. Sci.* **2018**, *133*, 502–509. [CrossRef]
5. Xue, X.Y. Develop an unmanned aerial vehicle based automatic aerial spraying system. *Comput. Electron. Agric.* **2016**, *128*, 58–66. [CrossRef]
6. He, X.K.; Bonds, J.; Herbst, A.; Langenakens, J. Recent development of unmanned aerial vehicle for plant protection in East Asia. *Int. J. Agric. Biol. Eng.* **2017**, *10*, 18–30.
7. Chen, Z.G.; Chen, M.X.; Wei, X.H.; Li, J.Y.; Li, L. Variable prescription pesticide spraying system for farmland based on the Beidou Navigation Satellite system. *J. Drain. Irrig. Mach. Eng.* **2015**, *33*, 965–970.
8. Perez-Ruiz, M.; Aguera, J.; Gil, A.; Slaughter, D.C. Optimization of agrochemical application in olive groves based on positioning sensor. *Precis. Agric.* **2011**, *12*, 564–575. [CrossRef]
9. Qiu, B.J.; Li, K.; Shen, C.J.; Xu, X.C.; Mao, H.P. Experiment on response characteristics of variable-rate continuous spraying system. *Trans. Chin. Soc. Agric. Mach.* **2010**, *41*, 32–35. (In Chinese)
10. Gonzalez, R.; Pawlowski, A.; Rodriguez, C.; Guzman, J.L.; Sanchez-Hermosilla, J. Design and implementation of an automatic pressure-control system for a mobile sprayer for greenhouse applications. *Span. J. Agric. Res.* **2012**, *10*, 939–949. [CrossRef]
11. Shahemabadi, A.R.; Moayed, M.J. An algorithm for pulsed activation of solenoid valves for variable rate application of agricultural chemical. *IEEE Int. Symp. Inf. Technol.* **2008**, *4*, 1–3.
12. Reyes, J.F.; Esquivel, W.; Cifuentes, D.; Ortega, R. Field testing of an automatic control system for variable rate fertilizer application. *Comput. Electron. Agric.* **2015**, *113*, 260–265. [CrossRef]

13. Farooque, A.A.; Chang, Y.K.; Zaman, Q.U.; Groulx, D.; Schumann, A.W.; Esau, T.J. Performance evaluation of multiple ground based sensors mounted on a commercial wild blueberry harvester to sense plant height, fruit yield and topographic features in real-time. *Comput. Electron. Agric.* **2013**, *91*, 135–144. [CrossRef]
14. Yalew, S.G.; Griensven, V.A.; Zaag, V.P. AgriSuit: A web-based GIS-MCDA framework for agricultural land suitability assessment. *Comput. Electron. Agric.* **2016**, *128*, 1–8. [CrossRef]
15. Jiang, H.; Zhang, L.; Shi, W. Effects of Operating Parameters for Dynamic PWM Variable Spray System on Spray Distribution Uniformity. *IFAC-PapersOnLine* **2016**, *49*, 216–220. [CrossRef]
16. Wang, D.S.; Zhang, J.X.; Li, W.; Xiong, B.; Zhang, S.L. Design and test of the dynamic variable spraying system of plant protection UAV. *Trans. Chin. Soc. Agric. Mach.* **2017**, *48*, 86–93. (In Chinese)
17. Chen, S.D.; Lan, Y.B.; Li, J.Y.; Zhou, Z.Y.; Liu, A.M.; Mao, Y.D. Effect of wind field below unmanned helicopter on droplet deposition distribution of aerial spraying. *Int. J. Agric. Biol. Eng.* **2017**, *10*, 67–77.
18. Gonzalez-de-Soto, M.; Emmi, L.; Perez-Ruiz, M.; Aguera, J.; Gonzalez-de-Santos, P. Autonomous systems for precise spraying–Evaluation of a robotised patch sprayer. *Biosyst. Eng.* **2016**, *146*, 165–182. [CrossRef]
19. El Aissaoui, A. *A Feasibility Study of Direct Injection Spraying Technology for Small Scale Farms: Modeling and Design of a Process Control System*; Universite de Liege: Liege, Belgique, 2015.
20. Yin, D.F.; Chen, S.R.; Pei, W.C.; Shen, B.G. Design of map-based indoor variable weed spraying system. *Trans. Chin. Soc. Agric. Eng.* **2011**, *27*, 131–135. (In Chinese)
21. Llorens, J.; Gil, E.; Llop, J.; Escola, A. Ultrasonic and LIDAR sensors for electronic canopy characterization in vineyards: Advances to improve pesticide application methods. *Sensors* **2011**, *11*, 2177–2194. [CrossRef]
22. Cai, X.; Walgenbach, M.; Doerpmond, M.; Schulze, L.P.; Sun, Y. Closed-Loop Control of Chemical Injection Rate for a Direct Nozzle Injection System. *Sensors* **2016**, *16*, 127. [CrossRef] [PubMed]
23. Deng, W.; Zhao, C.; Chen, L.; Wang, X. Constant pressure control for variable-rate spray using closed-loop proportion integration differentiation regulation. *J. Agric. Eng.* **2016**, *47*, 148–156. [CrossRef]
24. Qin, Y.; Sun, L.; Hua, Q.; Liu, P. A Fuzzy Adaptive PID Controller Design for Fuel Cell Power Plant. *Sustainability* **2018**, *10*, 2438. [CrossRef]
25. Xue, X.Y.; Tu, K.; Qin, W.C.; Lan, Y.B.; Zhang, H.H. Drift and deposition of ultra-low altitude and low volume application in paddy field. *Int. J. Agric. Biol. Eng.* **2014**, *7*, 23–28.
26. Salyani, M.; Fox, R.D. Performance of Image Analysis for Assessment of Simulated Spray Droplet Distribution. *Trans. ASAE* **1994**, *37*, 1083–1089. [CrossRef]
27. Kesterson, M.A.; Luck, J.D.; Sama, M.P. Development and preliminary evaluation of a spray deposition sensing system for improving pesticide application. *Sensors* **2015**, *15*, 31965–31972. [CrossRef]
28. Gil, E.; Llorens, J.; Llop, J.; Fabregas, X.; Gallart, M. Use of a terrestrial LIDAR sensor for drift detection in vineyard spraying. *Sensors* **2013**, *13*, 516–534. [CrossRef]
29. Zhu, H.; Salyani, M.; Fox, R.D. A portable scanning system for evaluation of spray deposit distribution. *Comput. Electron. Agric.* **2011**, *76*, 38–43. [CrossRef]
30. Yao, W.X.; Lan, Y.B.; Wang, J.; Wen, S.; Wang, G.B. Droplet drift characteristics of aerial spraying of AS350B3e helicopter. *Trans. Chin. Soc. Agric. Eng.* **2017**, *33*, 75–83. (In Chinese)
31. Zhai, C.Y.; Wang, X.; Liu, D.Y.; Ma, W.; Mao, Y.J. Nozzle flow model of high pressure variable-rate spraying based on PWM technology. *Adv. Mater. Res.* **2011**, *422*, 208–217. [CrossRef]

11

Navigating a Service Robot for Indoor Complex Environments

Jong-Chih Chien [1], Zih-Yang Dang [2] and Jiann-Der Lee [2,3,4,*]

1 Degree Program of Digital Space and Product Design, Kainan University, Taoyuan 33857, Taiwan; jcchien@mail.knu.edu.tw
2 Department of Electrical Engineering, Chang Gung University, Taoyuan 33302, Taiwan; yang1227@hotmail.com
3 Department of Neurosurgery, Chang Gung Memorial Hospital at LinKou, Taoyuan 33302, Taiwan
4 Department of Electrical Engineering, Ming Chi University of Technology, New Taipei City 24301, Taiwan
* Correspondence: jdlee@mail.cgu.edu.tw

Featured Application: Navigation for a service robot with facial and gender recognition capabilities in an indoor environment with static and dynamic obstacles.

Abstract: This paper investigates the use of an autonomous service robot in an indoor complex environment, such as a hospital ward or a retirement home. This type of service robot not only needs to plan and find paths around obstacles, but must also interact with caregivers or patients. This study presents a type of service robot that combines the image from a 3D depth camera with infrared sensors, and the inputs from multiple sonar sensors in an Adaptive Neuro-Fuzzy Inference System (ANFIS)-based approach in path planning. In personal contacts, facial features are used to perform person recognition in order to discriminate between staff, patients, or a stranger. In the case of staff, the service robot can perform a follow-me function if requested. The robot can also use an additional feature which is to classify the person's gender. The purpose of facial and gender recognition includes helping to present choices for suitable destinations to the user. Experiments were done in cramped but open spaces, as well as confined passages scenarios, and in almost all cases, the autonomous robots were able to reach their destinations.

Keywords: robot; obstacle avoidance; facial and gender recognition

1. Introduction

Today's workers in care-taking facilities have their hands full taking care of patients, thus require help for everyday routines. There are many branches of investigation into using machines to help out with daily routines in the health-care industry, such as gesture recognition [1], or pedestrian movement prediction [2]. We choose to investigation helpful tasks, such as delivering proper medicines to target patients, or helping to carry heavy loads in follow-me mode, performed by an autonomous robot with face and gender recognition abilities. For these purposes, this paper presents a tri-wheeled autonomous service robot equipped with a camera, an RGB-Depth (RGB is acronym for Red, Green, and Blue) sensor, and sonar sensors. It has built-in facial recognition ability, ability to separate staff or patients from visitors, and it can also distinguish the gender of a visitor for record-keeping purposes. By recognizing staff, it can offer the staff functions not available to patients or visitors, such as the follow-me function. By recognizing patients, it can help dispense the proper medication to each individual patient. A possible scenario for this robot could be using it to dispense medicine to selected patients who may be walking around the hallways. In navigation, it can avoid static as well as dynamic obstacles while moving toward its objective, using a dual-level Adaptive Neuro-Fuzzy Inference

System (ANFIS)-based fuzzy controller. It uses a depth-map to detect obstacles ahead, then uses image processing techniques to extract information as input into an ANFIS-based fuzzy system for analysis in order for the service to be able to avoid obstacles. The robot also incorporates sonar information using another ANFIS-based fuzzy system when it determines that depth-map information is insufficient.

In 1998, Yamauchi and Schultz [3] proposed an idea to include extra distance sensors on robots, such as laser distance sensors and ultrasound distance sensors. The distance sensors can aid in detecting surrounding objects or obstacles in real time so that the robot can locate or correct itself using known map data. However, the high costs of the precision sensors make them unsuitable for popular use. Even though ultrasonic sensors are relatively cheap by comparison, their deviation of errors is wider and they have detection blind spots; the detection distance is only proportional to the volume of the sound generator. Prahlad [4] proposed a driver-less car that has face detection and tracking capabilities. Similarly, we added onto our architecture the ability to perform face and gender recognition. Correa [5] proposed the development of a sensing system in an indoor environment, allowing the robot to have autonomous navigation and the ability for identification of its environment. His proposed system consisted of two parts: The first part is a reactive navigation system where the robot uses the RGB-Depth sensor to receive depth information and uses the arrangement of obstacles as the basis to determine the path to avoid the obstacles indoors; the second part uses an artificial neural network to identify different configurations of the environment. Csaba [6] introduced an improved version based on fuzzy rules. His system uses 16 rules, three inputs and one output, one Mamdani-type fuzzy controller, and obtained acceptable results in real-time experiments. In 2016, Algredo-Badillo [7] presented the possibility of a fuzzy control system, with the output from depth sensor as its input, for an autonomous wheelchair as a possible design. In 2018, de Silva [8] and Jin et al. [9] show that fusion of data from multiple sensors can perform better than a single sensor for a driver-less vehicle.

Our study builds an autonomous robot with a camera, a RGB-depth sensor, and sonar sensors. Prior to moving, the robot scans the entire room with a pre-trained pedestrian classifier in order to determine if a targetable person or persons exist that need to be tracked. If the robot locates its target and moves toward the person, the person's face is located and facial and gender recognitions would be performed simultaneously in order to determine if the target person's identity is in the database or if the person is a stranger. While moving, the dual sensor inputs to the fuzzy-based real-time obstacle avoidance system are used for path planning. The target person can then input target location and follows the robot or activates the robot's follow-me function instead. Using the follow-me function, the robot can help carry heavy loads for the user. The flowchart of the system is shown below in Figure 1.

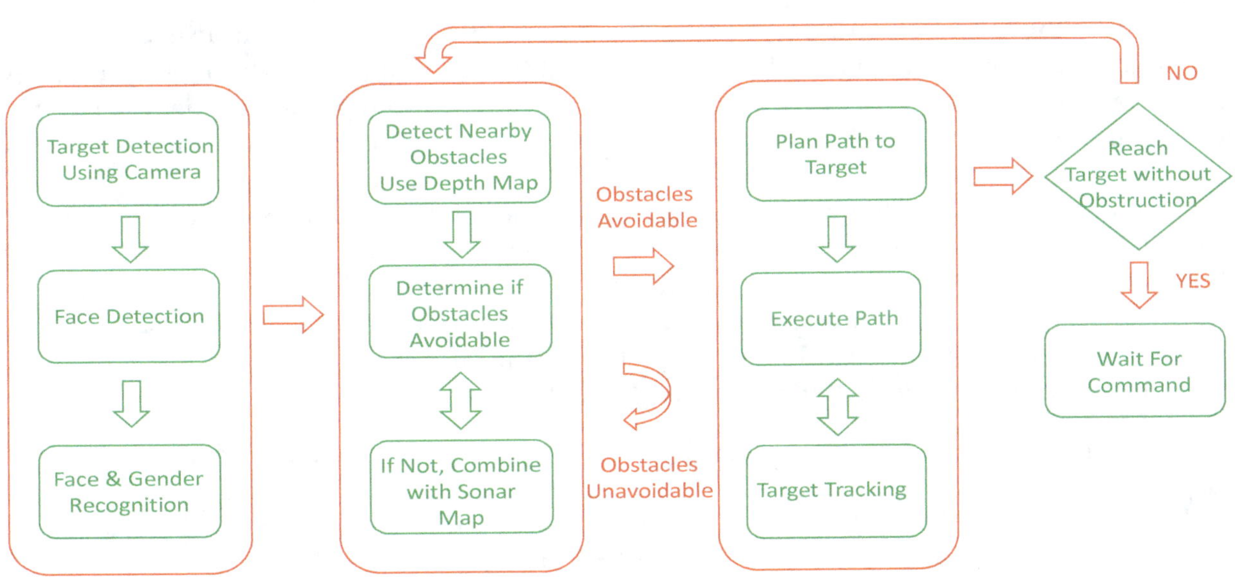

Figure 1. The system flowchart.

2. Method

2.1. Facial Recognition

We used OpenCV's LPBH (Local Binary Patterns Histogram) face recognizer method [10], which uses Local Binary Patterns Histogram as descriptors. It has shown to work better than the EigenFaces [11] or the FisherFaces [12] face recognizer methods under different environments and lighting conditions, and can reach an accuracy rate of more than 94% when 10 or more faces per person are used during its training [13]. The training and testing datasets were taken from Aberdeen [14], GUFD [15], and Utrecht ECVP [16]. Pictures of laboratory personnel were later insert into the training dataset. The purpose of adding face and gender recognition is so that the robot can locate an assigned target or targets, then performs its assigned tasks accordingly. The flowchart for face and gender recognition is shown in Figure 2.

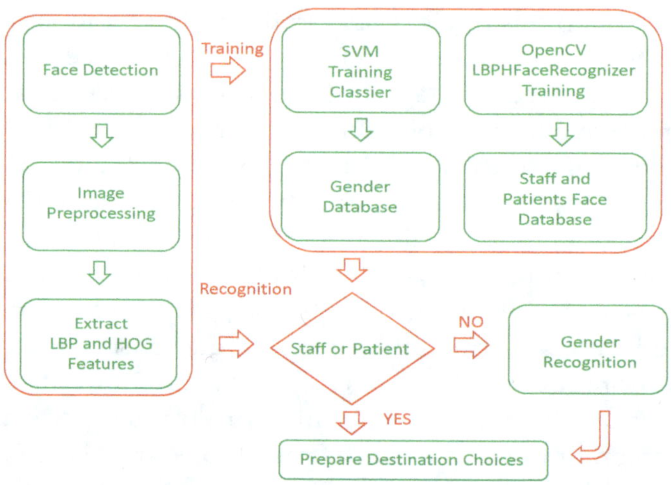

Figure 2. Facial and gender recognition flowchart.

2.2. Gender Recognition

In the gender identification part, the features with higher classification accuracy are selected from LBP [17] and HOG (Histogram of Oriented Gradients) [18] of different scales, and then screened using *p*-values before being combined. The advantage of this method is that it uses the differences between LBP and HOG methods of calculation to increase the classification accuracy. Simple experiments show that the results of combining both is higher than using either one alone. In addition, using the *p*-value to filter can be used to pick out features that are more prominent, so that the number of classification features required would be greatly reduced, and the time required for training or testing could also be improved.

After *p*-value is used to filter the statistical values of the texture features of HOG and LBP, the more robust features in the training samples are used to train a SVM (Support Vector Machine) [19] model, which is used as the basis for the classification of the test samples. The training and testing data were taken from the same online databases for facial recognition. We compared the accuracy rates of LBP, HOG, and LPB + HOG + *p*-value filtering. The results are shown below in Table 1. The rate of success of HOG + LBP + *p*-value filtering has reached 92.6%.

Table 1. Comparison of Gender Classification Accuracy Using Different Features.

Feature (s)	Classification Accuracy	Feature Points
LBP	91.8862%	2891
HOG	90.0464%	1548
LBP + HOG + *p*-value	92.6829%	1365

2.3. Object Segmentation

Object segmentation is performed using map from the depth sensor. An example of depth map vs. standard RGB camera is shown below in Figure 3.

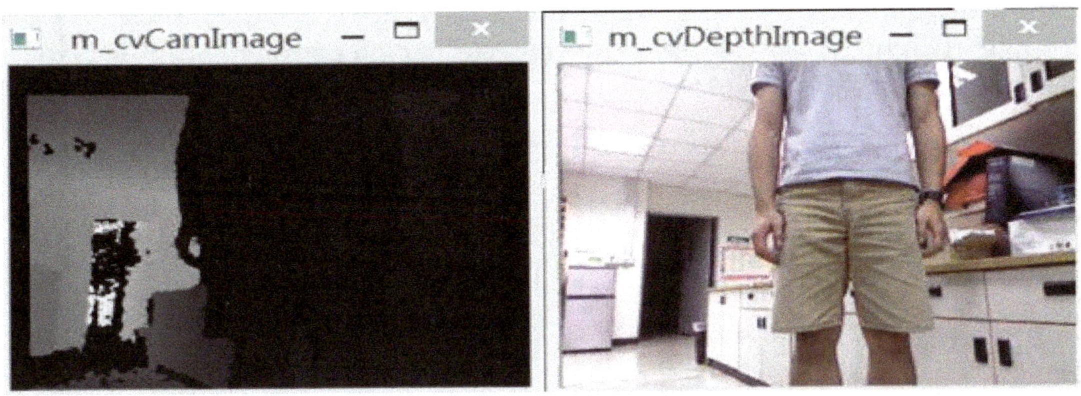

Figure 3. Example of depth map vs. standard camera image.

A depth map includes all objects nears and far, and contains too much information for accurate processing, so we decided that the threshold of the depth map should first be limited by a depth value less or equal to 1.5 m. An example result is shown below in Figure 4:

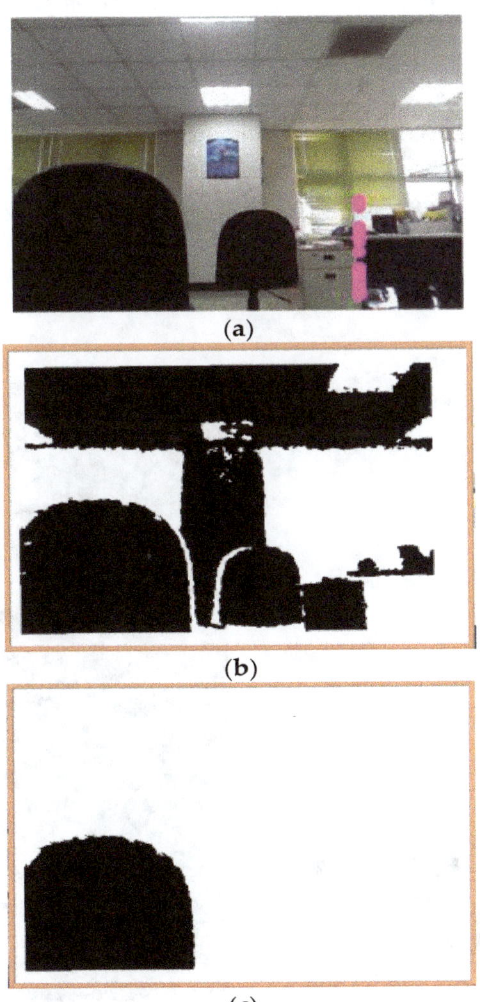

Figure 4. (**a**) Camera image, (**b**) depth map, (**c**) depth map threshold of 1.5 m.

At this range, generally the obstacles tend to be at the bottom of the depth image. In order to speed up the process, we extract the regions of interest (ROI) from the entire image by disregarding the depth information from above the middle of the height of the image. This is also to avoid the effects of lighting from above. In Correa's paper [5], he listed eight possible scenarios for obstacle arrangement; where the scene is divided into five parts, and if any of the space is occupied by an obstacle then it is marked, as shown in Figure 5. For example, the obstacle arrangement in Figure 4c is the scenario of Figure 5d.

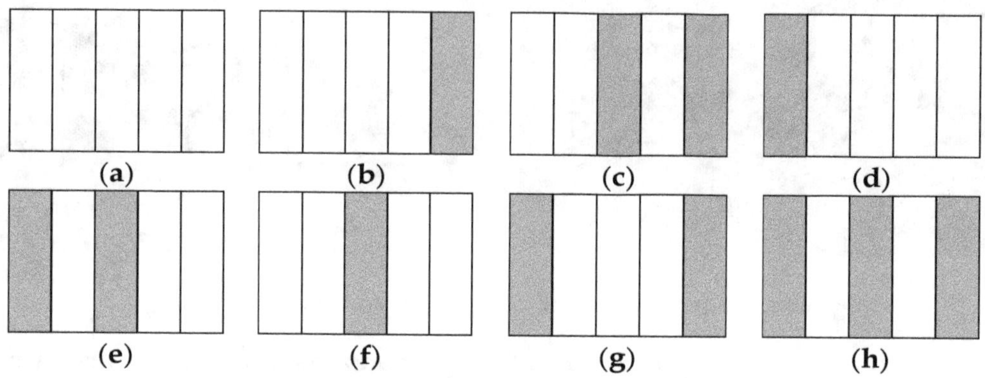

Figure 5. Correa's eight obstacles scenarios. (**a**) No obstacle; (**b**) One obstacle at the far right; (**c**) Two obstacles at the right, separated by a gap; (**d**) One obstacle at the far left; (**e**) Two obstacles at the left, separated by a gap; (**f**) An obstacle at the middle; (**g**) Two obstacles at the far right and far left; (**h**) Three obstacles separated by two gaps.

In the most likely cases are cases (f–h), scenarios where the robot uses Correa's method randomly move left or right, and can easily make the wrong decision. So instead of detecting obstacle arrangement first before moving, we decided that the robot should actively seek gaps between obstacles while moving in order to find the largest gap and judge if there is a chance to pass through. Figure 6 shows the various gaps detected using this method. The proposed method would be more proactive than that proposed by Correa.

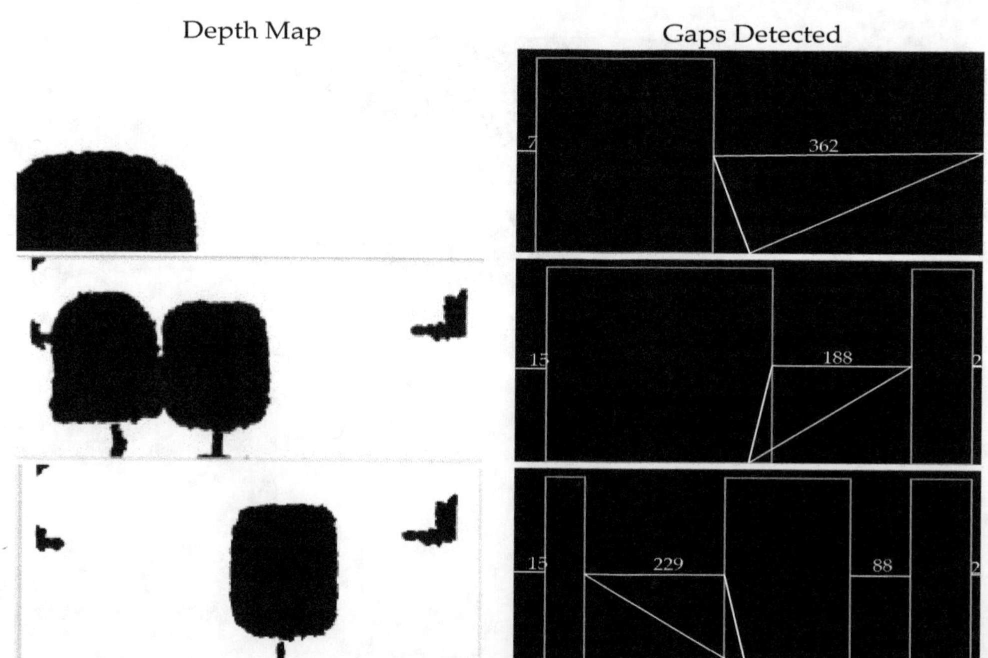

Figure 6. Threshold depth maps and detected gaps.

2.4. Robot Movements in the Presence of Obstacles

There are five possible commands for the robots to move: forward, toward left, toward right, pause, and turn around. Forward command is issued when fuzzy analysis of the obstacles shows that there is a gap in the middle that is passable. There is a necessary initial condition for the robot to move forward: that within its depth sensor's field-of-view, at least one passable gap exists. If this initial condition does not exist, then the robot can turn around, move forward a little, then turn around again in order to increase its field-of-view. The turn towards left command is issued if the fuzzy analysis of the obstacles determines that there is a gap wide enough to pass and the center of the gap is toward the left of the depth map, but it is not a hard right, rather at an angle that causes the robot to move toward the center of the gap; similarly for the turn towards right command. The turnaround command is issued when fuzzy analysis determines that there is no gap wide enough to pass through. The pause command is issued if additional information from the sonar is required. The flowchart for obstacle avoidance is shown below in Figure 7.

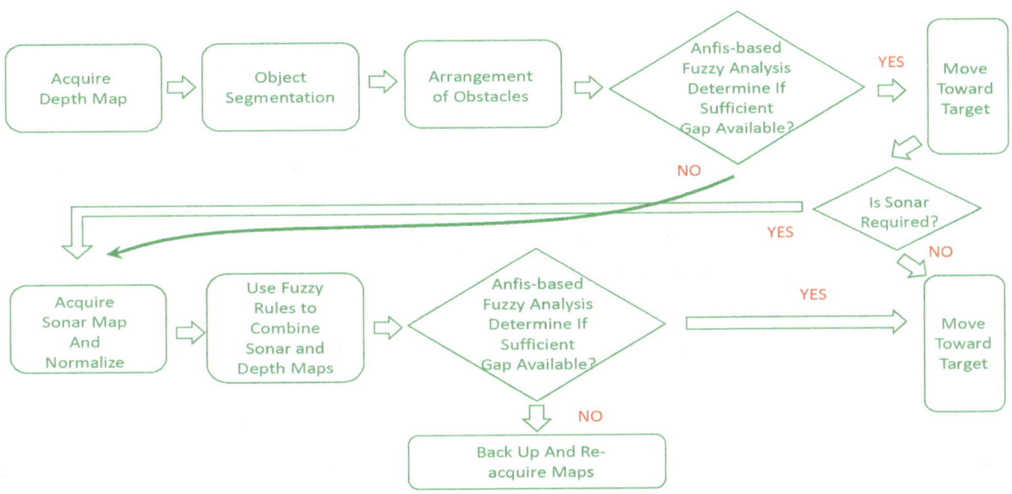

Figure 7. The flowchart for robot's obstacle avoidance.

2.5. ANFIS-Based Fuzzy System

ANFIS is the acronym for Adaptive Neuro-Fuzzy Inference System. An ANFIS [20] system structure for two variables and two rules is shown in Figure 8, where

Rule 1: If input x is A_1 and input y is B_1,
Then $f_1 = p_1x + q_1y + r_1$.
Rule 2: If input x is A_2 and input Y is B_2,
Then $f_2 = p_2x + q_2y + r_2$.

Then the output f is a linear combine of weighted f_1 and f_2. In Figure 8, the first layer is input layer, which contains the membership functions of variables. The second layer is the rule layer, which gets fuzzy rules from the combinations of the membership function of each variable. The third layer is the normalization layer, which normalizes the results from the previous layer. The fourth layer is the inference layer. The fifth layer is output layer, which calculates the sum from previous layer's output values. De-fuzzification is then performed on the output.

In our system, there are two fuzzy systems, as shown in Figure 7, where the first system uses depth map alone in determining whether a crossable gap exists between obstacles. In the second system, where depth map alone is determined to be insufficient for accurate judgement of the closeness of obstacles, the input from sonars are then used in addition to the depth map as inputs. In the following subsections, we will discuss these two fuzzy systems.

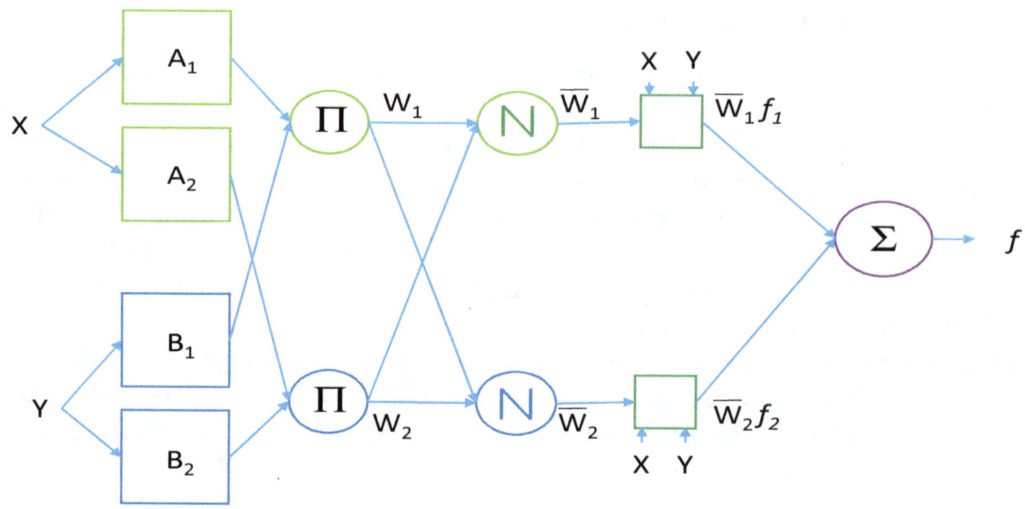

Figure 8. The ANFIS-based fuzzy system for two rules.

2.5.1. Fuzzy System Using Depth Map Alone

Because our RGB-Depth sensor is only accurate for obstacles up to 1.5 m away, there is a possibility of mis-judging gap width, and that a crossable gap may be judged as not crossable. Therefore, this is where the first fuzzy system is used to judge whether a gap is crossable. We use as input the value of the depth map, and the absolute difference between the depth values of neighboring obstacles. COG (Center of Gravity) is used as the de-fuzzification method. The 9 rules are shown in Table 2. The threshold values of 400 and 700 were determined experimentally first.

Table 2. Fuzzy rules for using depth map alone.

1	When gap width is less than robot width, and the absolute depth difference between neighbors is less than 400, then output LOW.
2	When gap value is less than robot width, and the absolute depth difference between neighbors is between 400 and 700, then output LOW.
3	When gap value is less than robot width, and the absolute depth difference between neighbors is greater than 700, then output LOW.
4	When gap value is around robot width, and the absolute depth difference between neighbors is less than 400, then output LOW.
5	When gap value is around robot width, and the absolute depth difference between neighbors is between 400 and 700, then output LOW.
6	When gap value is around robot width, and the absolute depth difference between neighbors is greater than 700, then output HIGH.
7	When gap value is greater than robot width, and the absolute depth difference between neighbors is less than 400, then output HIGH.
8	When gap value is greater than robot width, and the absolute depth difference between neighbors is between 400 and 700, then output HIGH.
9	When gap value is greater than robot value, and the absolute depth difference between neighbors is greater than 700, then output HIGH.

An example using these rules is shown in Figure 9, where the gap value is 241, around the robot's width, and the depth difference between neighbors is 663; the 9 rows in Figure 9 represent inputs and outputs for each of the 9 fuzzy rules. The bottom right graph is the visualization for the final output. The red lines for the input columns are visual representation of the input values.

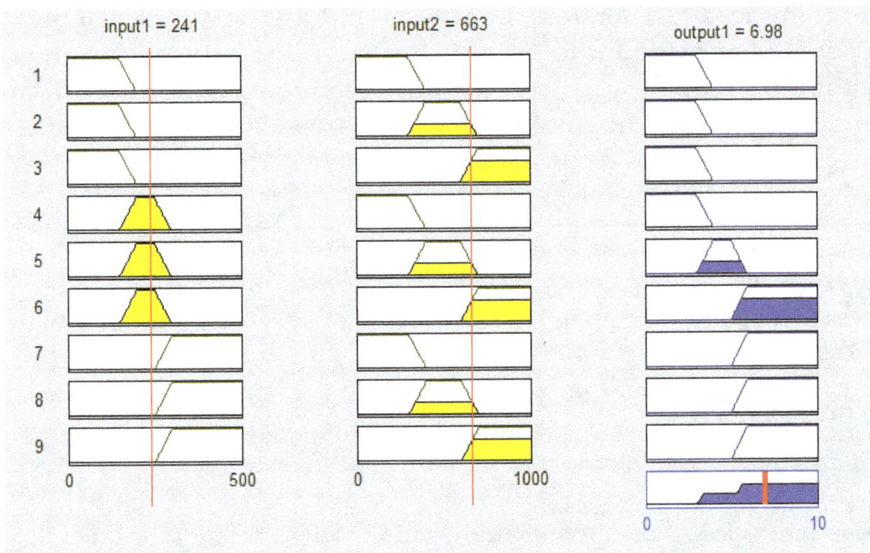

Figure 9. An example of fuzzy-rules-based decision.

2.5.2. Combining Sensors

One drawback using the depth map is the problem that if the obstacles are too low and too close to the robot, then they are undetectable by using the depth map alone. In this case, the sonars will be activated as aid in obstacle-avoidance. However, judging how much weight should be given to the depth map or the sonar map in order to yield the optimal map so as to find the largest available gap is a problem. It requires solving the weights in Equation (1) at each instance of decision making:

$$(\text{Result Map}) = W_d * (\text{Depth Map}) + W_s * (\text{Normalized Sonar Map}) \quad (1)$$

The field of research of combining sonar map with other sensors, such as depth map, is still an area that requires exploration [21,22]. The sonar map can only be used for obstacles that are close, so using the sonar map alone may cause the robot to spin continuously in order to acquire more information. Because of this, our research uses the depth map to move the robot until it moves too close to the gap or obstacles and determines that additional information would be required, then it activates the sonar system in order to acquire the sonar map. There are eight sonar sensors installed in a semi-circular fashion, as illustrated in Figure 10.

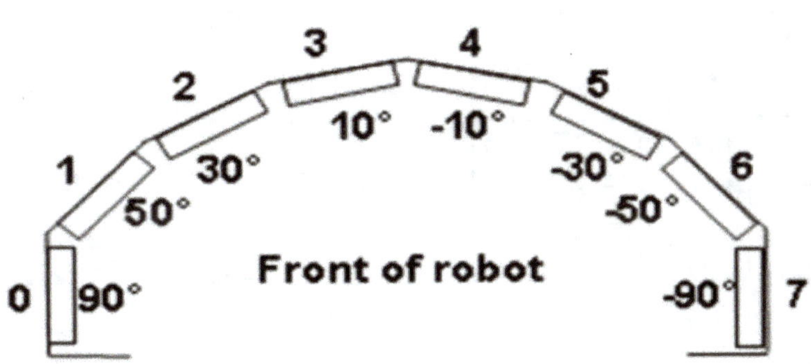

Figure 10. Placement of sonar sensors near our robot's base.

However, the leftmost and rightmost sensors are not used because they seem to cause erroneous decisions. The sonar signals are first normalized between 0 and 1.0, and it is determined that signal strength less than 0.4 indicates that obstacles are close, while greater than 0.6 indicates that obstacles are far away. If each of the six sonar signals are used as individual fuzzy inputs, then the system would become overly complicated. Therefore, in order to reduce the number of fuzzy rules, the signals for the left three sensors are averaged into a single input, and the signals for the right three sensors are averaged into another single input. This would effectively reduce the number of fuzzy rules. We designed a total of 8 fuzzy rules for combining the sensors. They are listed in Table 3.

Table 3. Fuzzy rules for combining sensors.

1	If the gap within depth map is greater than robot width, and the left sonar input is high, and the right sonar input is high, then $W_d = 1.0$ $W_s = 0.0$.
2	If the gap within depth map is greater than robot width, and the left sonar input is high, and the right sonar input is low, then $W_d = W_s = 0.5$.
3	If the gap within depth map is greater than robot width, and the left sonar input is low, and the right sonar input is high, then $W_d = W_s = 0.5$.
4	If the gap within depth map is greater than robot width, and the left sonar input is low, and the right sonar input is low, then $W_d = W_s = 0.5$.
5	If the gap within depth map is less than robot width, and the left sonar input is low, and the right sonar input is low then, $W_d = 0.0$, $W_s = 0.5$.
6	If the gap within depth map is less than robot width, and the left sonar input is high, and the right sonar input is low then, $W_d = W_s = 0.5$.
7	If the gap within depth map is less than robot width, and the left sonar input is low, and the right sonar input is low then, $W_d = W_s = 0.5$.
8	If the gap within depth map is less than robot width, and the left sonar input is high, and the right sonar input is high then, $W_d = W_s = 0.5$.

An example of using these rules as shown in Figure 11, where gap width is around the robot's width, the normalized and averaged value for left sonar is 0.259, and the normalized and averaged value for right sonar is 0.729. The eight rows in Figure 11 represents the inputs and outputs for the 8 rules. The red lines are symbolic representations of the input values.

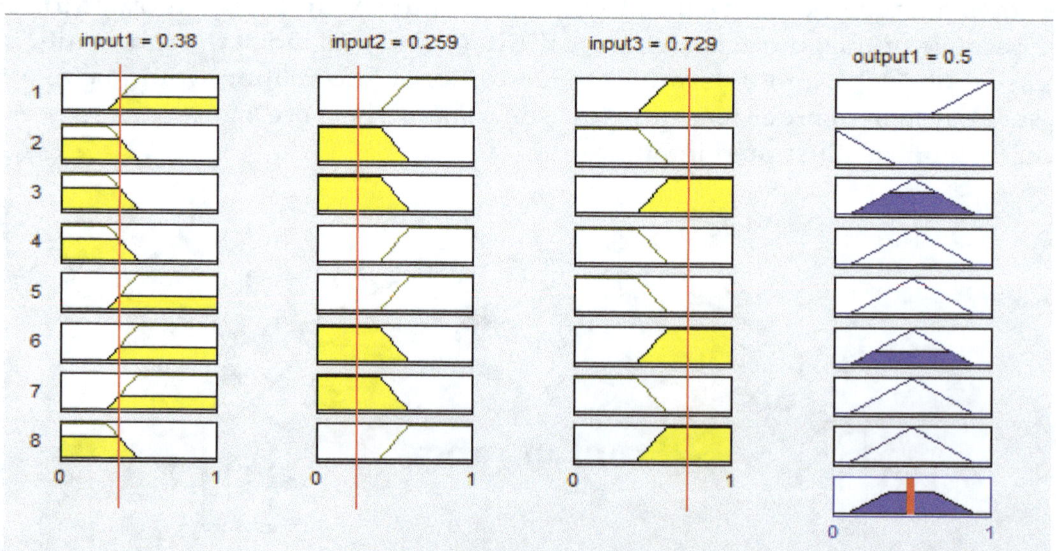

Figure 11. An example of using fuzzy rules to combine sensors input.

Figure 12 shows the paths the robot takes when decisions are made based on using fuzzy rules to combine sonar map and depth map versus using only the sonar map alone. In situations like this, we can see the advantage of combining the inputs of sensors.

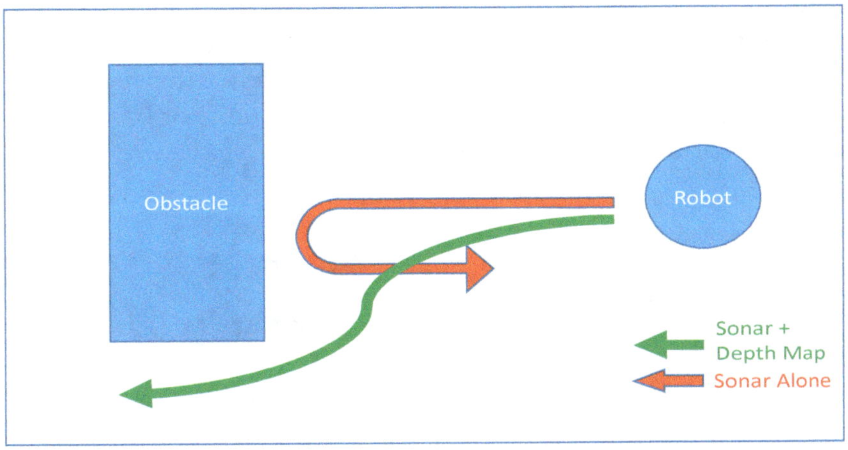

Figure 12. The path our robot takes when using sonars alone vs. combining sensors.

3. Experiment

3.1. Setup

The experimental robot is shown in Figure 13, with camera and RGB-Depth sensor on its top and middle, respectively. The sonar sensors are located below. Its width is 381 mm, and it has a swing radius of 26.7 cm. Its software is running on a laptop located in the middle.

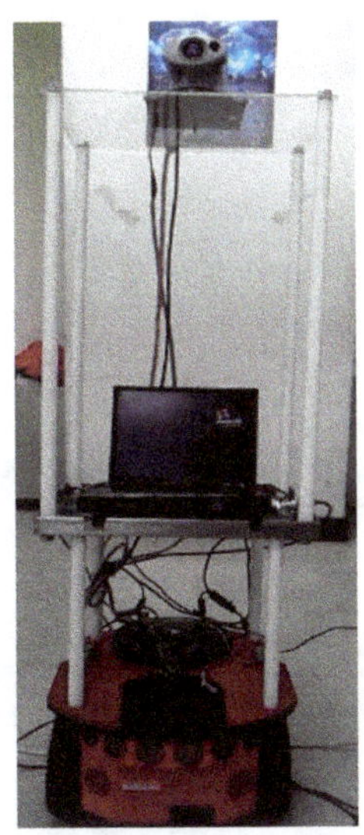

Figure 13. The experimental robot.

The experiments are divided into 2 subsections. The first is the face recognition plus gender determination. The second subsection is to test the obstacle avoidance capability of the robot. This particular subsection is further divided into situations when the obstacles are static, and when the obstacles are moving (e.g., pedestrians). These experiments were performed in real-time. These experiments were performed in an area of about 7 m by 3 m, with office chairs as obstacles.

3.2. Face and Gender Recognition

In the first experiment, the robot is placed a little distance away from the human person. The robot would adjust its orientation and distance to the human person so as to place the human face, detected using Viola's method [23], square in the middle of its field-of-view, and tracked using the KLT feature tracker [24]. The training sample included 400 male photos and 300 female photos. The test samples included 265 male photos and 227 female photos. The correct gender recognition rate was 92.6829%. Later, 10 photos of each laboratory personnel were added to the training set, and the robot was able to perform on-line face recognition during the test, and recognize the human persons in front as laboratory personnel and correctly classified their gender. After recognition and classification, choices for destinations are then presented to the user via the monitor based on the result, including a Follow-Me choice for lab workers. Examples of classification results are shown in Figure 14.

(**a**) Female Lab Worker (**b**) Male Lab Worker

Figure 14. Examples of facial recognition and gender classification.

3.3. Obstacles Avoidance

3.3.1. Static Obstacle Avoidance

In the static obstacles experiment, the testing site is a hallway. The obstacles are placed along the hallway, as shown in Figure 15.

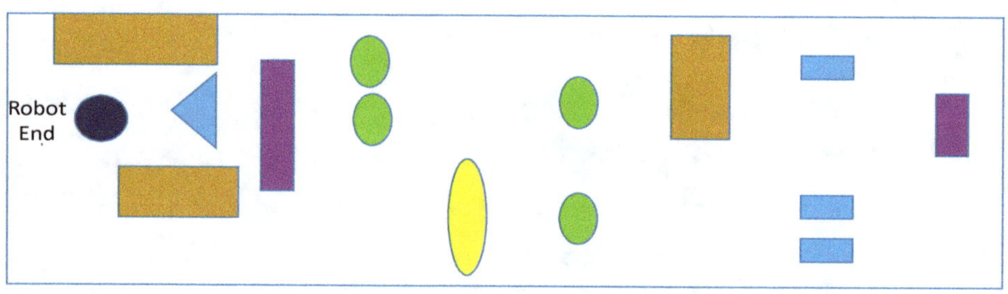

Figure 15. Static obstacles for the robot.

Sufficient gaps are left between the obstacles so that the robot should be able to reach its final destination. The following plot, Figure 16, shows the path the robot took. The robot's position was recorded at an interval of every 10 s.

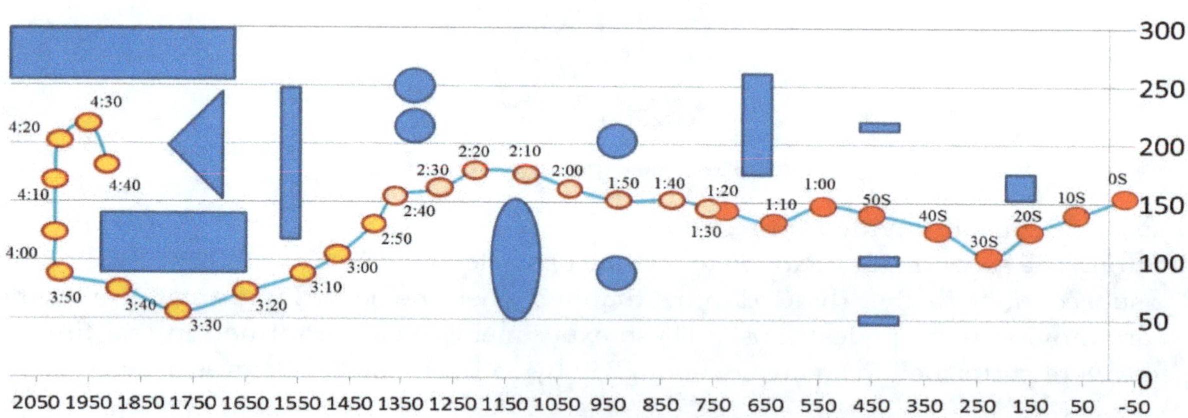

Figure 16. The path the robot took with positions taken at 10 s interval.

There are a total of 17 possible scenarios designed along the path that we think that the robot would need to make critical decisions about whether to incorporate sonar sensors in its decisions. The following figure, Figure 17, shows where each of the 17 scenarios, T_1–T_{17}, took place, and Table 4 shows the input values and the robot's decisions for the 17 scenarios.

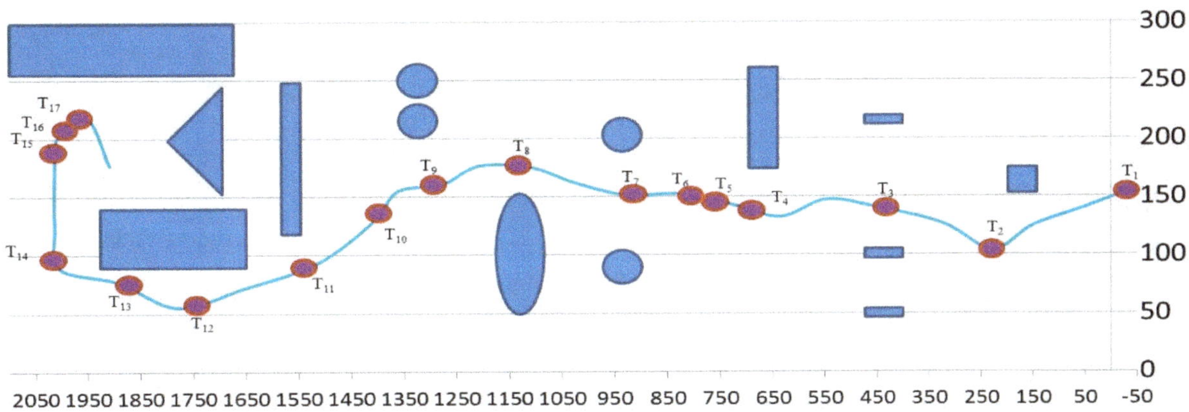

Figure 17. The seventeen scenarios (T_1–T_{17}) where decisions of the robot were recorded.

Table 4. Sensors Inputs and Decision for Scenarios T_1–T_{17}.

Scenario	Left Obstacle Depth Value	Right Obstacle Depth Value	Gap Normalized to Robot Width	Left Sonar Normalized	Right Sonar Normalized	Final Decision
T_1	N/A	317	1.41	1	1	Forward
T_2	97.6	10,000	1.67	0.2	1	Right Turn
T_3	N/A	106.3	1.20	0.8	0.44	Left Turn
T_4	N/A	N/A	0.0	0.22	0.22	Forward
T_5	128.9	N/A	0.0	0.2	0.62	Right Turn
T_6	122.1	125.6	1.64	0.4	0.4	Forward
T_7	139.9	176.3	1.56	0.58	0.66	Forward
T_8	272.2	N/A	2.24	0.4	0.36	Forward
T_9	N/A	N/A	0.0	0.3	0.22	Forward
T_{10}	N/A	188.8	1.67	1	0.5	Forward
T_{11}	163.0	160.3	1.05	0.5	0.38	Forward
T_{12}	100.1	N/A	1.33	0.08	0.13	Right Turn
T_{13}	193.1	1000.0	0.84	0.4	1	Right Turn
T_{14}	163.5	196.4	0.83	0.54	0.4	Forward
T_{15}	N/A	N/A	0.0	0.04	0.22	Forward
T_{16}	N/A	N/A	0.0	0.22	0.22	Forward
T_{17}	115.9	N/A	1.13	0.22	0.42	Right Turn

Figure 18 shows the fuzzy inputs and outputs at T_1–T_{17}.

Table 5 below illustrates the comparisons between our method, Correa's method [3], and Csaba's method [4], using single-point analysis at each scenario point. Success is defined as being able to avoid collision.

Table 5. Comparisons between Correa, Csaba, and our proposed method on T_1–T_{17} scenarios.

Scenario	Correa	Csaba	Our Method
T_1	Fail!	Success	Success
T_2	Fail!	Fail!	Success
T_3	Fail!	Success	Success
T_4	Success	Success	Success
T_5	Fail!	Fail!	Success
T_6	Fail!	Fail!	Success
T_7	Success	Success	Success
T_8	Success	Success	Success
T_9	Success	Success	Success
T_{10}	Fail!	Success	Success
T_{11}	Success	Success	Success
T_{12}	Fail!	Success	Success

Table 5. *Cont.*

Scenario	Correa	Csaba	Our Method
T_{13}	Fail!	Fail!	Success
T_{14}	Fail!	Success	Success
T_{15}	Fail!	Success	Success
T_{16}	Success	Success	Success
T_{17}	Fail!	Success	Success
T_1	Success	Success	Success

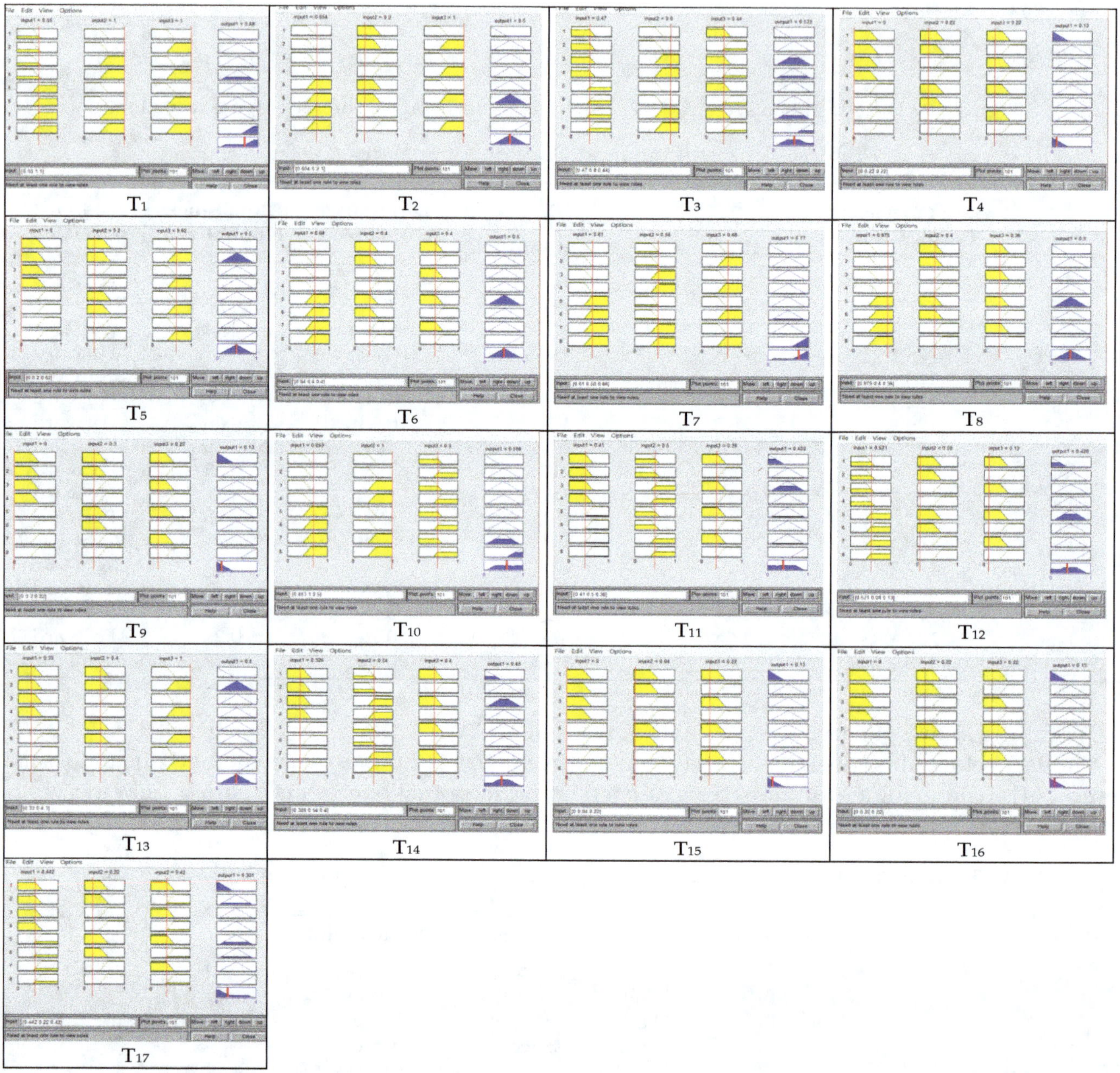

Figure 18. Fuzzy inputs and decisions for combining sensor maps at T_1–T_{17}.

3.3.2. Dynamic Obstacle Avoidance

Eleven scenarios were setup using one or two moving obstacles for this experiment. The obstacles are to simulate pedestrians in a service environment, and none was the robot's target. Table 6 illustrates these scenarios.

Table 6. Eleven scenarios for dynamic moving obstacles.

Scenario #	Scenario
1	Single Obstacle Moving Towards the Robot
2	Single Obstacle Moving Fast from Right of the Robot
3	Single Obstacle Moving Slow from Right of the Robot
4	Single Obstacle Moving Fast from Left of the Robot
5	Single Obstacle Moving Slow from Left of the Robot
6	Dual Obstacles Moving Towards the Robot Then Separates
7	Dual Obstacles Moving From Left and Right of the Robot Then Crisscross
8	Dual Obstacles Moving From Left of the Robot At The Same Speed
9	Dual Obstacles Moving From Left of the Robot At Different Speeds
10	Dual Obstacles Moving From Right of the Robot At The Same Speed
11	Dual Obstacles Moving From Right of the Robot At Different Speeds

Figure 19 illustrates how the robot responded at each of these scenarios. The red line represents the path of the robot, the blue line represents the path of the first obstacle, and orange line represents the path of the second obstacle. Each dot represents sampled locations taken between fixed time intervals. In each of these scenarios, except for the third, the robot was able to pass through the obstacles and reach the other side. In the third scenario, the robot was able to avoid collision by turning around.

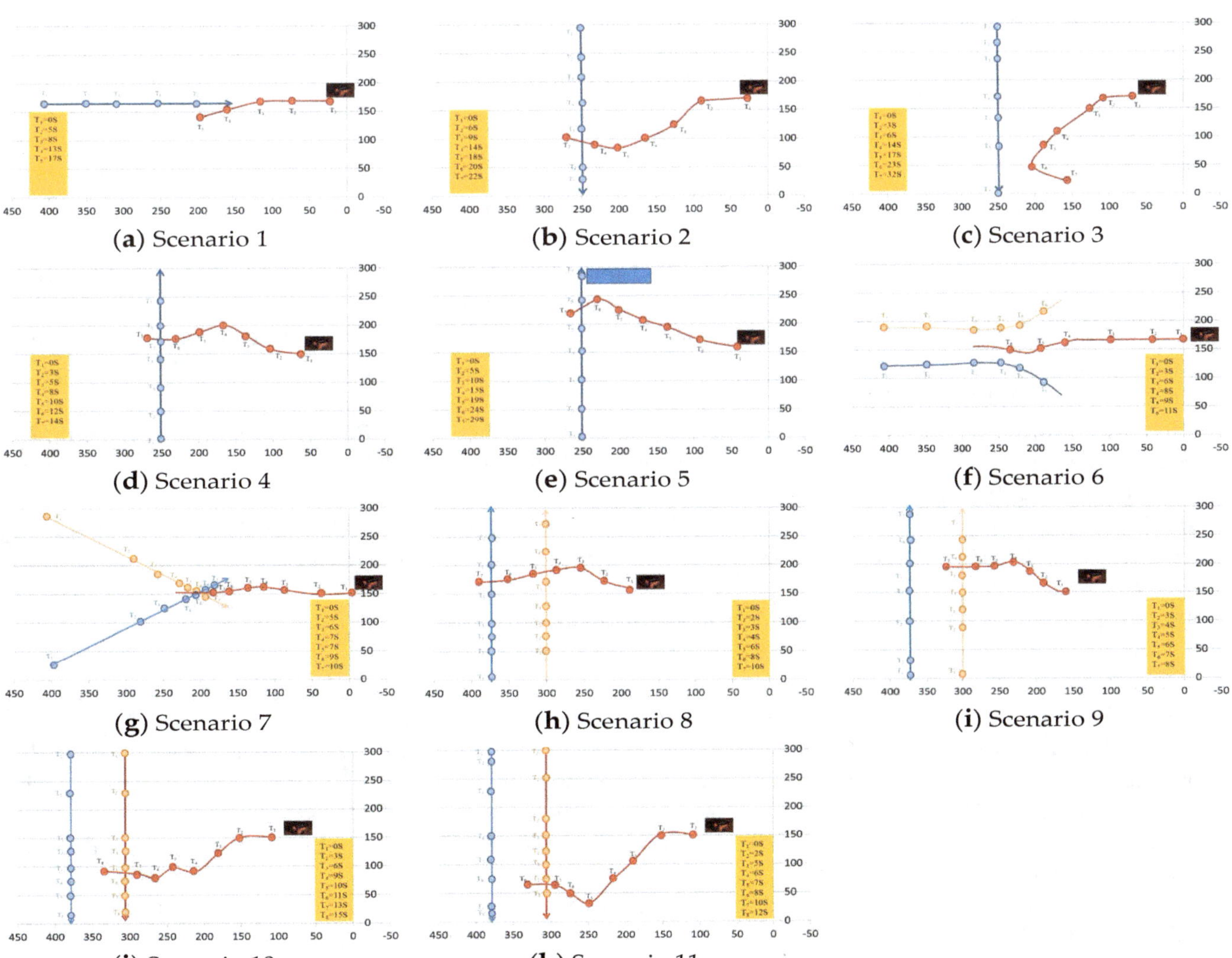

(**a**) Scenario 1 (**b**) Scenario 2 (**c**) Scenario 3 (**d**) Scenario 4 (**e**) Scenario 5 (**f**) Scenario 6 (**g**) Scenario 7 (**h**) Scenario 8 (**i**) Scenario 9 (**j**) Scenario 10 (**k**) Scenario 11

Figure 19. The paths of the moving obstacles (blue, orange) and the robot (red) for the 11 scenarios.

4. Discussion and Conclusions

In this paper, we presented a service robot with face recognition and gender classification abilities, with accuracy reaching 92.68%. The service robot can perform different tasks based on the classification results, such as activating the Follow-Me function for laboratory staff only. Possible uses of the robots including delivering medicines to target patients, assist visitors to find patients, and other tasks. We also developed a dual-level ANFIS-based fuzzy obstacle-avoidance system based on the inputs from two different types of sensors: RGB-Depth and sonars. It is found that the obstacle-avoidance capability using both types of sensors surpasses the performance of using just a single type sensor. We performed the experiment testing the obstacle avoidance capability of the robot under both static and dynamic environments and found that the robot can successfully maneuver around obstacles in almost all cases. In the future, we hope to improve the robot's navigation abilities by including advanced path planning using indoor maps of the environment, or different hardware configurations for other purposes.

Author Contributions: Conceptualization, methodology, analysis, writing, J.-C.C. and J.-D.L.; Experiments, Z.-Y.D.

References

1. Yang, G.; Lv, H.; Chen, F.; Pang, Z.; Wang, J.; Yang, H.; Zhang, J. A Novel Gesture Recognition System for Intelligent Interaction with a Nursing-Care Assistant Robot. *Appl. Sci.* **2018**, *8*, 2349. [CrossRef]
2. Chen, Z.; Song, C.; Yang, Y.; Zhao, B.; Hu, Y.; Liu, S.; Zhang, J. Robot Navigation Based on Human Trajectory Prediction and Multiple Travel Modes. *Appl. Sci.* **2018**, *8*, 2205. [CrossRef]
3. Yamauchi, B.; Shultz, A.; Adams, W. Mobile robot exploration and map-building with continuous localization. In Proceedings of the Internet Content Rating Association (ICRA), Leuven, Belgium, 20 May 1998; Volume 4, pp. 3175–3720.
4. Vadakkepat, P.; Lim, P.; de Silva, L.C.; Jing, L.; Ling, L.L. Multimodal Approach to Human-Face Detection and Tracking. *IEEE Trans. Ind. Electron.* **2008**, *55*, 1385–1393. [CrossRef]
5. Correa, D.S.O.; Sciotti, D.F.; Prado, M.G.; Sales, D.O.; Wolf, D.F.; Osorio, F.S. Mobile Robots Navigation in Indoor Environments Using Kinect Sensor. In Proceedings of the 2012 Second Brazilian Conference on Critical Embedded Systems, São Paulo, Brazil, 21–25 May 2012; pp. 36–41.
6. Csaba, G. Fuzzy Based Obstacle Avoidance for Mobil Robots with Kinect Sensor. In Proceedings of the 2012 4th IEEE International Symposium on Logistics and Industrial Informatics (LINDI), Smolenice, Slovakia, 5–7 September 2012; pp. 135–144.
7. Algredo-Badillo, I.; Morales-Raales, L.A.; Hernández-Gracidas, C.A.; Cortés-Pérez, E.; Pimentel, J.J.A. Self-navigating Robot based on Fuzzy Rules Designed for Autonomous Wheelchair Mobility. *Int. J. Comput. Sci. Inf. Secur.* **2016**, *14*, 11–19.
8. de Silva, V.; Roche, J.; Kondoz, A. Fusion of LiDAR and Camera Sensor Data for Environment Sensing in Driverless Vehicles. 2018. Available online: https://dspace.lboro.ac.uk/2134/33170 (accessed on 10 November 2018).
9. Jin, X.-B.; Su, T.-L.; Kong, J.-L.; Bai, Y.-T.; Miao, B.-B.; Dou, C. State-of-the-Art Mobile Intelligence: Enabling Robots to Move Like Humans by Estimating Mobility with Artificial Intelligence. *Appl. Sci.* **2018**, *8*, 379. [CrossRef]
10. OpenCV. Available online: http://www.opencv.org/ (accessed on 2 September 2018).
11. Turk, M.; Pentland, A. Eigenfaces for recognition. *J. Cogn. Neurosci.* **1991**, *3*, 71–86. [CrossRef] [PubMed]
12. Belhumeur, P.N.; Hespanha, J.P.; Kriengman, D.J. Eigenfaces vs. fisherfaces: Recognition using class specific linear projection. *IEEE Trans. Pattern Anal. Mach. Intell.* **1997**, *19*, 711–720. [CrossRef]
13. Chien, L.W.; Ho, Y.F.; Tsai, M.F. Instant Social Networking with Startup Time Minimization Based on Mobile Cloud Computing. *Sustainability* **2018**, *10*, 1195. [CrossRef]
14. Aberdeen Facial Database. Available online: http://pics.psych.stir.ac.uk/zips/Aberdeen.zip (accessed on 10 February 2017).
15. GUFD Facial Database. Available online: http://homepages.abdn.ac.uk/m.burton/pages/gfmt/Glasgow%20Face%20Recognition%20Group.html (accessed on 10 February 2017).

16. Utrecht ECVP Facial Database. Available online: http://pics.psych.stir.ac.uk/zips/utrecht.zip (accessed on 10 February 2017).
17. Ullah, I.; Aboalsamh, H.; Hussain, M.; Muhammad, G.; Mirza, A.; Bebis, G. Gender Recognition from Face Images with Local LBP Descriptor. *Arch. Sci. J.* **2012**, *65*, 353–360.
18. Ren, H.; Li, Z. Gender Recognition Using Complexity-Aware Local Features. In Proceedings of the 2014 22nd International Conference on Pattern Recognition, Stockholm, Sweden, 24–28 August 2014; pp. 2389–2394.
19. Moghaddam, B.; Yang, M.-H. Gender Classification with Support Vector Machines. In Proceedings of the Fourth IEEE International Conference on Automatic Face and Gesture Recognition, Grenoble, France, 28–30 March 2000; p. 306.
20. Jang, J.-S.R. ANFIS: Adaptive-network-based fuzzy inference system. *IEEE Trans. Syst. Man Cybern.* **1993**, *23*, 665–685. [CrossRef]
21. Flynn, A.M. Combining Sonar and Infrared Sensors for Mobile Robot Navigation. *Int. J. Robot. Res.* **1988**, *7*, 5–14. [CrossRef]
22. Elfes, A.; Matthies, L. Sensor integration for robot navigation: Combining sonar and stereo range data in a grid-based representataion. In Proceedings of the 26th IEEE Conference on Decision and Control, Los Angeles, CA, USA, 9–11 December 1987; pp. 1802–1807.
23. Viola, P.; Jones, M.J. Rapid Object Detection using a Boosted Cascade of Simple Features. In Proceedings of the IEEE Computer Society International Conference on Computer Vision and Pattern Recognition, Kauai, HI, USA, 8–14 December 2001; Volume 1, pp. 511–518.
24. Tomasi, C.; Kanade, T. *Detection and Tracking of Point Features*; Technical Report CMU-CS_91-132; Carnegie Mellon University: Pittsburgh, PA, USA, 1991.

12

A Master-Slave Separate Parallel Intelligent Mobile Robot used for Autonomous Pallet Transportation

Guo Li [1], Rui Lin [2,*], Maohai Li [2], Rongchuan Sun [2] and Songhao Piao [1,*]

[1] Department of Computer Science and Technology, Harbin Institute of Technology, Harbin 150001, China; 14b903020@hit.edu.cn

[2] School of Mechanical and Electrical Engineering, Soochow University, Suzhou 215000, China; limaohai@suda.edu.cn (M.L.); sunrongchuan@suda.edu.cn (R.S.)

* Correspondence: linrui@suda.edu.cn (R.L.); piaosh@hit.edu.cn (S.P.)

Abstract: This work reports a master-slave separate parallel intelligent mobile robot for the fully autonomous transportation of pallets in the smart factory logistics. This separate parallel intelligent mobile robot consists of two independent sub robots, one master robot and one slave robot. It is similar to two forks of the forklift, but the slave robot does not have any physical or mechanical connection with the master robot. A compact driving unit was designed and used to ensure access to the narrow free entry under the pallets. It was also possible for the mobile robot to perform a synchronous pallet lifting action. In order to ensure the consistency and synchronization of the motions of the two sub robots, high-gain observer was used to synchronize the moving speed, the lifting speed and the relative position. Compared with the traditional forklift AGV (Automated Guided Vehicle), the mobile robot has the advantages of more compact structure, higher expandability and safety. It can move flexibly and take zero-radius turn. Therefore, the intelligent mobile robot is quite suitable for the standardized logistics factory with small working space.

Keywords: intelligent mobile robot; pallet transportation; master-slave; compact driving unit; high-gain observer

---

1. Introduction

Logistics is an important part of the manufacturing enterprise [1,2], and the pallet is one of the logistics carriers in the factory [3]. The automation of pallet transportation is an important factor which affects the smart factory logistics [4]. Pallet handling equipment mainly refers to manually operated hydraulic trucks and electric or fuel forklifts, both of which require a large working space for operation. When the goods are delivered, it is necessary for the operator to handle the loading and unloading of the goods. A large amount of labor is required, which does not meet the requirements of factory automation.

In recent years, with the development of factory intelligence, flexible manufacturing systems and automated warehousing, the autonomous navigation forklift or forklift AGV (Automated Guided Vehicle) has gradually become an important device for solving the internal logistics of pallet transportation in the factory [5,6]. Compared with the traditional forklift, the autonomous navigation forklift AGV is improved with intelligent components such as external sensors, navigation and positioning modules, software scheduling, and power management system. Forklift AGV can transport the pallets placed on the ground to meet the unmanned logistics requirements [7–9]. As early as 2015, the forklift company, for example, Dematic or Jungheinrich, launched a forklift AGV based on the traditional forklift. For AGV products based on forklift, a single steering wheel or a differential fixed double steering wheel with a front integrated steering motor was generally used as the driving wheel, and the two-forked casters with two auxiliary wheels. Forklift AGV could move freely by controlling a single steering wheel or a differential double steering wheel. Forklift AGV used the traditional forklift structure and had the function of stacking high pallets. However, due to its own structural characteristics, the overall size and its self-weight are both large. And the arrangement of the driving wheels and the attached wheels determined that this type of AGV cannot achieve zero-radius rotation. It requires a rather large working space [10].

In order to adapt to small working space of factory logistics, another type of pallet transportation robot was designed and used. The pallet transportation robot moved underneath the pallet rack to lift and delivered the pallet to a destination, which was placed on a pallet rack with a certain height in advance. In 2007, Professor Manuel Weber of the University of Stuttgart, Germany, proposed and designed the double-pronged pallet handling robot, the Doppelkufen system [11], to minimize the required transportation space. The Doppelkufen system used a visual recognition ribbon to guide between different production lines. In order to realize autonomous navigation, the two separate double forks connected with each other physically, and peripheral sensors were assembled to develop a compact forklift robot capable of zero-radius rotation and omnidirectional motion, for example, Agumos G130 [12] of Melkus Mechatronic GmbH and Nipper [13] of Dutch F3 Design. Combining with the advantages of both the traditional forklift AGV and the compact forklift AGV, INTREST Services GmbH of Germany introduced the Agilox product [14]. Agilox could perform palletizing automatically. It had a heavy self-weight and high height and cannot move underneath the pallet directly like a forklift. However the pallet needed to be placed on a custom-height pallet rack in advance.

In order to solve the problems, this paper mainly introduces a master-slave separate parallel intelligent mobile robot for autonomous pallet transportation in the smart factory logistics. The separate parallel intelligent mobile robot consists of two independent sub robots, which is similar to two forks of the forklift, but these sub robots do not have any physical connection with each other to achieve the pallet transportation. It can move underneath the narrow free entry of the pallet and has a load capacity of up to 1 ton. It was designed with a compact driving unit, in addition to actual motions, such as linear motion, oblique linear motion, and zero-radius turn. It could also perform synchronous lifting and laying of the pallet. The slave robot follows the master robot synchronously and consistently. Each robot had two driving units. In order to synchronize the motion and trajectory of these two sub robots, a nonlinear control system was used [15]. The application of nonlinear control systems in robotics control was very common, from the simplest two-wheel driving self-balancing robot dynamics [16,17], the robotic arm joint PD control [18], to the cluster control of UAV formations [19–21]. In the design we used the synchronic control strategy of slave robot dynamic adaptive following master robot. By analyzing the practical effects of sliding mode control [22] and high-gain observer [23] of the nonlinear control tools, we selected high-gain observer to obtain the following speed and state control of slave robot which was to ensure that the following errors converged and achieved synchronization of the master-slave robot motion.

Compared with the manual and forklift truck, the separate parallel intelligent mobile robot can fulfill an autonomous pallet transportation. Compared with the traditional intelligent forklift AGV, it has the advantages of compact structure, small self-weight and large payload, high expandability and safety. It can implement zero-radius turn and requires small channel width. It can also be combined with automation equipment or storage system seamlessly. So it is very suitable for small and standardized logistics warehouses.

The rest of this paper is organized as follows. Section 2 describes the mechanical structure of master-slave separate parallel intelligent mobile robot, such as compact driving units, communication modules, and safety protection modules. High-gain observer is adopted to ensure the consistency and synchronization of actual motions of two sub robots of intelligent mobile robot in Section 3. Experimental results and related analyses are given in Section 4, followed by conclusions in Section 5.

2. Separate Parallel Intelligent Mobile Robot

2.1. A Compact Driving Unit

As the EPAL Euro pallet is widely used in the factory logistics (PALETTE EUR-EPAL, dimension: L1200 mm × W800 mm and L1200 mm × W1000 mm), a compact driving unit was designed ensure that the two sub robots could move into the free entry of the standard pallet and lift the pallet up. Each sub robot has front and rear driving units. Separate parallel intelligent mobile robot implemented synchronous motions such as linear motion, oblique linear motion, and zero-radius turn. It could also perform synchronous lifting and laying of the pallet.

2.1.1. Motor Performance Calculation

In the actual motions of the master-slave separate parallel robot during the pallet transportation, it is necessary to overcome some resistance, such as the rolling resistance from the ground and the air resistance from the air. The resistance expressions $\sum F$ are:

$$\sum F = F_f + F_w + F_i + F_j \tag{1}$$

where, F_f is tolling resistance; F_w is air resistance; F_i is climbing resistance; F_j is acceleration resistance.
Tolling resistance for a sub robot is

$$F_f = \mu m g \tag{2}$$

where, μ is coefficient of tolling resistance. m represents the its own weight and maximum payload.
The resistance during acceleration of the sub robot is:

$$F_j = ma \tag{3}$$

Due to surface roughness, climbing resistance of the sub robot is

$$F_i = mg\sin\alpha \tag{4}$$

The rolling diameter of sub robot's driving wheel is set to D. The total motion-resistance force is $\sum F$. Then total resistance moment $\sum M$ is:

$$\sum M = \sum F \cdot \frac{D}{2} \tag{5}$$

As each unit has two driving wheels, the driving moment of every wheel M_d is

$$M_d = \frac{\sum M}{4} \tag{6}$$

According to the design requirements, a sub-robot's self-weight and maximum payload m = 600 kg. The normal velocity is 0.7 m/s. The acceleration is $a = 0.3$ m/s^2. The surface slope of working ground is $\alpha = 2°$. The ground friction coefficient is $\mu = 0.02$.

2.1.2. Mechanism Design

Each sub robot uses front and rear driving units and one driving unit consists of two DC motors, reducers and driving wheels. The power transmission sequence is: DC motor, planetary reducer, small timing pulley, timing belt, large timing pulley, and driving wheel. The slip ring is assembled to simplify wiring during the zero-radius turn of the driving units. The slip ring is divided into upper and lower parts: the upper part is the stator, and the lower part is the rotor. The stator is fixed with the integral support plate, and the rotor rotates together with the driving mechanism. Slip ring not only solves the problem of wiring of power and cable during the lifting and laying of the pallet, but also provides direction angle value for the front and rear driving units through the built-in multi-turn absolute encoder, which provides input for the nonlinear synchronization control in Section 3. It is known that the pitch, length and other parameters of the screw can lift the pallet up precisely to a certain height by controlling the number of wheel rotations of the driving units. Since the screw has a self-holding characteristic, the pallet does not fall down by itself even though power outage may occur unexpectedly. The compact driving unit is shown in Figure 1.

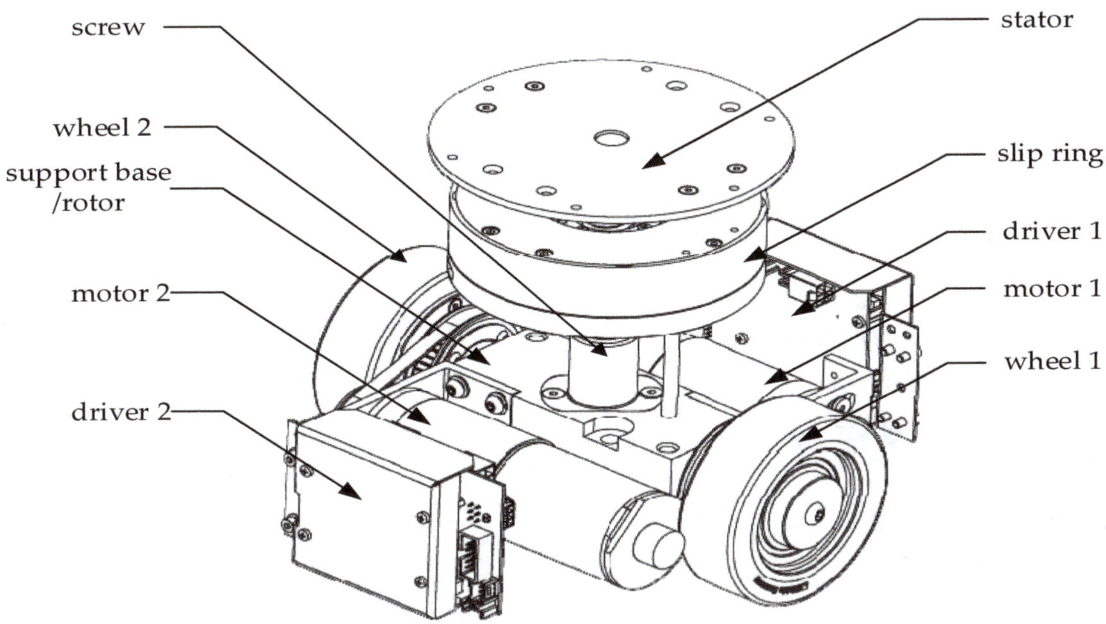

Figure 1. Three dimensional sketches of the compact driving unit.

2.1.3. Analysis of Motion Models

Motion models of separate parallel intelligent mobile robot are mainly based on the master robot's motion, such as visual navigation, and laser-based navigation. Slave robot follows master robot to dynamically adjust the relative speed and position. The two sub robots implement synchronous actual motions, such as linear motion, oblique linear motion, zero-radius turn, and lifting and laying. The motion model is shown in Figure 2.

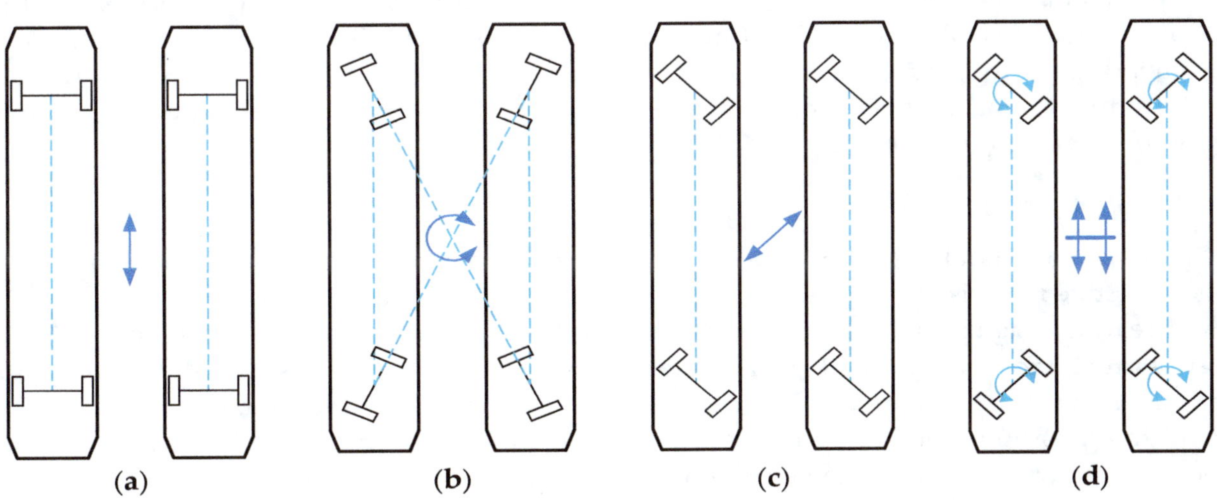

Figure 2. Analysis of motion model. (**a**) Forwards/Backwards; (**b**) Rotation; (**c**) Sideways/crab motion; (**d**) Lifting and laying.

2.2. Schematic Diagram of Separate Parallel Intelligent Mobile Robot

The main components of master-slave separate parallel intelligent mobile robot are: driving units, control and power supply modules, sensor modules, shell cover and safety protection modules, as shown in Figure 3. The individual modules of the sub robot are mechanically independent but electrically connected to each other. Each sub robot consists of two driving units, two safety protection modules, two sensor modules, and control and power supply modules with the cover. Safety sensors and infrared data transmission module are installed around the sides of each sub robot. The shell cover is the mounting base, and each module is mounted on the cover by a cross recessed countersunk head screw.

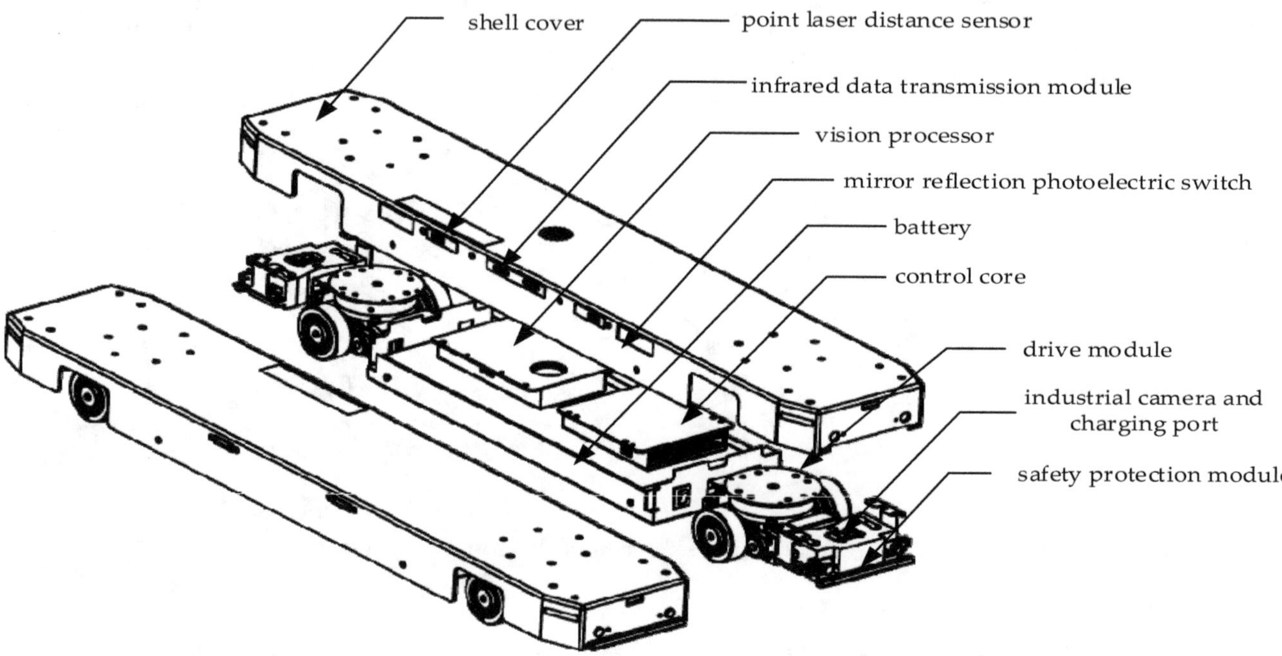

Figure 3. Schematic diagram of separate parallel intelligent mobile robot.

3. Synchronization Nonlinear Control

The most significant and complex problem in the real-time control system of the master-slave separate parallel intelligent mobile robot is the synchronization control of actual motions of the two sub robots, which is also a very difficult nonlinear control system. To get good performance of the synchronization control, complex nonlinear control strategy should be used. Therefore, this section mainly introduces some nonlinear control strategies, such as the input of nonlinear control system and high-gain observer control tool.

3.1. Control Strategies of Actual Motions

Different control strategies should be used in master robot and slave robot, while separate parallel intelligent mobile robot is moving. As the master robot, the only control principle is ensuring that the front and rear driving units work precisely and move as ordered. But for the slave robot, its control strategy is much more complex. (1) Self-motion conforms to self-physical model. (2) Following the master robot on front-rear axis and left-right axis, which are two different axes (X-axis and Y-axis). The repetitive position accuracy of master robot is normally distributed and related to the control accuracy of the driving units, the smoothness of the ground and the operation cycle time of the feedback control system. While the precision of driving wheel is higher and the smoothness of ground is better or the operation cycle time of control system is shorter, the variance of repetitive position accuracy will get smaller. As to the slave robot, the factors affecting repetitive position accuracy are much more complex, which include master robot's position error, slave robot's following error, inherent errors, etc.

3.1.1. Control Strategy of Master Robot Motions

As mentioned above, master robot only needs to move according to the driving task demands. Therefore, PID control strategy is suitable for master robot to move to the destination or lift up and lay down the pallet by controlling its driving motors. What is more, in simulation diagram, module "Constant" means setting trajectory, and "Integrator" means that the odometer is the integral result of velocity and time. In "Control section", the speeds of the two driving units are adjusted dynamically based on the position feedback. Because the system is an open loop including integral part, a proportional part in feedback control can make master robot motion system gradually stable. Control strategy of master robot motion is shown in Figure 4.

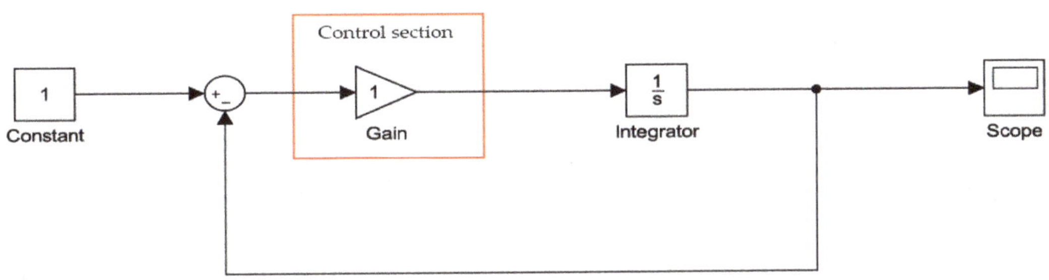

Figure 4. Control strategy of master robot motion.

3.1.2. Control Strategy of Slave Robot Motions

In fact, slave robot control system is a very complex nonlinear control strategy. Here, the main control ideas of slave robot are described in this section. The controlled object for slave robot are driving units. Control goal is to eliminate position errors on X-axis and Y-axis. The feedback signal includes the distance data between master robot and slave robot, and the SIGN signal indicates if master robot position leads or lags on Y-axis.

In order to keep the system in a steady stable state, the system's energy derivative should be non-positive definite according to Lyapunov's second stability theorem. Therefore, in this control system the driving motors' speed should tend to be stable in the process of system. Because slave robot cannot know the accurate error in real time between master robot and slave robot on Y-axis, which means it is impossible for slave robot to follow master robot without position error on Y-axis. Therefore, high-gain observer is used for keep slave robot stable within a certain error value while following master robot on Y-axis. Actually, when high-gain observer is used, it will affect the motions on both X-axis and Y-axis. Because when driving units move in the oblique linear direction, slave robot motion on X-axis and Y-axis are coupling. The slave robot moving on Y-axis will also result in moving on X-axis. Therefore, high-gain observer should be used when slave robot motion on X-axis and-Y axis are in the positive correlation.

In order to assure the sub robot motion on X-axis and Y-axis are in positive correlation, the moving direction of driving units should be adjusted accordingly. Because the control system energy will not be changed if only directions of driving units are adjusted. In fact, by adjusting driving motors' direction, the motions on X-axis and Y-axis are able to be changed as ordered. By using the above control strategy, the following errors between master robot and slave robot will be converged, and slave robot is able to keep stable within a certain error range. The control strategy of slave robot is shown in Figure 5.

Given the initial position offset and angle offset, the errors on X-axis and Y-axis will approach zero but not reaches an asymptotically stability using high-gain observer control strategy. In the actual motions, the following errors between master robot and slave robot can be observed by industrial camera. And jitters may occur during slave robot motion, which also verifies the accuracy of high-gain observer control tool.

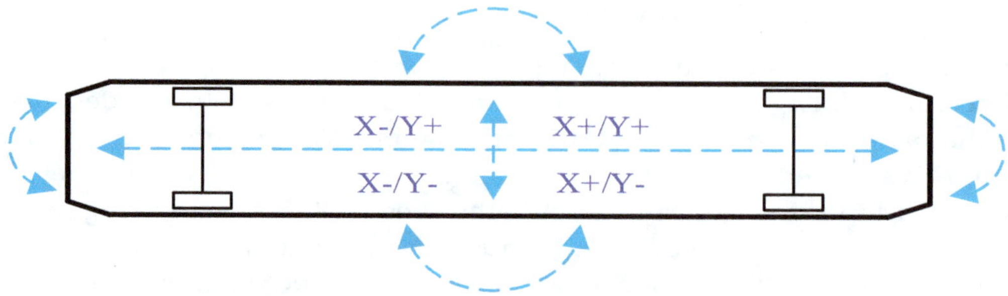

Figure 5. Control strategy of slave robot.

3.2. Input of Nonlinear Control System

Separate parallel intelligent mobile robot uses master-slave robot structure. In order to achieve high performance of real-time control and response, the chip with over 600 MHZ main frequency should be selected as main control chip. What is more, RTOS system should be embedded for using time slices efficiently. And infrared data transmission module is used for real-time communication between master robot and slave robot. In this way, the operation cycle time of the control system can be within 5 milliseconds, which is an important factor of affecting master-slave robot real-time synchronization control.

In order to reduce the following errors and achieve high repetitive position accuracy, high-precision sensors should be used, for example distance sensors. In this way, sensors will get relative distance data, which are also the input parameters of slave robot nonlinear control system. Here are the input parameters of slave robot nonlinear control system:

(1) Distance between master robot and slave robot on X-axis: the precise distance between slave robot and master robot can be obtained by two sets of high-precision point laser distance sensors which are installed on the relative side of the front and rear of slave robot.

(2) Distance between master robot and slave robot on Y-axis: position error on Y-axis can be roughly measured by tree groups of mirror reflection photoelectric switches, which return a set of SIGN function signal to the control system. That is to say, if the difference between slave robot and master robot is greater than a preset value, the Y-axis distance sensor returns +1 signal to the control system. If the difference is within a preset range, the Y-axis distance sensor returns 0 signal to the control system. Otherwise, the Y-axis distance sensor returns -1 signal.

Obviously, the procedure of slave robot following master robot is a nonlinear control process. The input parameters of the nonlinear control system include the distance on X-axis and position error SIGN signal on Y-axis. And the output is the moving speed of each driving motor. The goal of the slave robot control system is to make the following process stable and keep the following errors between master robot on X-axis and Y-axis within a preset value.

3.3. High-Gain Observer

High-gain observer is a good tool for the nonlinear control process in which some state variables cannot be measured accurately due to technical or economic reasons. In some practical cases, the tool can be modified to produce an output feedback controller [14].

Suppose that a nonlinear system can be transformed into the form:

$$\begin{cases} \dot{x} = Ax + g(y,u) \\ \hat{y} = Cx \end{cases} \tag{7}$$

where (A, C) is observable. This form is special because the nonlinear function g depends only on the output y and the control input u. Taking the observer as:

$$\dot{\hat{x}} = A\hat{x} + g(y, u) + H(y - C\hat{x}) \tag{8}$$

it can be easily seen that the estimation error $\tilde{x} = x - \hat{x}$ satisfies the linear equation:

$$\dot{\tilde{x}} = (A - HC)\tilde{x} \tag{9}$$

Hence, designing H such that $A - HC$ is Hurwitz guarantees asymptotic error convergence, that is $\lim_{x\to\infty} \tilde{x}(t) = 0$. Aside from the fact that the observer works only for a special class of nonlinear systems, its main drawback is the assumption that the nonlinear function g is perfectly known. Any error in modeling g will be reflected in the estimation error equation.

The upper nonlinear observer is suitable for most system. But for the sub-style parallel intelligent mobile robot, slave robot should follow master robot synchronously and timely. There are also many measurement deviations. We should give a special design of the observer gain that makes the observer robust to uncertainties in modeling the nonlinear functions. High-gain observer can guarantee that the output feedback controller recovers the performance of the state feedback controller when the observer gain is sufficiently high.

We use the observer in output feedback stabilization. The main results in Section 4 are separation principles that allow us to separate the design into two tasks. First, we design a state feedback controller that stabilizes the system and meets other specifications. Then, an output feedback controller is obtained by replacing the state x by its estimate $\hat{x}$ provided by high-gain observer. A key property that makes this separation possible is the design of the state feedback controller to be globally bounded in x. High-gain observer can be used in a wide range of control problems.

Consider the second-order nonlinear system:

$$\begin{cases} \dot{x}_1 = x_2 \\ \dot{x}_2 = \phi(x, u) \\ y = x_1 \end{cases} \tag{10}$$

where $x = [x_1, x_2]^T$. Suppose $u = y(x)$ is a locally Lipschitz state feedback control law that stabilizes the origin $x = 0$ of the closed-loop system.

To implement this feedback control using only measurements of the output y, we use the observer:

$$\begin{cases} \dot{\hat{x}}_1 = \hat{x}_2 + h_1(y - \hat{x}_1) \\ \dot{\hat{x}}_2 = \phi_0(\hat{x}, u) + h_2(y - \hat{x}_1) \end{cases} \tag{11}$$

We set $\tilde{x}_1(0) = 0$ and $\tilde{x}_2(0) = 0$. Where $\phi_0(x, u)$ is a nominal model of the nonlinear function $\phi(x, u)$. The estimation error:

$$\tilde{x} = \begin{bmatrix} \tilde{x}_1 \\ \tilde{x}_2 \end{bmatrix} = \begin{bmatrix} x_1 - \hat{x}_1 \\ x_2 - \hat{x}_2 \end{bmatrix} \tag{12}$$

Satisfies the equation:

$$\begin{cases} \dot{\tilde{x}}_1 = -h_1\tilde{x}_1 + \tilde{x}_2 \\ \dot{\tilde{x}}_2 = -h_2\tilde{x}_1 + \delta(x, \tilde{x}) \end{cases} \tag{13}$$

where $\delta(x, \tilde{x} = \phi(x, \gamma(\hat{x})) - \phi_0(\hat{x}, \gamma(\hat{x})))$. We want to design the observer gain $H = [h_1, h_2]^T$ such that $\lim_{t \to \infty} \tilde{x}(t) = 0$. In the absence of the disturbance term δ, asymptotic error convergence is achieved by designing H such that:

$$A_o = \begin{bmatrix} -h_1 & 1 \\ -h_2 & 0 \end{bmatrix} \tag{14}$$

is Hurwitz. For this second-order system. A_o is Hurwitz for any positive constants h_1 and h_2. In the presence of δ, we need to design H with the additional goal of rejecting the effect of δ on x. This is ideally achieved, for any δ, if the transfer function:

$$G_o(s) = \frac{1}{s^2 + h_1 s + h_2} \begin{bmatrix} 1 \\ s + h_1 \end{bmatrix} \tag{15}$$

From δ to x is identically zero. While this is not possible, we can make $\sup_{\omega \in R} |G_o(j\omega)$ arbitrarily small by choosing $h_2 >> h_1 >> 1$. In particular, taking

$$h_1 = \frac{\alpha_1}{\varepsilon}, h_2 = \frac{\alpha_2}{\varepsilon^2} \tag{16}$$

for some positive constant α_1, α_2 and ε, with $\varepsilon << 1$, it can be shown that:

$$G_o(s) = \frac{\varepsilon}{(\varepsilon s)^2 + \alpha_1 \varepsilon s + \alpha_2} \begin{bmatrix} \varepsilon \\ \varepsilon s + \alpha_1 \end{bmatrix} \tag{17}$$

Hence, $\lim_{\varepsilon \to 0} G_o(s) = 0$. The disturbance rejection property of high-gain observer can be also seen in the time domain by representing the error equation in the singularly perturbed form.

The initial conditions are $\varepsilon = 0.01$, $h_1 = 100$, and $h_2 = 1000$. We can obtain the best high-gain observer for synchronization control of these two sub robots of separate parallel intelligent mobile robot.

4. Experiment and Analysis

The separate parallel intelligent mobile robot consists of two independent sub robots, a master robot and a slave robot, which are similar to two forks of the forklift, but the sub robot does not have any physical connection with master robot. Each robot has front and rear driving units. These two sub robots collaborate to deliver the pallet synchronously. This section will analyze the steady state error in order to synchronize the actual motions of these two sub robots. The center of gravity distribution and offset error during flexible docking to the free entry of the pallet will be also analyzed.

4.1. Steady State Error Analysis of Actual Motions

This section will summarize the steady state error of actual motions based on both simulation in Simulink and convergence of results.

4.1.1. Linear Motion

When the master-slave mobile robot is moving linearly frontwards or backwards, slave robot should follow master robot synchronously. In order to get the value of front or rear deviation, three non-uniformly distributed are used to get relative position, which is set to be SIGN (+1, −1, 0). Slave robot should align its velocity of every driving motor by high-gain observer in real time. Otherwise, jitter of slave robot will be intolerable. Then slave robot cannot follow master robot synchronously. The procedure of adjusting relative velocity and position will be convergent and easy to reach a steady state after a period of time. The jitter of slave robot following master robot will also be reduced. The simulation in Simulink and convergence of results are shown in Figure 6. As shown in Figure 6b, there are following errors at the beginning of linear motion. In Figure 6b, the abscissa unit is second and ordinate unit is millimeter. Self-adjusting process during the following procedure will then eliminate.

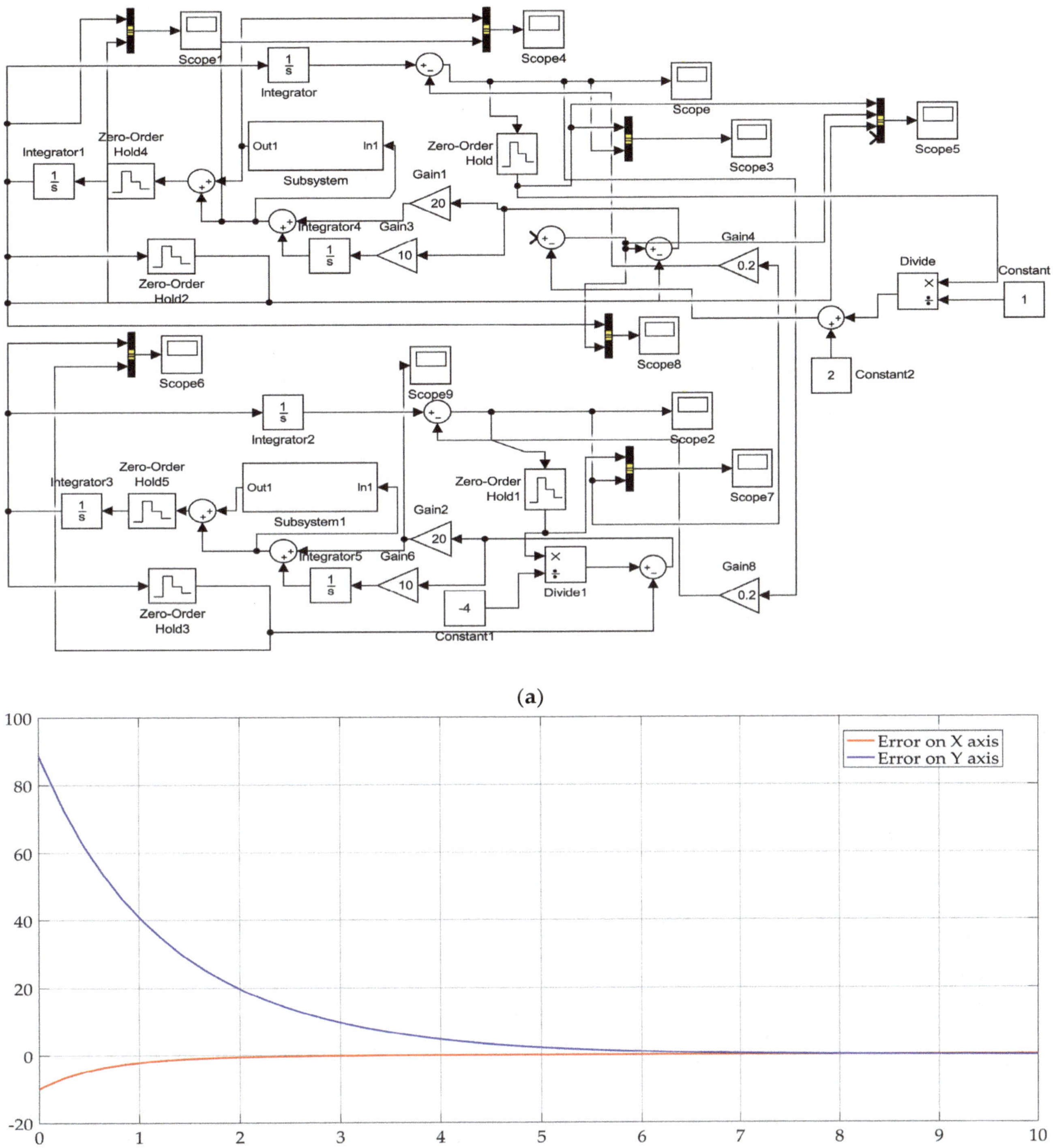

(a)

(b)

Figure 6. The simulation in Simulink and convergence of results for linear motion. (**a**) Simulink simulative model graph; (**b**) The model reaches a steady convergent state.

4.1.2. Oblique Linear Motion

Integrator and *integrator1* are two integrations of the model which describe the driving motor of the sub robot. Also this represents that the speed difference of driving motors would not directly cause the offset of robot motion. Simple Zero-Order-Hold is sample hold. Due to the integral element could lead to time-delay and vibration, the system would not use integral element in system offset reducing. So only *integrator4* is added to *I* in outer-loop in Simulink to reduce steady state error in the actual following motions. Theoretically, when the initial offset of following motion is 15 mm and angle is $-5°$, the following errors on X-axis and Y-axis will approximate 0 after reaching the steady state. As shown in Figure 7, the model reaches a steady convergent state. In Figure 7b, the abscissa unit is second and ordinate unit is millimeter.

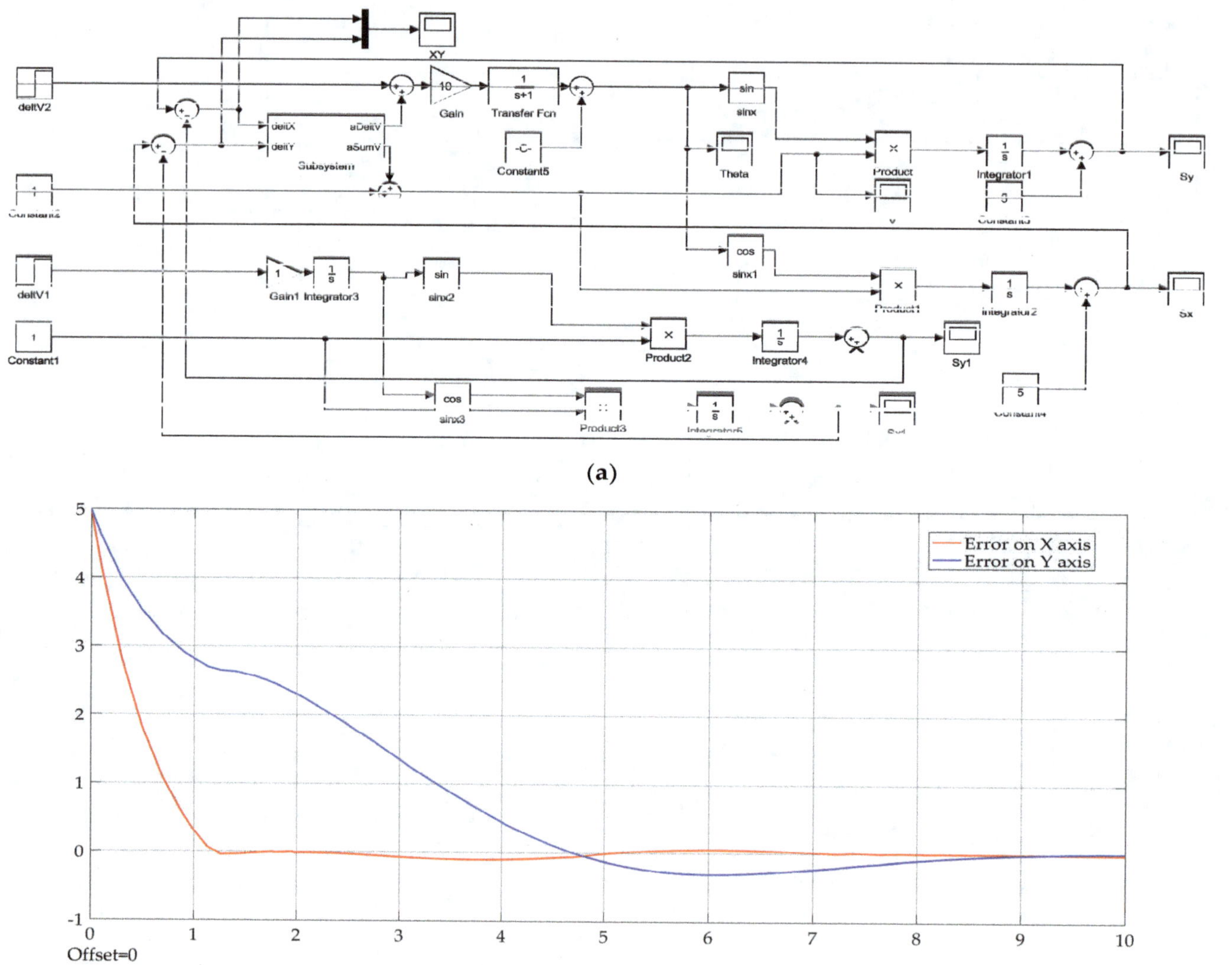

(a)

(b)

Figure 7. The simulation in Simulink and convergence of results for oblique linear motion. (**a**) Simulink simulative model graph; (**b**) The model reaches a steady convergent state.

4.1.3. Zero-radius Turn

When the master-slave separate parallel intelligent mobile robot takes a zero-radius turn, slave robot should follow master robot's position synchronously and timely on both X-axis and Y-axis. While rotating, the motion analysis of driving units should have a unique solution in the robot motion model. Then the velocities of slave robot's front and rear driving motors will cancel out each other on Y-axis. And the velocity on X-axis of the sub robot's driving motors is taken as the linear velocity while rotating. When the position following errors on X-axis and Y-axis occur on slave robot, the velocity and direction of driving units should be adjusted in real time, by using Lyapunov second stability theorem. It should be noted that when slave robot is adjusted, it is not guaranteed that both the position following errors on X-axis and Y-axis converge simultaneously. However the energy derivative of control system is always non-position define. The simulation in Simulink and convergence of results for zero-radius turn are shown in Figure 8. In Figure 8b, the abscissa unit is second and ordinate unit is millimeter.

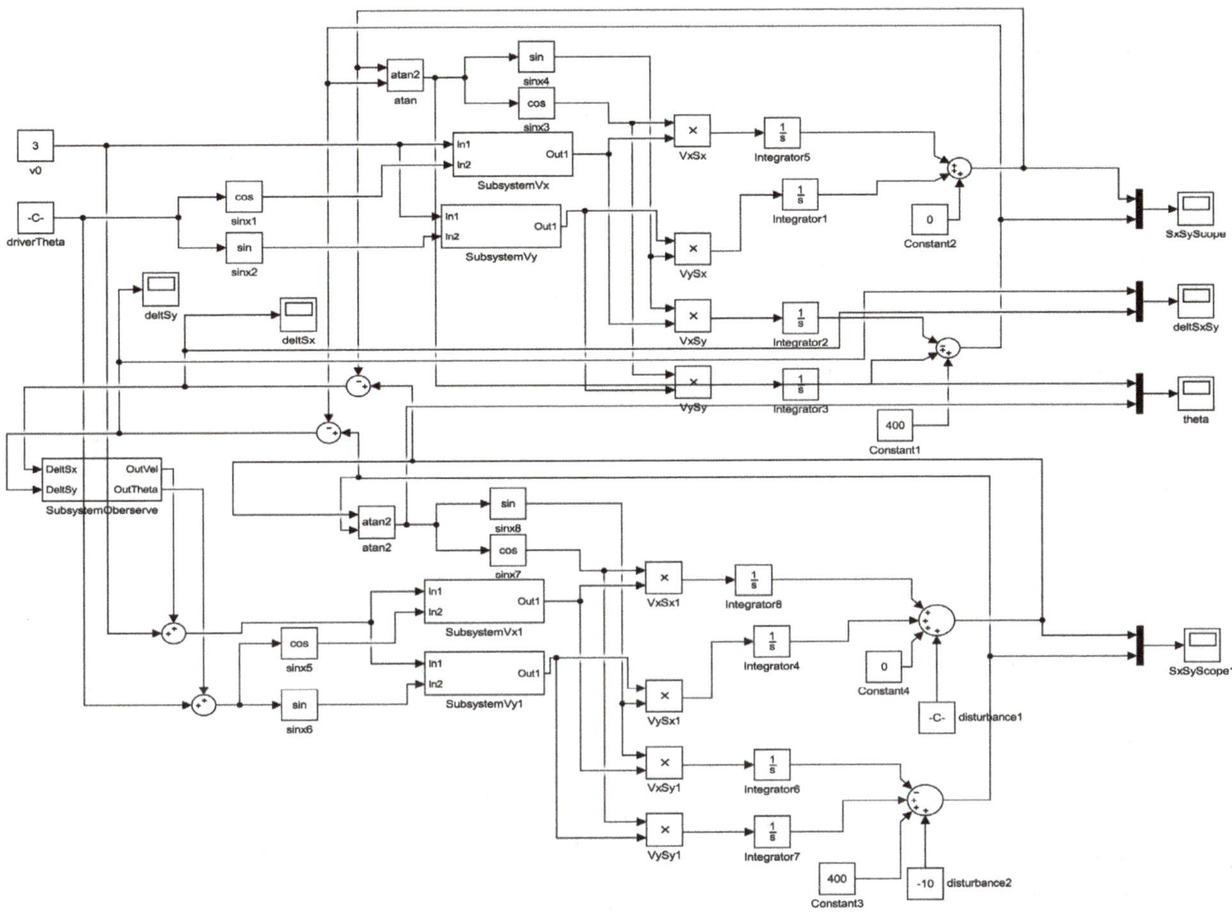

(a)

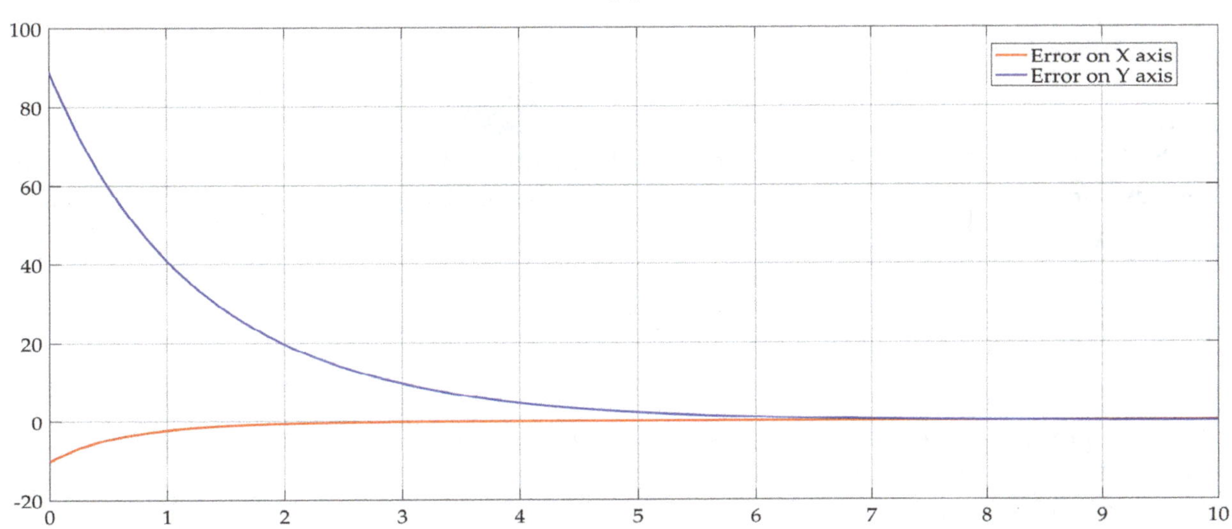

(b)

Figure 8. The simulation in Simulink and convergence of results for zero-radius turn. (**a**) Simulink simulative model graph; (**b**) The model reaches a steady convergent state.

4.1.4. Steady State Error Factors

As we know, in the ideal case, the zero offset error means that slave robot follows master robot absolutely according to calibrated data, and there is no error. So the most significant factor that affects the stability of the system is the accuracy of calibrated data. (1) Camera installation error: The installation angle deviation of camera and driving unit is unavoidable. The camera should be calibrated first to reduce the initial angle deviation. (2) Encoder calibration offset: The error of encoder deviation would be transferred from inner-loop to the outer-loop. Thus the encoder calibration offset would directly lead to the error of system stability. (3) Accuracy of driving control: In actual motions, the precision of each driving motor is up to 1 rpm. As the result, the precision of the following velocity and position control could not be well guaranteed. (4) Mechanical installation error: The mechanical installation error represents installation tolerance error such as driving unit and other mechanical parts. All these errors would result in the difference between calibrated zero position and actual zero position.

As we known, the errors of actual motions of these two sub robots cannot reach zero. This section analyzes most of the factors that may affect the steady state errors of the following motions. The two driving units of slave robot may jitter during the adjusting control. Surface roughness of the road and the curve degree of vision-based color bar on the road can both have significant effect on the stability of the master-slave parallel robot' motion. As discussed above, various factors affect the steady state error. The steady state error for the master-slave mobile robot measured in the experiment is about 3mm, which is not easy to eliminate.

4.2. *Accurate Flexible Docking to the Free Entry of the Pallet*

To ensure that the two sub robots can move underneath the pallet accurately and lift up and lay down the pallet synchronously, an evaluation of docking to the free entry is required. After arriving at the pre-docking position, the mobile robot should recognize and compute the accurate position the free entry of pallet by laser sensor or vision sensor, and then move into the free entry smoothly. In this section, the maximum deviation angle error $\beta_{\max}$ during docking to the free entry is analyzed through synthetical consideration of the following error and positioning error.

4.2.1. The Ideal Case (No Direct Collision)

In order to lift the pallet up, the sub robot should move underneath the pallets accurately and safely. In the ideal case, the shell cover of the sub robot won't collide with the edge of the free entry of the pallets, as shown in Figure 9a. $\beta_1 = (90° - \theta_1)$ represents the intersection angle of the sub robot and the free entry of the pallet.

$$\begin{cases} AC = l_r - \frac{w_r - w_h}{2} \tan\delta \\ BC = \frac{w_p - w_r}{2} \pm \varepsilon \\ \cos\theta_1 = \frac{BC}{AC} \end{cases} \tag{18}$$

where, $l_r = 1180$ mm is the length of the basis of the sub robot. $w_r = 212$ mm is the width of the basis of the sub robot. $w_h = 170$ mm is the width of the head of the sub robot. $\delta = 70°$ is the conical chamfer of the head of the sub robot. $w_p = 382.5$ mm represents the width of free entry of the pallet. $\varepsilon = 10$ mm is the maximum following error of the sub robot.

Relative deviation angle β_1:

$$3.84° < \beta_1 = (90° - \theta_1) < 4.87° \tag{19}$$

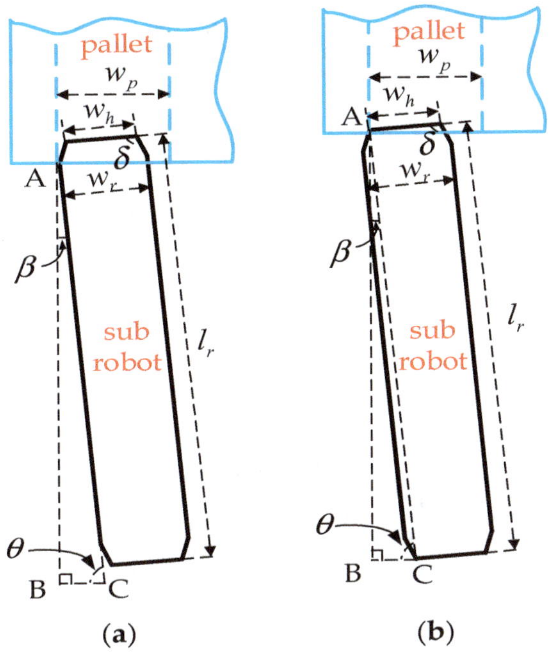

Figure 9. The maximum deviation angle error of a sub robot docking to the free entry of the pallet. (**a**) The ideal case; (**b**) The extreme case.

4.2.2. The Extreme Case (Contact Guidance)

Although the shell cover of the sub robot will collide with the free entry slightly in some conditions, the sub robot can move underneath the pallets yieldingly, as shown in Figure 9b. $\beta_2 = (90^\circ - \theta_2)$ represents the included angle of the sub robot and the free entry of the pallet.

$$\begin{cases} AC = l_r \\ BC \approx \frac{l_r - w_r}{2} + \frac{w_r - w_h}{2} \pm \varepsilon \\ \cos\theta_2 = \frac{BC}{AC} \end{cases} \tag{20}$$

Relative deviation angle β_2:

$$4.70^\circ < \beta_2 = (90^\circ - \theta_2) < 5.65^\circ \tag{21}$$

By combing Equation (19) with (21), we have the the maximum allowable angle error $\beta_{\max}$:

$$-\max(\beta_1, \beta_2) < \beta_{\max} < \max(\beta_1, \beta_2) \tag{22}$$

Therefore, $\beta_{\max} \in \left(-5.65^\circ \quad 5.65^\circ \right)$.

4.3. *Analysis of Lifting and Laying*

The main procedures of pallet transportation for intelligent mobile robot are that it firstly moves underneath the pallet, then lifts the pallet up and delivers the pallet to the destination, and finally lays it down. In order to control motions of the two sub robots synchronously, this section analyzes the pallet offset error, including longitudinal and transverse offset error during lifting and laying.

4.3.1. Analysis of Center of Gravity

As we know, the sub robot may roll over while the four driving motors are at two parallel lines at a certain moment. Relative velocity and position control loops are used during the lifting and laying of the pallet. The front and rear driving wheels are always vertical through position feedback. The initial relative angle of front and rear driving units is shown in Figure 10. Thus the center of gravity of robot will be within the base of the sub robot during the procedure of lifting and laying.

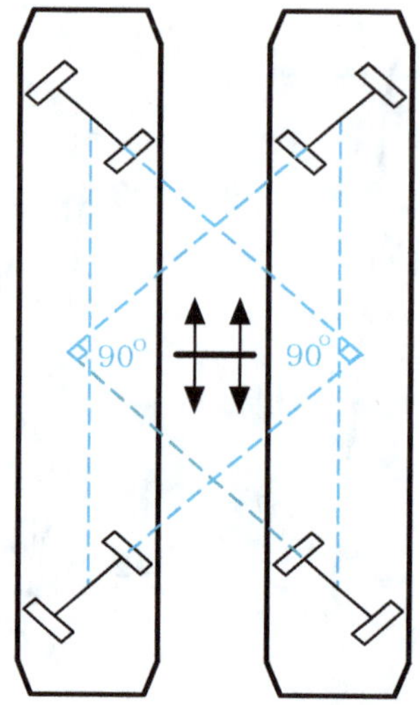

Figure 10. The initial relative angle of front and rear driving units before lifting and laying.

4.3.2. Analysis of Pallet Offset Error

During the lifting process, the offset errors of the position of the pallet include longitudinal offset error and transverse offset error. They are both caused by uneven ground and jittering of sub-robot's motion. It is difficult to guarantee the level of the ground, so changing or optimizing the robot motion is the most effective way.

- Analysis of pallet longitudinal offset errors

The longitudinal offset errors of the pallet cannot be completely eliminated. It is unable to measure the absolute accurate deviation between pallet position and robot position, so no correction can be made. There must be cumulative error between them. What we should do is to ensure the absolute stop position between lifting and laying as consistent as possible. The two sub robots should be controlled to stop accurately and synchronously while height position of the pallet changes. The control strategy of accurate stop is shown in Figure 11.

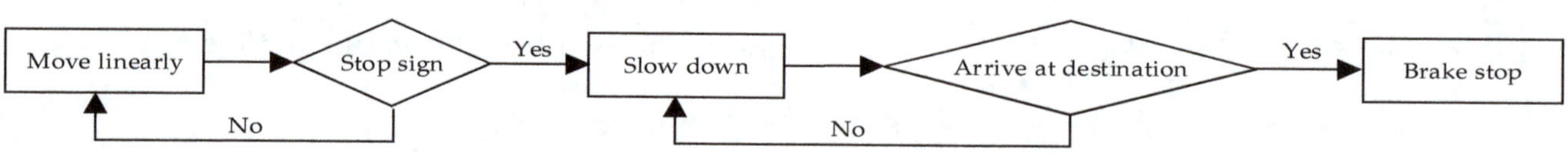

Figure 11. Control strategy of accurate stop.

- Analysis of pallet transverse offset errors

The transverse offset of the pallet refer to errors in the left and right direction. In general, the mobile robot moves underneath of the pallet and then lift the pallet up. But once there are some transverse offset errors, the mobile robot may be stuck under the free entry of the pallet. It may be caused by uneven ground or jitters of the sub robot's motions. In order to reduce the transverse offset errors, dynamic position adjustment is required during lifting and laying. The dynamic adjustment strategy is shown in Figure 12. The rotation angle of each driving unit is divided into 2 areas. (1) Angle adjustment areas. If the angle of rear driving unit is in the range of angle that needs to be adjusted, the rear driving unit will follow the front one, and they will be perpendicular. Within the angle adjustment areas, the center of gravity of the sub robot can always maintain a relatively large area, which will prevent the sub robot to roll over. (2) Position adjustment areas. If the angle of driving unit is in the range of position that needs to be adjusted, the speed of the driving motor will be changed according to the position error collected by the camera or navigation sensor. It ensures that the sub robot performs left and right position feedback during the procedure of lifting and laying. According to the position adjustment areas, it is possible to optimize the transverse offset errors between the sub robot and the pallet.

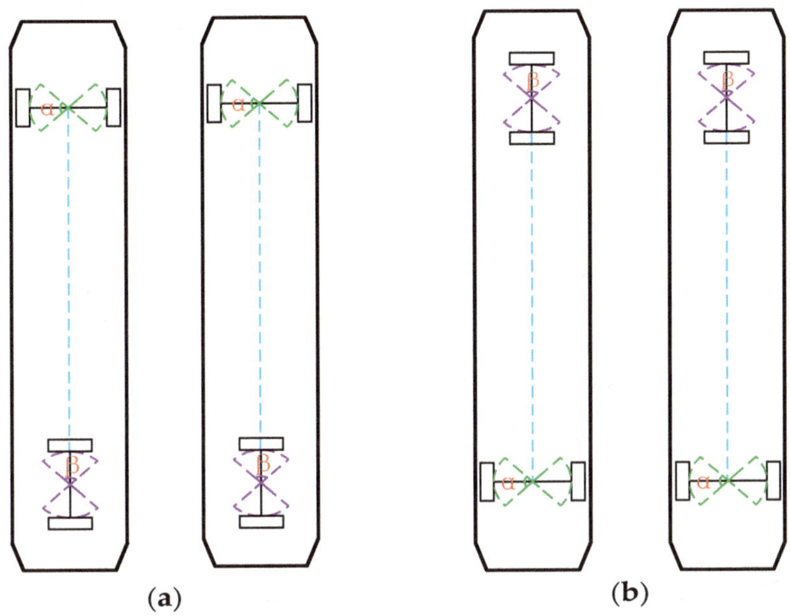

Figure 12. Dynamic adjustment strategy of master-slave robot during lifting and laying. (**a**) Angle adjustment areas; (**b**) Position adjustment areas.

When the front or rear driving unit lies in the angle adjustment area, we adjust the speeds of the two motors of this driving unit synchronously. The two motors may accelerate or decelerate at the same speeds. But when the front or rear driving unit lies in the position adjustment area, we adjust the speeds of the two motors of this driving unit asynchronously. The two motors may accelerate or decelerate at the different speeds.

4.4. Actual Running Performance

The master-slave separate parallel intelligent mobile robot was designed and assembled. The performance was tested by actual delivering the pallets in the factory. The two sub robots could also lift up and lay down the pallet synchronously. The mobile robot moved flexibly at the crossroads. And detailed specifications of the master-slave separate parallel intelligent mobile robot are listed in Table 1.

Table 1. The specifications of the master-slave separate parallel intelligent mobile robot.

Feature	Specification	Feature	Specification
max payload	1000 kg	swing radius	>730 mm
max speed	0.7 m/s	navigation	color bar/vision
outline dimension	L1180 × W212 × H98 mm	running time	10 h
weight	40 kg	charging time	2 h
uplift height	28 mm	turning radius	0 mm
uplift time	16 s	pallets	PALETTE EUR-EPAL

5. Conclusions

To pick up and transport pallets autonomously in the factory, a master-slave separate parallel intelligent mobile robot was designed. The mobile robot consists of two independent sub robots, which are similar to two forks of the forklift, but master robot does not have any physical connection with slave robot. In order to adapt to the compact space of the entry of pallet, four compact motion driving units are designed for these two sub robots. High-gain observer is used to control the following speed timely. The state control of slave robot is to ensure that the following error converges and achieve synchronization of the two sub robots' motions. The experiment results demonstrate that the separate parallel intelligent mobile robot can transport the pallet autonomously. The two sub robots can fulfill synchronous motions, such as linear motion, oblique linear motion, and zero-radius turn. They can also lift up and lay down the pallet synchronously. The mobile robot moves flexibly and is quite suitable for the standardized logistics factory with small working space. In our future research, the autonomous navigation module based on the 2D laser will be assembled with the intelligent mobile robot. Then the mobile robot can move freely for pallet transportation in the factory.

Author Contributions: Data curation, G.L. and R.L.; Formal analysis, R.S.; Investigation, M.L. and R.S.; Methodology, R.L., M.L. and S.P.; Project administration, S.P.; Software, G.L.; Writing—original draft, R.L.; Writing—review & editing, R.L. and S.P.

Acknowledgments: Thanks associate Professor Hongmiao Zhang for giving suggestions to improve this manuscript. We would like to take the opportunity to thank the reviewers for their thoughtful and meaningful comments.

References

1. Behzad, E.; Sara, B.; Ben, W. The evolution and future of manufacturing: A review. *J. Manuf. Syst.* **2016**, *39*, 79–100.
2. Jeremie, C. *A Look into Logistics Automation. ROBO Global Defining the Universe of Robotics and Automation*; ETF Securities: London, UK, 2017; pp. 1–18.
3. Paweł, Z. Model of Forklift Truck Work Efficiency in Logistic Warehouse System. In *Logistics Operations, Supply Chain Management and Sustainability*; Springer: Cham, Switzerland, 2014; pp. 467–479.
4. Armesto, L.; Tornero, J.; Torres, J. Transport process automation with industrial forklifts. *IFAC Proc.* **2003**, *36*, 33453–33455. [CrossRef]
5. Kelen, C.T.; Marcelo, B.; Glauco, A.P. Automatic routing of forklift robots in warehouse applications. *ABCM Symp. S. Mechatron.* **2010**, *4*, 335–344.
6. Wang, J.; Zhang, N.; He, Q. Application of Automated Warehouse Logistics in Manufacturing Industry. In Proceedings of the IEEE 2009 ISECS International Colloquium on Computing, Communication, Control, and Management, Sanya, China, 8–9 August 2009.
7. Li, L.Y.; Liu, Y.H.; Fang, M.; Zheng, Z.-Z.; Tang, H.-B. Vision-Based Intelligent Forklift Automatic Guided Vehicle (AGV). In Proceedings of the 2015 IEEE International Conference on Automation Science and Engineering (CASE), Fort Worth, TX, USA, 21–24 August 2015; pp. 264–265.
8. Seelinger, M.; Yoder, J.D. Automatic Pallet Engagment by a Vision Guided Forklift. In Proceedings of the 2005 IEEE International Conference on Robotics and Automation, Barcelona, Spain, 18–22 April 2005.
9. Qian, J.; Zi, B.; Wang, D.; Ma, Y.; Zhang, D. The Design and Development of an Omni-Directional Mobile Robot Oriented to an Intelligent Manufacturing System. *Sensors* **2017**, *17*, 2073. [CrossRef] [PubMed]
10. Patel, P.; Parekh, R.; Panchal, R.; Solanki, V. Material Handling System Based on Automated Guided Vehicle: Scope, Limitation and Application. In Proceedings of the International Conference on Emerging Trends in Mechanical Engineering, Gujarat, India, 6–7 July 2017.
11. Weber, M.; Vorwerk, C. Neuartiges antriebskonzept zum fahren, lenken und heben. *Log. J. Proc.* **2010**, *6*. [CrossRef]
12. Melkus, A. Fahreloses transportfarzeug: EP, No. EP3216747A1. 7 March 2016.
13. F3-Design. Available online: https://www.dinostretchhood.com/nipper/ (accessed on 21 November 2018).
14. AGILOX SYSTEM. Available online: http://agilox.net/ (accessed on 19 July 2018).
15. Hassan, K.K. *Nonlinear Systems*, 3rd ed.; Prentice Hall: Englewood, IL, USA, 2002; pp. 551–625.

16. Kim, S.; Kwon, S. Nonlinear Control Design for a Two-Wheeled Balancing Robot. In Proceedings of the 2013 10th International Conference on Ubiquitous Robots and Ambient Intelligence, Jeju, Korea, 30 October–2 November 2013; pp. 486–487.
17. Osama, J.; Mohsin, J.; Yasar, A.; Khubab, A. Modeling, Control of a two-Wheeled Self-Balancing Robot. In Proceedings of the IEEE International Conference on Robotics and Emerging Allied Technologies in Engineering (iCREATE), Islamabad, Pakistan, 22–24 April 2014.
18. Antonio, H.J.; Yu, W. High-Gain Observer-Based PD COntrol for Robot Manipulator. In Proceedings of the IEEE Xplore Conference on American Control Conference, Chicago, IL, USA, 28–30 June 2000; Volume 4, pp. 2518–2522.
19. Voos, H. Nonlinear Control of a Quadrotor micro-UAV Using Feedback-Linearization. In Proceedings of the IEEE International Conference on Mechatronics, Malaga, Spain, 14–17 April 2009; pp. 1–6.
20. Goodarzi, F.; Lee, D.; Lee, T. Geometric Nonlinear PID Control of a Quadrotor UAV on SE(3). In Proceedings of the 2013 European Control Conference (ECC), Zürich, Switzerland, 17–19 July 2013; pp. 3845–3850.
21. Hu, Q.; Fei, Q.; Wu, Q. Research and application of nonlinear control techniques for quad rotor UAV. *J. Univ. Sci. Technol. China* **2012**, *42*, 706–710.
22. Young, K.D.; Utkin, V.I.; Ozguner, U. A control engineer's guide to sliding mode control. *IEEE Trans. Control Syst. Technol.* **1999**, *7*, 328–342. [CrossRef]
23. Boizot, N.; Busvelle, E.; Gauthier, J.P. An adaptive high-gain observer for nonlinear systems. *Automatica* **2010**, *46*, 1483–1488. [CrossRef]

13

Design of Wing Root Rotation Mechanism for Dragonfly-Inspired Micro Air Vehicle

Jae Hyung Jang and Gi-Hun Yang *

Robotics Group, Korea Institute of Industrial Technology, Ansan-si 15588, Gyeonggi-do, Korea; jhjang7@kitech.re.kr

* Correspondence: yanggh@kitech.re.kr.

Abstract: This paper proposes a wing root control mechanism inspired by the drag-based system of a dragonfly. The previous mechanisms for generating wing rotations have high controllability of the angle of attack, but the structures are either too complex or too simple, and the control of the angle of attack is insufficient. In order to overcome these disadvantages, a wing root control mechanism was designed to improve the control of the angle of attack by controlling the mean angle of attack in a passive rotation mechanism implemented in a simple structure. Links between the proposed mechanism and a spatial four-bar link-based flapping mechanism were optimized for the design, and a prototype was produced by a 3D printer. The kinematics and aerodynamics were measured using the prototype, a high-speed camera, and an F/T sensor. In the measured kinematics, the flapping amplitude was found to be similar to the design value, and the mean angle of attack increased by approximately 30° at a wing root angle of 0°. In the aerodynamic analysis, the drag-based system implemented using the wing root control mechanism reduced the amplitude of the force in the horizontal direction to approximately 0.15 N and 0.1 N in the downstroke and upstroke, respectively, compared with the lift-based system. In addition, at an inclined stroke angle, the force in the horizontal direction increased greatly when the wing root angle was 0° at the inclined stroke angle, while the force in the vertical direction increased greatly at a wing root angle of 30°. This means that the flight mode can be controlled by controlling the wing root angle. As a result, it is shown that the wing root control mechanism can be applied to the MAV (micro air vehicle) to stabilize hovering better than the MAV using a lift-based system and can control the flight mode without changing the posture.

Keywords: biomimetic robot; micro air vehicle; flapping; drag-based system; dragonfly

---

1. Introduction

Small birds and insects are good objects to mimic for developing a micro air vehicle (MAV) for stable flight at low Reynolds fluid [1–7]. The RoboBee developed by researchers at Harvard [1], the robot hummingbird of DARPA [2], the beetle robot of Konkuk University [3], and the tailless aerial robot inspired by the flies of Delft [7] are representative robots developed by mimicking the flight of small birds and insects. Among the insects mimicked, the dragonfly has the most stable hovering ability, the ability to switch flight modes without changing posture, and ability to fly backwards [8,9]. In order to achieve this high maneuverability, the dragonfly uses characteristics such as the phase difference between the forewing and hind wing [8,10–12], independent control of each wing [13], and a drag-based system in hovering flight [14,15].

Among these characteristics, we focus on the drag-based system. Most insects, except the dragonfly, use a lift-based system in which they stroke their wings in the horizontal direction of the body during hovering. On the other hand, dragonflies use a drag-based system in which they stroke their wings in an inclined direction to the body. This drag-based system not only provides more stable flight but also improves maneuverability by maintaining posture when changing flight mode [14,15]. Dragonflies that use a drag-based system rely on drag force for 76% of the force required for hovering, and most vertical forces for hovering flight are obtained on the downstroke [15]. For this reason, in order to maximize their vertical force, dragonflies increase the drag force through a large angle of attack during the downstroke, while minimizing the drag force through a small angle of attack in the upstroke [16,17].

The aforementioned angle of attack is an important factor that enables most insects, including dragonflies, to control the magnitude and direction of the force generated when flapping their wings [18]. Most small birds and insects produce an angle of attack by rotating their wings in the longitudinal direction when the direction of the flapping stroke is reversed, such as from the upstroke to the downstroke, or from the downstroke to the upstroke [19,20]. These wing rotations are divided into passive rotation using aerodynamic force and inertial force and active rotation that is directly controlled through the muscles. It is not known exactly what type of wing rotation insects use, but it has been proven that passive rotation explicitly occurs during flapping [21]. A variety of mechanisms for generating wing rotation for the FW-MAV (flapping wing micro air vehicle) have been proposed in studies mimicking the flight characteristics of insects.

These mechanisms are divided into passive [22] and active rotation mechanisms [23,24] like insect wing rotation. A passive rotation mechanism has been produced by creating a wing frame using a flexible material or loose membrane. For the active rotation mechanism, a method of artificially generating rotation using a spring, and a method of simultaneously implementing flapping and rotation with a single actuator using a spatial four-bar link mechanism have been presented. Various attempts have been made to develop a wing rotation mechanism, but there is a clear limit to mimicking the drag-based system of dragonflies. In the case of the active rotation mechanism, since wing rotation is directly generated using the power of an actuator, it is suitable to implement the desired movement through mechanical design. Moreover, the flapping pattern is easily switched from an asymmetric flapping pattern to a symmetric flapping pattern or vice versa by adding degrees of freedom to the MAV. However, this mechanism is difficult to apply in practical MAV development. The reason is because not only is the structure complex, such that the weight of the airframe increases, and friction and torsion occur to a large extent, but additional energy is also required to generate wing rotation. On the other hand, the passive rotation mechanism has a relatively simple structure and is widely used in the development of MAV [1–5,25,26]. A passive rotation mechanism has also been developed to control the magnitude of the angle of attack by adding degrees of freedom to the wing roots in a passive rotation mechanism in which the wing's membrane is loosely constructed. The passive rotation mechanism, which depends on the proposed external force, is suitable for developing an MAV with the same angle of attack in the up-down stroke, such as a symmetric flapping pattern. However, it is not suitable for developing MAVs with asymmetric flapping patterns that require a different angle of attack in up-down strokes. Thus, both proposed wing rotation mechanisms have limitations in terms of mimicking the drag-based system of a dragonfly.

The study proposes a mechanism that mimics the flight characteristics of dragonflies, using a drag-based system for hovering flight and a lift-based system for forward flight. The mechanism can be applied to an MAV using a passive rotation mechanism and can implement an asymmetric flapping pattern by changing the mean angle of attack. The proposed mechanism consists of a slide-crank mechanism designed to operate independently of flapping. The flapping motion was implemented using a spatial four-bar link mechanism, while the passive rotation mechanism was fabricated using the design parameters of the wing, optimized for the symmetric flapping pattern from previous research [27]. A high-speed camera was used for kinematic analysis of the prototype, and force measurement tests were conducted to determine whether the horizontal force amplitude of the drag-based system implemented by the proposed mechanism was reduced compared with that of the lift-based system, as well as the force direction variations with respect to changes in the wing root angle.

2. Flight Characteristics of Dragonflies

2.1. Kinematics of Dragonfly Flight

The kinematics [28,29] and aerodynamics [14–16,20,30,31] of dragonflies have been studied extensively in the past. Wang analyzed the stroke amplitude and wing angle of attack placed at 70% from the wing root in forward flight using an ultrahigh-speed camera [28]. In the same way, Azuma measured the stroke amplitude and angle of attack in hovering flight of the dragonfly [29]. However, the maximum and minimum stroke angle and mean angle of attack must be clearly defined in order to proceed with the aerodynamic experiment on the MAV. In this study, these parameters were defined as listed in Table 1, and the coordinate system of the kinematics is shown in Figure 1. The figure shows the flapping motion in the relative coordinate system $x'y'z'$ rotated by α with respect to the absolute coordinate system xyz, where α is the stroke angle, β is the stroke amplitude, and δ is the angle of attack.

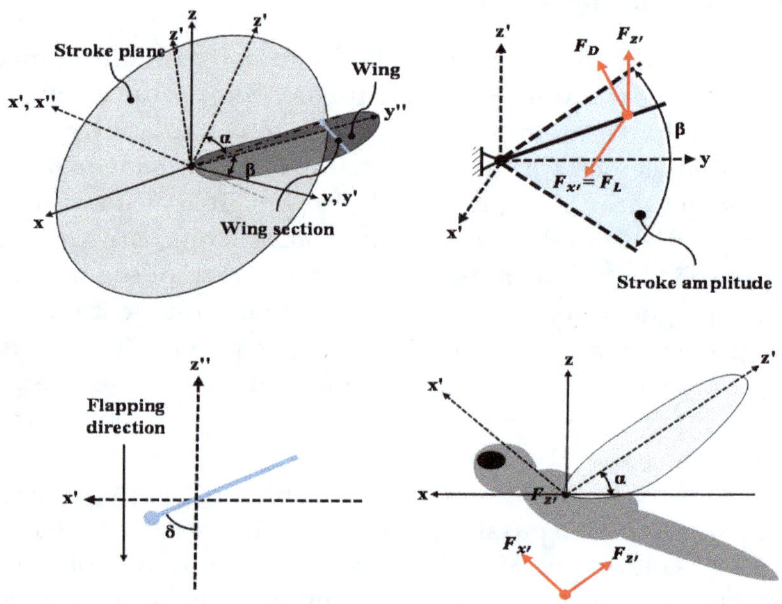

Figure 1. Coordinate system of flapping motion.

Table 1. Kinematics of dragonfly for experiment.

		Angle of Attack		Stroke Amplitude	Stroke Angle
		Hover	Forward		
Angle	Down stroke	75°	45°	−40°~40°	60°
	Up stroke	15°	45°		
	Mean angle of attack	30°	0°		

2.2. Aerodynamics of Dragonfly Flight

The systems used by organisms for swimming or flying are divided into drag-based systems and lift-based systems, depending on the type of force [14,15]. The lift-based system uses lift force in the manner of most fish or flying insects. Lift forces are obtained through up-down strokes, whereas drag force is generated in the opposite direction with the each up-down stroke, and the net force generated for one cycle is close to zero, as shown in Figure 2a. The drag-based system is based on the drag force, used in rowing locomotion or jellyfish swimming. In order to generate enough drag force to fly or swim, dragonflies flap their wings, and with each down-stroke maintain a large angle of attack to increase the drag force. In contrast, the upstroke minimizes the drag force by flapping with a small angle of attack, as shown in Figure 2b. This method allows the creature to generate force in the direction of movement through the drag force.

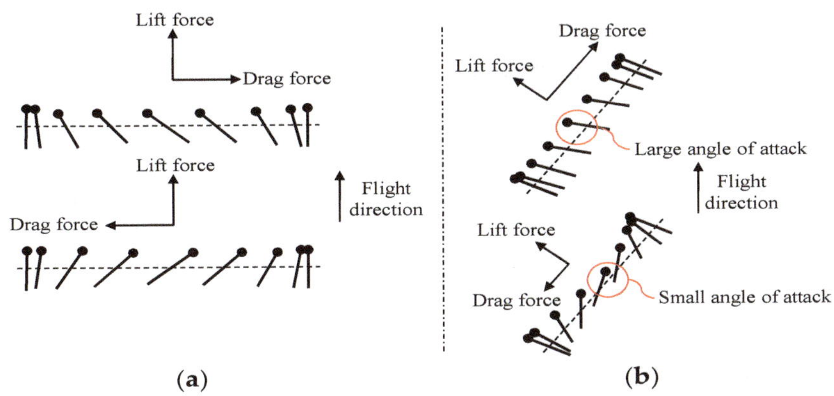

Figure 2. Stroke motion of a (**a**) lift-based system and (**b**) drag-based system.

Wang [14] mentioned that the drag-based system is more stable than the lift-based system in stationary flight, while the lift-based system has superior performance in forward flight. In addition, Vogel [15] experimentally proved that the drag-based system is more stable in stationary flight. Unlike most insects that use a lift-based system, dragonflies use both a drag-based system in hovering flight and lift-based system in forward flight. This system conversion is achieved by changing their angle of attack, which gives them more stable and superior flight abilities.

Dickinson experimentally defined lift and drag coefficients for when the thin wing moves in low Reynolds fluid as the following equations.

$$C_L = 0.225 + 1.58 \sin(2.13\alpha - 7.20), \quad (1)$$

$$C_D = 1.92 - 1.55 \sin(2.04\alpha - 9.82). \quad (2)$$

where C_L is the lift coefficient, C_D is the drag coefficient, and α is the angle of attack. The equation shows that the drag coefficient is a function of the angle of attack. In the relationship between the angle of attack and drag force, in order to use the drag-based system at hovering, the dragonfly increases the drag force using a large angle of attack at the downstroke and minimizes the drag force using a small angle of attack at the upstroke. On the other hand, in forward flight, the dragonfly flies by generating lift force using the same angle of attack in the up-down stroke like other insects. Thus, the rapid conversion from a drag-based system to a lift-based system without changing the stroke angle is achieved by changing the mean angle of attack.

3. Development of a Wing Root Rotation Mechanism

3.1. Passive Wing Rotation Mechanism

In the flight of insects, wing rotation is an essential factor for a positive angle of attack at flapping. To generate the wing rotation passively, this study used a wing rotation mechanism using a loose wing membrane, as shown in Figure 3 [27]. The proposed wing rotation mechanism was fabricated with a wing membrane larger than the designed wing membrane to generate wing rotation, and this changes the angle of attack by rotating the wing root. This wing rotation made by the loose wing membrane not only generates more lift force by the camber wing structure in flight but also increases the efficiency while gliding [32–35].

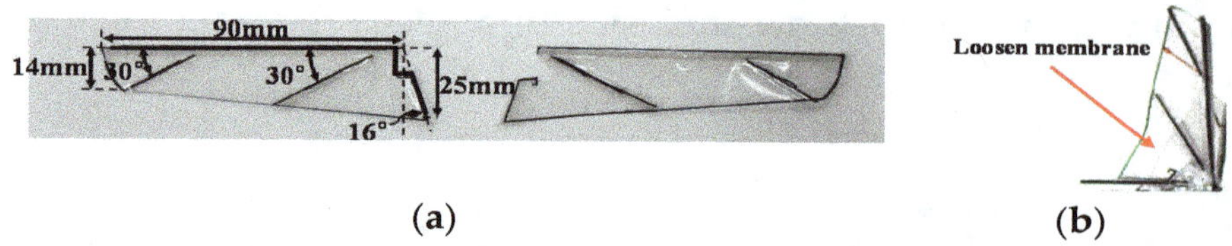

Figure 3. (**a**) Passive wing rotation mechanism using a loose wing membrane and (**b**) loosen membrane in moving wing.

The frame of the wing is made of carbon rods to reduce the weight, and the wing membrane is made of 25 μm thick polypropylene film. For the passive rotation of the wing, a polyolefin tube attached at the leading edge and root of the wings reduce the friction at wing rotation. In addition, two 0.6 mm carbon rods attached to the wing membrane maintain the rigidity of the wing after rotation, as shown Figure 3a.

3.2. Concept of Wing Root Rotation Mechanism

The purpose of this study is to develop a passive wing rotation mechanism that mimics the flight characteristics of dragonflies and can change the mean angle of attack by rotating the wing root. The mean angle of attack changes by wing root rotation with respect to the horizontal direction of the wing when the wing has the same angle of attack at each up-down stroke. Figure 4 shows the variation of the wing angle of attack range with this passive wing rotation mechanism. We developed the mechanism for changing the mean angle of attack based on a slide-crank link mechanism, where the crank is the wing root and rotates on the y″-axis.

To obtain a lift-based system with the same angle of attack at each up-down stroke, the wing root direction should be perpendicular to the stroke direction. However, the lift-based system of the wing root rotation mechanism has a singularity position problem when designed using a general slide-crank link mechanism. In order to solve this problem, we designed a slide-crank link mechanism with a three-bar link, as shown Figure 5. The wing root of the crank link with the three-bar link is located below the general slide-crank link mechanism's wing root. Therefore, the crank link is located above the singularity position when the wing root direction is perpendicular to the stroke direction. As a result, the wing root can rotate without singularity.

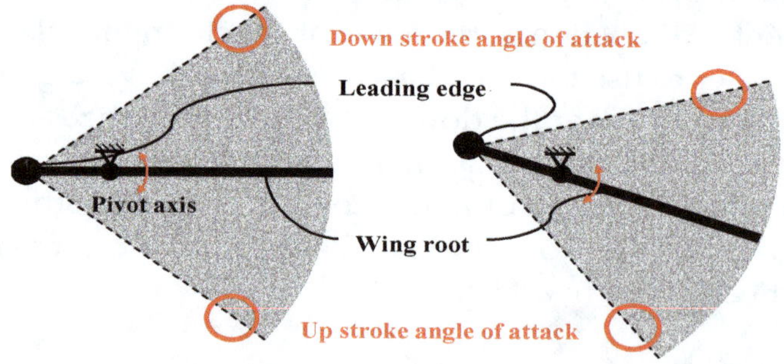

Figure 4. Range of angle of attack with respect to the wing root angle.

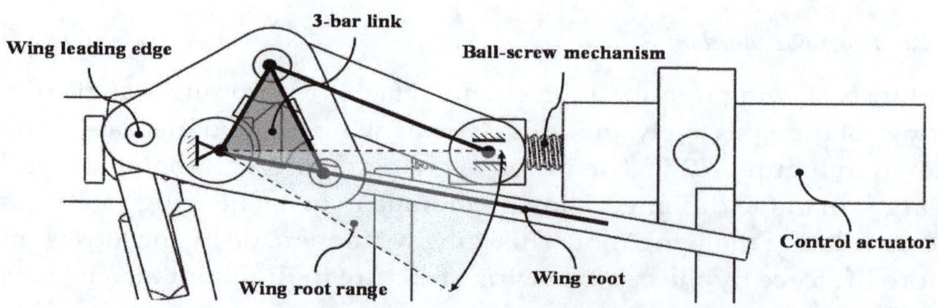

Figure 5. Slide-crank mechanism with three-bar link.

3.3. *Kinematics of Wing Root Rotation Mechanism*

Figure 6 shows a schematic of the proposed slide-crank mechanism with the three-bar link to define the relationship among variables, and the kinematics of this mechanism was analyzed by vector analysis.

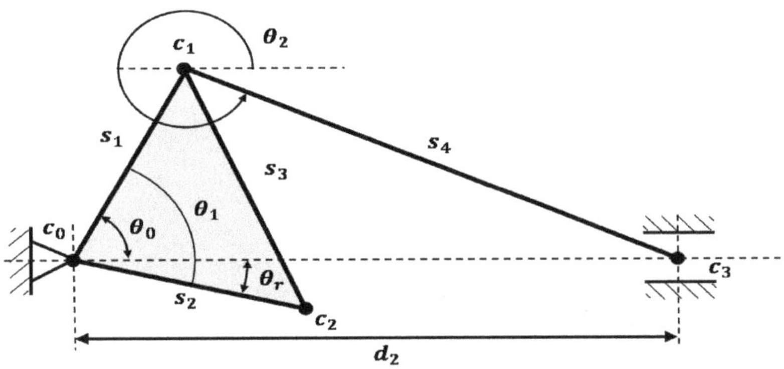

Figure 6. Schematic kinematics of wing root rotation mechanism.

The wing root angle θ_r is defined by θ_0 and θ_1, as in Equation (3).

$$\theta_r = \theta_1 - \theta_0 \tag{3}$$

where θ_1 is a constant value, and θ_0 is a dependent variable described by d_2, which is an independent variable.

$$\theta_0 = \cos^{-1}\left(\frac{d_2^2 + s_1^2 - s_4^2}{2s_1d_2}\right) \tag{4}$$

As a result, θ_r can be obtained as Equation (5) by substituting (4) into (3).

$$\theta_r = \theta_1 - \cos^{-1}\left(\frac{d_2^2 + s_1^2 - s_4^2}{2s_1d_2}\right) \tag{5}$$

After the kinematic analysis, each variable was defined for the design of the wing root control mechanism in Table 2. s_1, minimum value of d_2, and $\theta_2 - (\theta_0 + 180°)$ are the values to avoid physical collision. These design parameters were obtained by trial and error in 3D modeling. s_4 was obtained by applying the proposed values to (6).

$$s_4 = \sqrt{d_2^2 + s_1^2\cos^2(\theta_2 - (\theta_0 + 180°)) - s_1^2} + s_1\cos(\theta_2 - (\theta_1 + 180°)) \tag{6}$$

Lastly, θ_1, which makes θ_r equal to zero when d_2 is the minimum, is defined by (7).

$$\theta_1 = \cos^{-1}\left(\frac{s_1^2 + d_2^2 - s_2^2}{2s_2d_2}\right) \tag{7}$$

Figure 7 shows the results of the kinematics calculated by the defined design parameters. Figure 7a shows the simulation output motion (θ_r) with respect to input motion (d_2) and describes that the output motion is normally driven by input motion. The blue line is the s_4 link, the red line is the s_4 link, the blue dot is the input motion of linear motion, and the red dot is output motion of the wing root.

To implement the lift-based system and drag-based system of the dragonfly, each mean angle of attack should be 0° and 30°. Therefore, it is necessary to define the input variable for each system. The wing root angle for the lift-based system was defined as 0 at the above link optimization, but d_2 for the drag-based system, which is −30°, was not defined. Figure 7b shows relationship between d_2 and θ_r, and the desired value of d_2 when θ_r is −30°. As a result, we confirmed that the MAV implements a drag-based system when d_2 is 16.25 mm. Table 2 lists the design parameters of the wing root rotation mechanism.

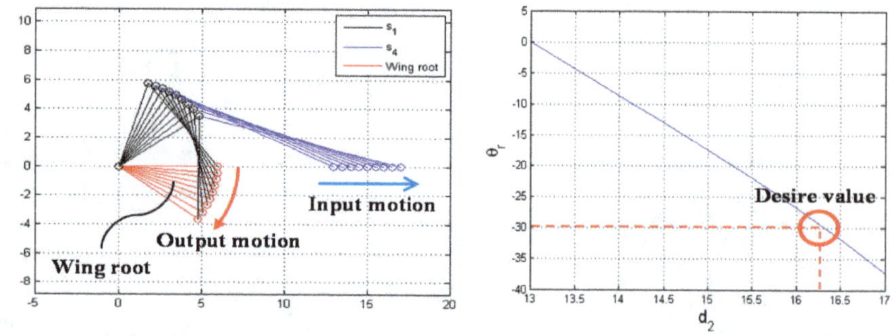

(a) (b)

Figure 7. (**a**) Kinematics of input motion (slide motion) and output motion (wing root rotation). (**b**) Graph of relationship between θ_r and d_2 to define d_2 with respect to desired θ_r (30°).

Table 2. Design parameters of wing root rotation mechanism.

Design Parameters	s_1	s_2	s_4	θ_1	$d_{2,min}$
Value	6 mm	6 mm	12.62 mm	72.97°	13 mm

4. Design of Flapping Mechanism

4.1. Analysis of Kinematics

In this work, we used the spatial four-bar-based flapping mechanism to convert from rotational motion to a couple of flapping motions using a rotary actuator. The spatial four-bar-based flapping mechanism consists of two spherical joints, a crank link, and a flapping link. In this mechanism, the factor considered for the design is the limited range of spherical joints. Figure 8 shows the angle limit of the spherical joint; when the y-axis is the joint axis, it has an angle limit of $-30°$ to $30°$ in the x-axis direction and z-axis direction.

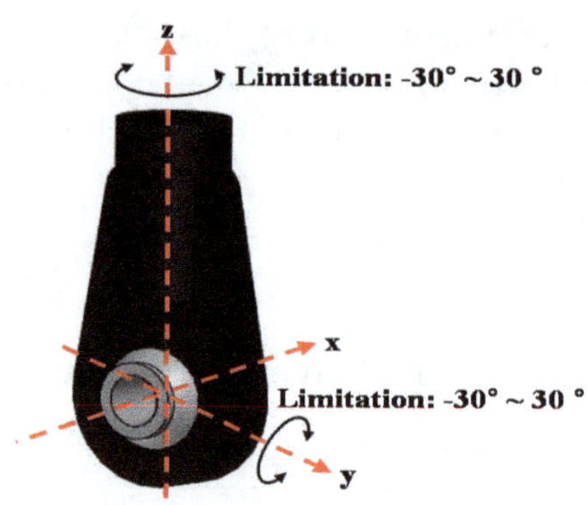

Figure 8. Limitation of spherical joint.

The schematic of the flapping mechanism is shown in Figure 9, where O_1, O_2 are the three-axis spherical joints, O_0, O_3 are one-axis joints; θ_1 is the crank angle; θ_3 is the flapping angle; and $\theta_{2,x}$, $\theta_{2,z}$ are the angles of the spherical joint. The kinematics of the flapping mechanism was analyzed by vector analysis; the input variable is θ_1, which is the crank angle. The flapping angle (θ_3) is defined by the crank angle and expressed as (8).

$$\theta_3 = \sin^{-1}\left(\frac{l_2^2 - (k_x^2 + k_y^2 + k_z^2 + l_3^2)}{2l_3\sqrt{k_x^2 + k_z^2}}\right) - \sin^{-1}\left(\frac{k_x}{\sqrt{k_x^2 + k_z^2}}\right) \tag{8}$$

In the equation, $k_x = l_{4,x}$, $k_y = l_{4,y} - l_1 \cos(\theta_1)$, and $k_z = l_{4,z} - l_1 \sin(\theta_1)$. Then, using the defined θ_3, the spherical joint angles $\theta_{2,x}$, and $\theta_{2,z}$ are described by (9) and (10), respectively.

$$\theta_{2,x} = \cos^{-1}\left(\frac{k_x + l_3 \cos(\theta_3)}{l_2}\right) \tag{9}$$

$$\theta_{2,z} = \cos^{-1}\left(\frac{k_y}{l_2 \sin(\theta_{2,z})}\right) \tag{10}$$

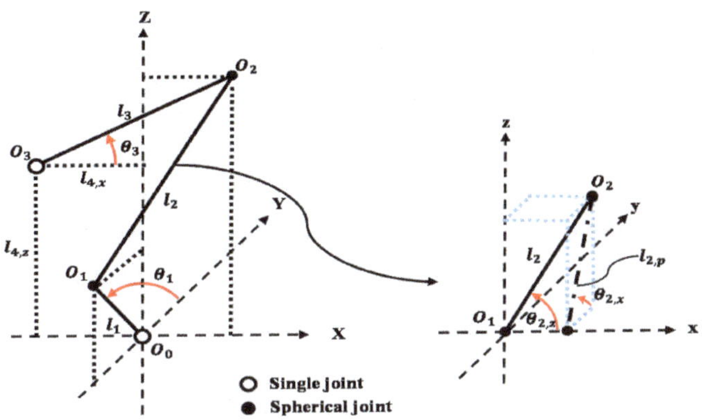

Figure 9. Schematic kinematics of spatial four-bar link-based flapping mechanism.

4.2. Define Link Length of Flapping Mechanism

We defined the link length of the flapping mechanism so that the stroke amplitude is within a given range, and $\theta_{2,z}$, $\theta_{2,x}$ are within the angle limits of the spherical joint. The mechanism was determined to have a stroke amplitude of $-40°$ to $40°$, an angle of the x-axis of the spherical joint within 60–120°, and an angle of the z-axis of the spherical joint within 60–90°. In the processing, the limit of the z-axis angle of the spherical joint was set as 90° rather than 120° to avoid physical collision at 90°. The link length was defined in three steps: (1) the determination of $l_{2,p}$ with respect to the x-axis limit of the spherical joint; (2) the calculation of l_3, $l_{4,z}$ with respect to the stroke amplitude; and (3) the determination of $l_{4,x}$ with respect to the z-axis limit of the spherical joint.

To determine the link length with respect to the $\theta_{2,x}$ range, $l_{2,p}$, which is projected on the y-z plane of l_2, as shown in Figure 10a, is defined as (11).

$$l_{2,p} = \frac{l_1}{\cos(\theta_{2,x})} \tag{11}$$

The maximum and minimum values of θ_2 for calculating $l_{2,p}$ are the same because the central axis of the crank link and flapping link are located on the same plane, as shown in Figure 10b. This means that if the link value is determined for one of the maximum and minimum values, then the other value is also defined. Therefore, $l_{2,p}$ was obtained by applying $\theta_{2,min}$ to (11), after $\theta_{2,min}$ was defined as 77°.

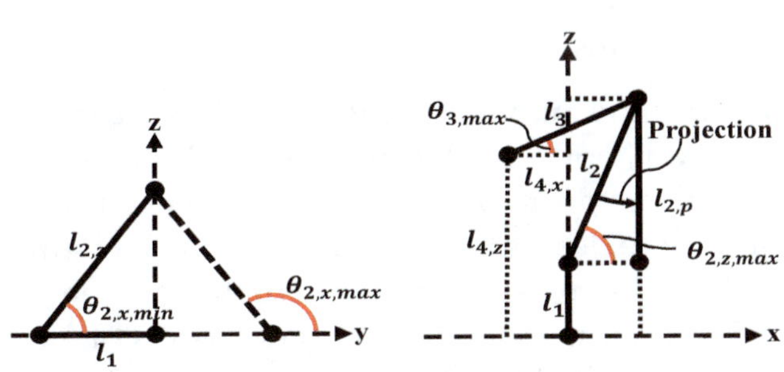

Figure 10. (**a**) Projected on the y-z plane; (**b**) projected on the x-z plane of flapping mechanism.

Second, in order to calculate l_3 and $l_{4,z}$, and having the stroke amplitude defined in Table 1, the relation between l_3 and $l_{4,z}$ is expressed as (12) and (13) when θ_3 is the maximum and minimum, respectively.

$$l_{3,\max} = \frac{(l_{2,p} + l_1 - l_{4,z})}{\sin(\theta_{3,\max})} \tag{12}$$

$$l_{3,\min} = \frac{(l_{2,p} - l_1 - l_{4,z})}{\sin(\theta_{3,\min})} \tag{13}$$

In addition, to determine l_3 and $l_{4,z}$ for the desired stroke amplitude, the object function was defined as (14), which is the absolute value of the difference between (12) and (13).

$$O_{(l_3,l_4)} = |l_{3,\max} - l_{3,\min}| \tag{14}$$

Afterwards, l_3 and $l_{4,z}$ were determined by the defined object function and line search method. $l_{2,p}$ is different from the previously defined value, but the defined $l_{2,p}$ was used because the difference in values is within the tolerance range. Finally, we defined the link length for the angle limit of $\theta_{2,z}$, as shown in Figure 10b; $\theta_{2,z}$ has the maximum value when θ_3 is the maximum. The maximum value of $\theta_{2,z}$ was defined as 80°, and $l_{4,x}$ was obtained by applying the variables to (15).

$$l_{4,x} = l_3 \cos(\theta_3) - \frac{l_{2,p}}{\tan(\theta_{2,z,\max})} \tag{15}$$

The values of θ_3, $\theta_{2,x}$, $\theta_{2,z}$, and the link lengths obtained from the kinematic equations are summarized in Table 3. Figure 11 shows each angular value according to the crank angle change. In the graph, we confirmed that $\theta_{2,x}$ and $\theta_{2,z}$ are within the angle limit of the spherical joint, and θ_3 is within the desired stroke amplitude.

Table 3. Value of design parameters of the flapping mechanism.

l_1	l_2	l_3	$l_{4,x}$	$l_{4,z}$	$\theta_{2,x}$	$\theta_{2,z}$	θ_3
6 mm	26.8 mm	9.331 mm	−4.43 mm	26.67 mm	76.85°~103.1°	79.46°~84.19°	−40°~39.91°

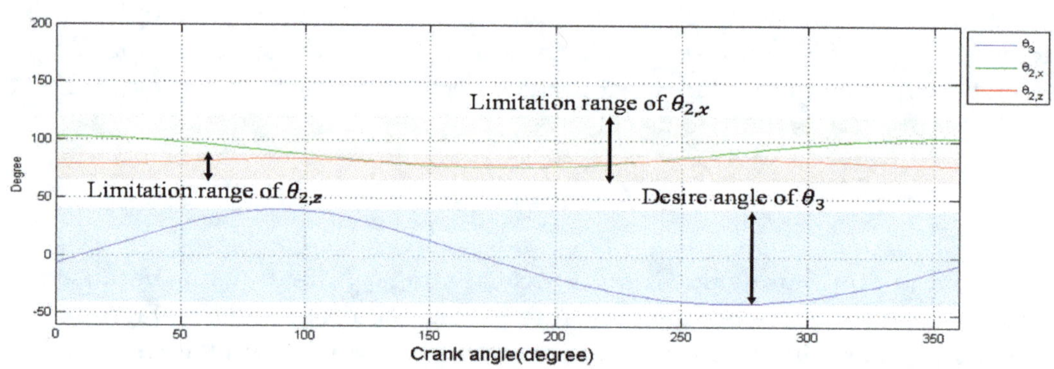

Figure 11. Values of $\theta_{2,x}$, $\theta_{2,z}$, and θ_3 with respect to the crank angle.

5. Design and Fabrication of Prototype

In order to combine the wing mechanism using a rotary actuator and the wing root rotation mechanism using a linear actuator, both mechanisms must be designed to move independently. Figure 12a shows the side view of the prototype design for the rotation mechanism, and Figure 12b shows the front view of the prototype design for the flapping mechanism. Figure 12a shows the main flapping link, slide flapping link, and wing root rotation link for independently driving two actuators. The wing root rotation link rotates with two flapping links as pivot axes and always moves in the same plane as the flapping link. The slide flapping link rotates with the spindle of the ball screw as the pivot axis and performs linear motion as the nut of the ball screw. Because the slide flapping link can simultaneously implement rotary motion and linear motion as the nut of the ball screw, the designed prototype has both independent flapping motion and wing root rotation motion.

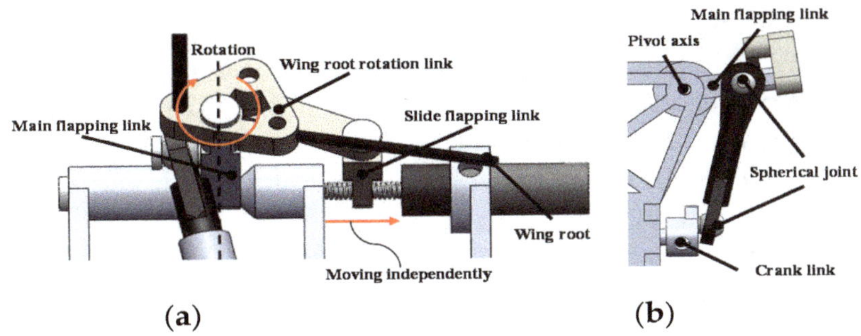

Figure 12. (**a**) Side view of the prototype design for the rotation mechanism; (**b**) front view of prototype design for the flapping mechanism.

Figure 13a,b shows the designed prototype and fabricated prototype, respectively. The prototype was designed using 3D CAD (Solidworks 2013, Dassault Systemes, France) software and design parameters determined for link optimization. The parts of the prototype, except a spherical joint link, were manufactured by an SLA (stereo lithography apparatus)-type 3D printer (ProJet 5000, resolution: 100 μm, 3D Systems, Rock Hill, SC, USA) using PC (polycarbonate). The joints were fixed with a 2 mm rivet for low friction and to affix without additional processing. The actuator for flapping is an EC6 (φ6, brushless, 2-watt, Hall sensors, maxon motor, Suisse, www.maxon-motor.com), the controller is a DEC 24/2, and a 37.5:1 gearbox ratio was used for the flapping mechanism. The LCP06-A03V-0026(D&J WITH Co., Geumcheon-Gu, Seoul, Korea) was used as the actuator of the wing root rotation mechanism, and 2 mm bolt was used for the spindle.

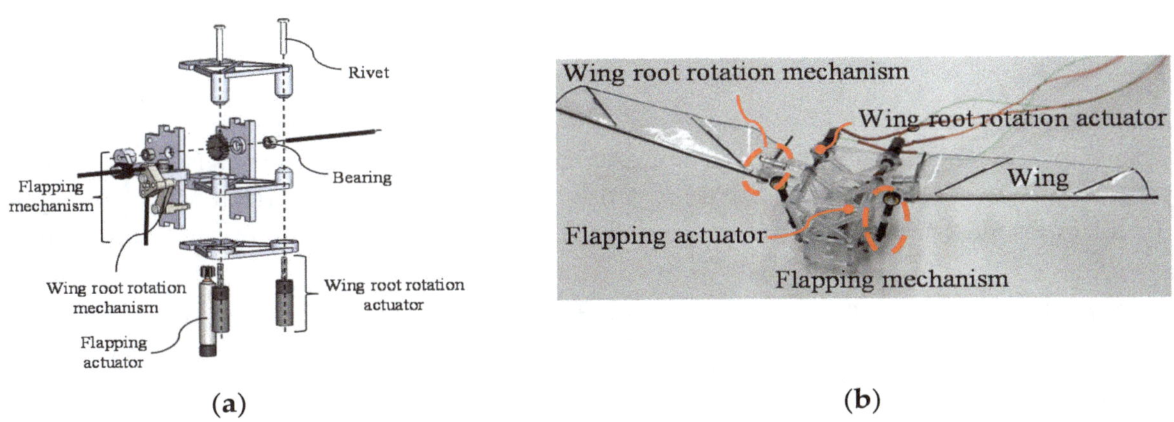

Figure 13. (**a**) Designed prototype and (**b**) front view of fabricated prototype.

6. Experiment and Analysis

6.1. Experimental Setup

This study proposes a mechanism for using the drag-based system in a hovering and lift-based system in forward flight by varying the mean angle of attack. We measured two factors to verify whether the drag-based and lift-based systems were implemented according to wing root rotation changes in the proposed mechanism. First, we took the front and side views of the MAV using a high-speed camera and measured the kinematics using the photographs. To obtain 30 images per cycle of flapping, the MAV was taken at a rate of 450 FPS (frames per second) at flapping frequency of 15 Hz. The angle of attack at flapping was obtained by measuring the length of the wing membrane at 70% from of wing root of the image.

Second, the aerodynamic force was measured by a force sensor. Figure 14 shows the force sensor, test stand, and signal line. High sampling rates and high resolution were required for force measurements during flapping. Therefore, the Nano 17 F/T sensor (resolution: 1/80 N, ATI Industrial Automation, Apex, NC, USA) was used for force measurement. The x-axis of the force sensor was installed at the front of the MAV and the z-axis was installed along the vertical direction of the MAV to measure data from each lift and drag. The tests proceeded at a flapping frequency of 10–15 Hz, and to obtain 500 data points per flapping cycle, the data was measured at 5000 RPS (rate per second) at 10 Hz and 7500 RPS (rate per second) at 15 Hz. Figure 15 shows the entire flowchart of the force and image data.

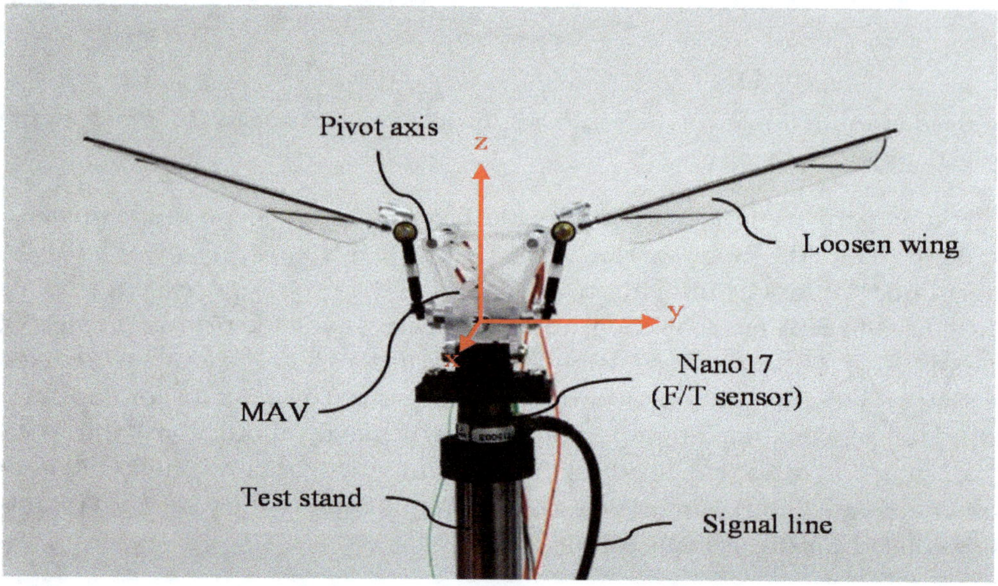

Figure 14. Experimental setup.

According to Caetano et al. [36], it has been determined that the frequency of the aerodynamic force is 2 times higher than flapping frequency, and the inertial force frequency is 3 times higher than the flapping frequency. Therefore, in this study, the measured data were filtered by a lowpass filter with a cutoff frequency 2 times larger than the flapping frequency to estimate the aerodynamic force.

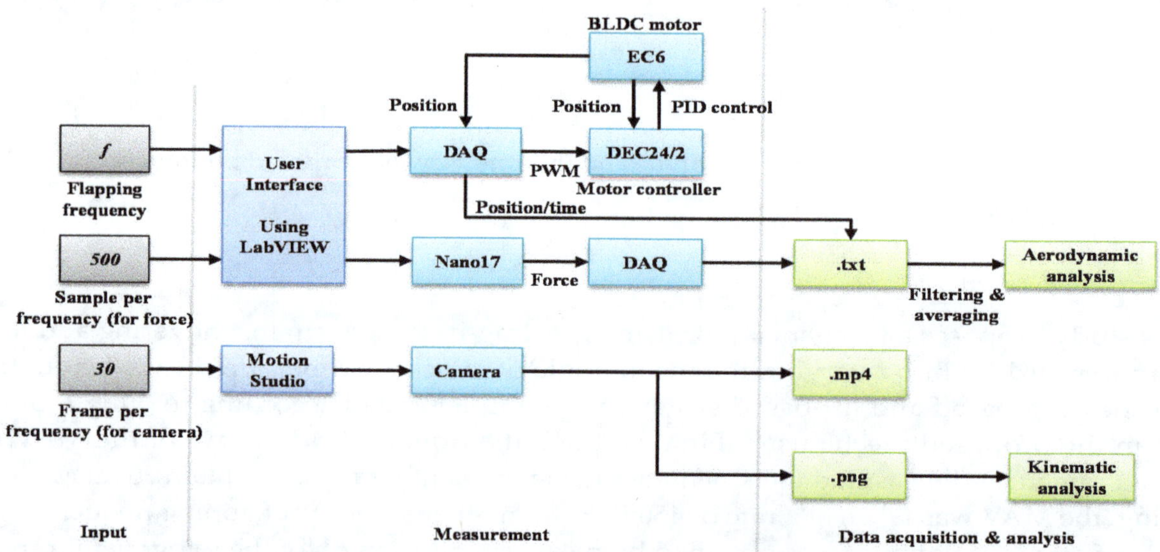

Figure 15. Flowchart of force and image data.

6.2. Analysis of Kinematics

6.2.1. Stroke Amplitude

Figure 16a shows the MAV at the minimum and maximum amplitudes taken with a high-speed camera. Figure 16b shows the measured amplitude during one cycle of the flapping link and wing by analyzing the image taken. In order to extract the kinematics from the image, the pivot, spherical joint tip, and wing tip were marked. Then, each mark was connected by a line, as shown in the left image in Figure 16a. The kinematics was extracted by measuring the angle between the connected lines. The amplitude of the flapping link has an error of approximately 4° with the designed amplitude by the joint's tolerance. On the other hand, the wing amplitude has an error of approximately 20° and is caused by the flexion of the wing frame.

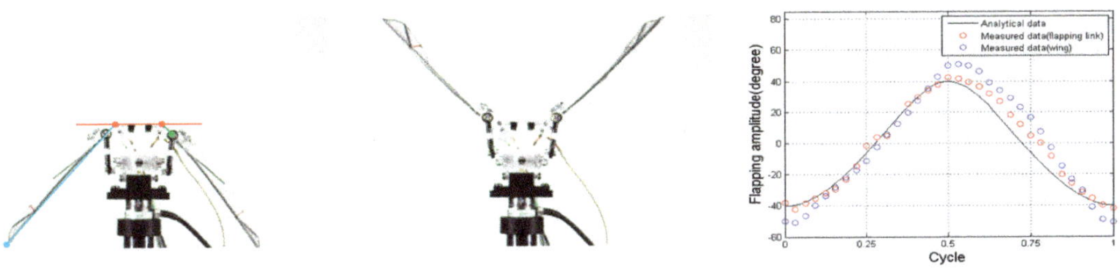

(a) (b)

Figure 16. (**a**) Flapping amplitude image of prototype; (**b**) flapping amplitude comparison between designed and measured value.

6.2.2. The Wing Angle of Attack

Figure 17 shows each angle of attack at upstroke and downstroke when the wing root angle is 0° and 30°.

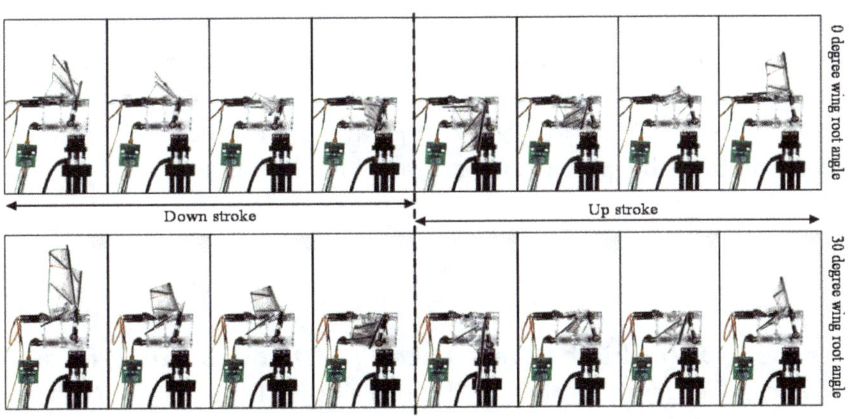

Figure 17. Angle of attack at each upstroke and downstroke at wing root angles of 0° and 30°.

Figure 18 shows the angle of attack of each wing at wing root angles of 0° and 30°. The red straight line is the wing root angle of 0°, the blue straight line is the wing root angle of 30°, and the dotted line is the mean of each angle of attack. Since the angle of attack is defined based on the wing root, it has a negative value in the upstroke and a positive value in the downstroke. The angle of attack at a wing root angle of 0° was symmetric with respect to the angle of attack of 0°, and the mean angle of attack was measured close to 0°. On the other hand, the angle of attack at the wing root angle of 30° had a large angle of attack in the downstroke and a small angle of attack in the upstroke, with mean angle of attack of 30°. These results indicate that the wing root angle and mean angle of attack have a linear relationship and that the angle of attack at the flapping of the MAV is similar to that of a real dragonfly.

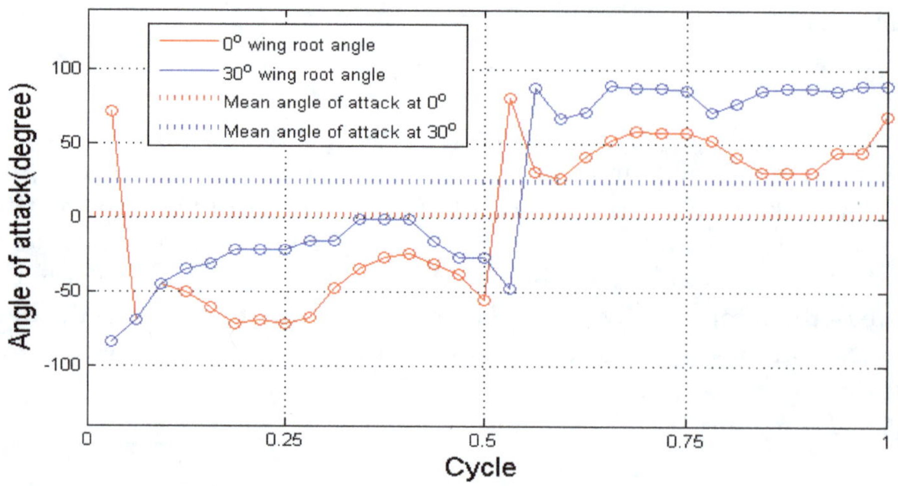

Figure 18. The measured angles of attack.

6.3. Analysis of Aerodynamics

The lift and drag forces acting on the object are given by (16) and (17), respectively, in the quasi-steady model.

$$F_L = \frac{1}{2}\rho C_L S U^2 \tag{16}$$

$$F_D = \frac{1}{2}\rho C_D S U^2 \tag{17}$$

where F_L and F_D are the lift force and drag force, respectively, ρ is the density of air, S is the area of the object, U is the velocity of the object, and C_L and C_D are the lift and drag coefficients, respectively. Lift force and drag force were defined with respect to wing, and the wing generates a drag force in the opposite direction of the moving direction and a lift force in the vertical direction of the moving direction. In the equation, the lift and drag are proportional to the velocity squared of the object because all parameters except the velocity are constant. $F_{x'}$ and $F_{z'}$, which were measured in the experiment, have the relationship indicated in (18) and (19) with F_L and F_D as shown Figure 1b. Therefore, $F_{x'}$ and $F_{z'}$ should increase with the flapping frequency squared of the object.

$$F_{x'} = F_L \tag{18}$$

$$F_{z'} = F_D \sin(\theta_s) \tag{19}$$

Figure 19a shows the relationship between each $F_{x'}$, $F_{z'}$ and flapping frequency of the MAV for wing root angles of 0° and 30° during flapping. The red dotted line (F_t) denotes the estimated value of $F_{x'}$ and $F_{z'}$ of the wing over the flapping frequency square through the characteristics of (16) and (17). The blue straight line is the measured lift force, and the black straight line is $F_{z'}$. Figure 19b shows the difference between the estimated and measured values. The difference between the measured force and estimated value is not over 0.006 N regardless of the flapping frequency, as shown in the figure. Therefore, $F_{x'}$ and $F_{z'}$ in the proposed MAV wing are proportional to the flapping frequency squared, and it is possible to estimate the force generated by increasing the flapping frequency.

In the left graph in Figure 19a, $F_{x'}$ at a 0° wing root angle increases in the positive direction with flapping frequency, while F_z increases in the negative direction or appears near 0. On the other hand, $F_{x'}$ at a wing root angle of 30°, shown in the right graph, is lower than that of the wing root angle of 0° but has a similar tendency, and $F_{z'}$ increases with increasing flapping frequency. A comparison of the two results shows that the magnitude of $F_{x'}$ and $F_{z'}$ that occurs in the wing changes when the wing root angle changes.

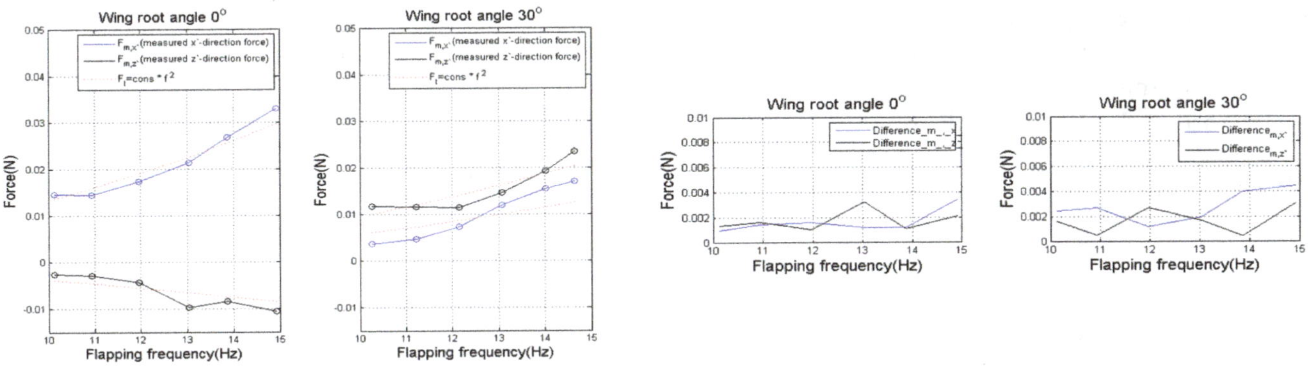

(a) (b)

Figure 19. (**a**) Measured $F_{x'}$ and $F_{z'}$ and (**b**) difference with respect to flapping frequency.

The top graph in Figure 20 shows $F_{z'}$ at a 0° wing root angle when the MAV flaps at a frequency of 15 Hz, while the bottom graph shows $F_{z'}$ at a 30° wing root angle. The top and bottom image in Figure 20 show the angle of attack when the wing is horizontal to the ground. The top two images indicate the angle of attack at a 0° wing root angle, while the bottom two images indicate the angle of attack at a 30° wing root angle. The force generated is an aerodynamic force, which is $F_{z'}$ of the MAV wing. The aerodynamic force at a 0° wing root angle does not show a significant difference when up-down strokes are compared. In contrast, the peak value after vibration due to the inertial force in the downstroke at a 30° the wing root angle means that the wing receives a large average $F_{z'}$. The decreasing tendency after the vibration due to the inertial force in the upstroke means that the wing receives a small average $F_{z'}$. The experimental results demonstrate that flapping at a wing root angle of 30° produces a reasonable average $F_{z'}$ in the up-down stroke.

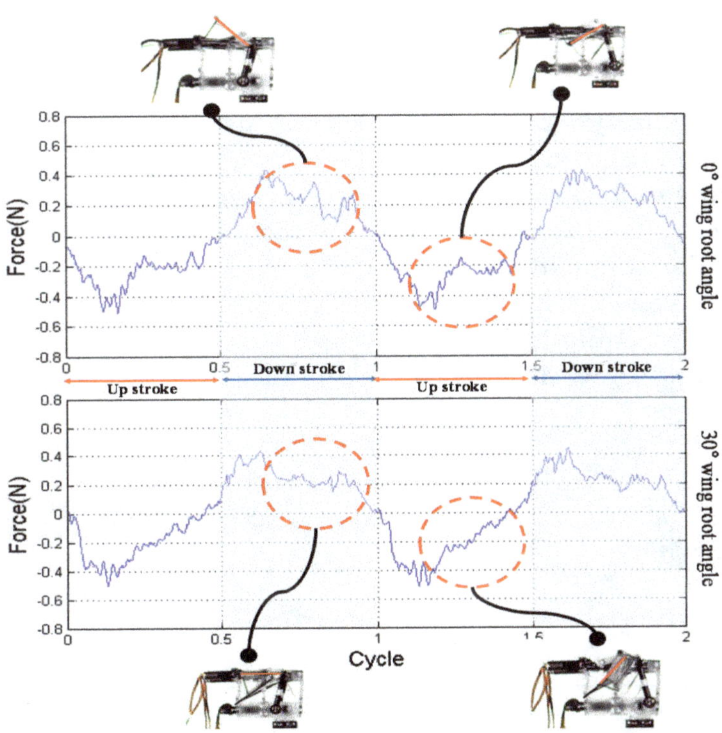

Figure 20. Drag force comparison between 0° and 30° wing root angle at 15 Hz flapping frequency.

6.3.1. Comparison between Horizontal Force Amplitude

The drag-based system used in hovering by a real dragonfly flaps in the inclined stroke angle direction of 60°, as shown in Figure 21. On the other hand, the direction of $F_{x'}$ measured in the study was generated toward the vertical direction of the stroke angle, and the direction of the drag force was generated toward the horizontal direction of the stroke angle. To measure the horizontal force generated by the wing root rotation mechanism at the same stroke angle as the drag-based system of a real dragonfly, the horizontal force was calculated by applying $F_{x'}$ and $F_{z'}$ measured at the wing root angle of 30° to (21).

$$F_V = F_{x'} \cos(\theta_s) + F_{z'} \sin(\theta_s) \tag{20}$$

$$F_H = F_{x'} \sin(\theta_s) - F_{z'} \cos(\theta_s) \tag{21}$$

where F_V is the vertical force, F_H is the horizontal force, and θ_s is the stroke angle. Since the lift-based system used for hovering has a horizontal directional stroke angle, the horizontal force generated during flapping of the MAV was defined as the $F_{z'}$ of the 0° wing root angle.

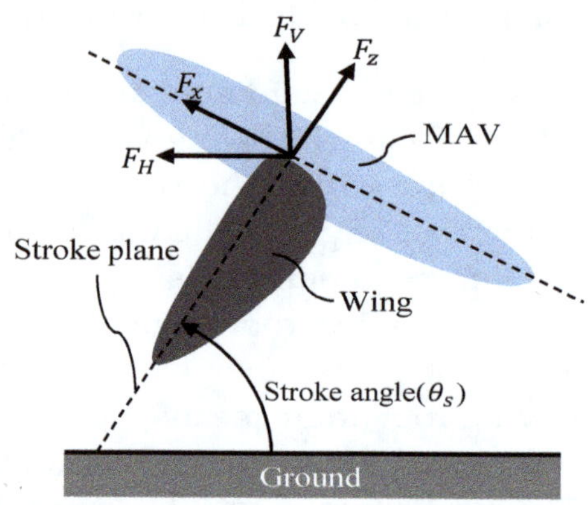

Figure 21. Force direction at an inclined stroke angle.

Figure 22 shows the horizontal forces of the two systems defined by the measured force. The black line is the horizontal force of the lift-based system implemented at a 0° the wing root angle, and the blue line is the horizontal force of the drag-based system implemented at a 30° wing root angle. The dotted line is the raw data and the straight line is the graph after applying the moving average filter, which averages a window of 30 data points. In drag-based systems, the MAV generated a large horizontal force in the upstroke because the lift force has more effect on horizontal force, and lift force was generated more at the upstroke. In contrast, the lift-based system has the same horizontal force amplitude in both the upstroke and downstroke. Comparison of the amplitudes of the two forces indicates that the horizontal force amplitude of the lift-based system is approximately 1.8 times larger than the horizontal force amplitude of the drag-based system. The difference between the two amplitudes is 0.15 N and 0.1 N in each downstroke and upstroke, respectively.

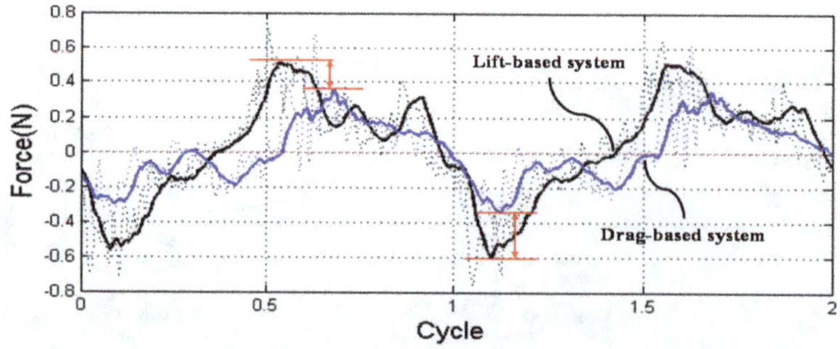

Figure 22. Horizontal force comparison between drag-based system and lift-based system at hovering.

6.3.2. Force Direction Change with Respect to Wing Root Angle

Figure 23 shows the vertical and horizontal forces generated in the wing during flapping at a stroke angle of 60°. The vertical and horizontal forces were calculated by Equations (17) and (18), respectively. The graph on the left shows the force generated at a wing root angle of 0° and the graph on the right shows the force generated at a wing root angle of 30°. At a wing root angle of 0°, the horizontal force increased with increasing flapping frequency, and the vertical force is 0.005 N, which is lower than the horizontal force. This means that the MAV generates force in the appropriate direction for forward flight when it has a wing root angle of 0° at flapping. Comparison with a 0° wing root angle shows that the vertical force at a wing root angle of 30° increases with increasing flapping frequency, and the horizontal force has a value close to 0 N. Therefore, when MAV flaps with a wing root angle of 30°, the wing generates a force in the direction for hovering. Thus, the wing root angle is an important factor in determining the flight mode of the MAV, and the flight mode can be changed by changing the wing root angle.

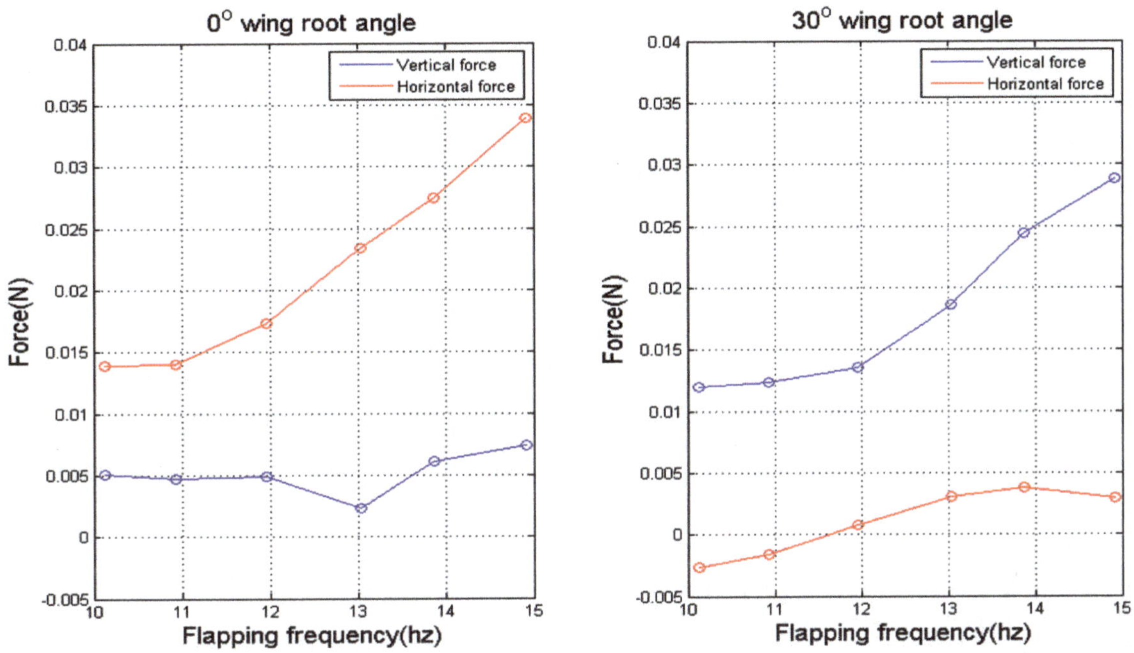

Figure 23. Vertical force and horizontal force with respect to flapping frequency at an inclined stroke angle.

7. Conclusions

In this study, we propose a mechanism to change the mean angle of attack of MAVs by rotating the wing roots. The links used in the flapping mechanism and the wing root rotation mechanism were optimized, and the graphs show the motion of the optimized mechanism. To confirm whether the proposed mechanism has stable hovering ability and can change flight mode without a posture change, a prototype incorporating the flapping mechanism and wing root rotation mechanism was fabricated, and the force generated by the MAV was measured. The prototype was designed using a ball-screw mechanism to independently control the wing root rotation mechanism and flapping mechanism. The measured force showed that the MAV with the wing root rotation mechanism can fly more stably when hovering and can change the flight mode by changing only the wing root angle. In the future, the wing root rotation mechanism is expected to fly more stably than the MAV with a lift-based system, and it will be possible to develop an MAV that mimics the interaction between the forewing and hindwing of an actual dragonfly.

Author Contributions: Supervision and Funding Acquisition, G.-H.Y.; Writing-Review & Editing, G.-H.Y.; Conceptualization, J.J.; Writing-Original Draft Preparation, J.J.; Design and Fabrication J.J.; Data acquisition and analysis, J.J.

References

1. Wood, R.J. Design, fabrication, and analysis of a 3DOF, 3 cm flapping wing mav. In Proceedings of the 2007 International Conference on Intelligent Robots and Systems, San Diego, CA, USA, 29 October–2 November 2007; pp. 1576–1581.
2. Keennon, M.; Klingebiel, K.; Won, H.; Andriukov, A. Development of the Nano Hummingbird: A tailless flapping wing micro air vehicle. In Proceedings of the 50th AIAA Aerospace Sciences Meeting, Nashville, TN, USA, 9–12 January 2012; p. 588.
3. Phan, H.V.; Park, H.C. Generation of Control Moments in an Insect-link Tailless Flapping-wing Micro Air Vehicle by Changing the Stroke-plane Angle. *J. Bionic Eng.* **2016**, *13*, 449–457. [CrossRef]
4. Rosen, M.H.; Pivain, G.L.; Sahai, R.; Jafferis, N.T.; Wood, R.J. Development of a 3.2 g Untethered Flapping-Wing Platform for Flight Energetics and Control Experiments. In Proceedings of the 2016 IEEE International Conference on Robotics and Automation (ICRA), Stockholm, Sweden, 16–21 May 2016; pp. 3227–3233.
5. Ramezani, A.; Shi, X.; Chung, S.J.; Hutchinson, S. Bat Bot (B2), A Biologically Inspired Flying Machine. In Proceedings of the 2016 IEEE International Conference on Robotics and Automation (ICRA), Stockholm, Sweden, 16–21 May 2016; pp. 3219–3226.
6. Hsiao, F.Y.; Yang, L.J.; Lin, S.H.; Lin, S.H.; Chen, C.L.; Shen, J.F. Autopilots for Ultra Lightweight Robotic Birds: Automatic Altitude Control and System integration of a Sub-10g Weight Flapping-Wing Micro Air Vehicle. *IEEE Control Syst. Mag.* **2012**, *14*, 35–48.
7. Karásek, M.; Muijres, F.T.; Wagter, C.D.; Remes, B.D.W.; de Croon, G.C.H.E. A tailless aerial robotic flapper reveals that flies use torque coupling in rapid banked turns. *Science* **2018**, *361*, 1089–1094. [CrossRef] [PubMed]
8. Alexander, D. Unusual phase relationships between the forewing and hindwings in flying dragonflies. *J. Exp. Biol.* **1984**, *109*, 379–383.
9. Norberg, R. Hovering flight of the dragonfly: *Aeschna juncea* L. Kinematics and aerodynamics. *Swim. Fly. Nat.* **1975**, *2*, 763–780.
10. Wang, J.K.; Sun, M. A computational study of the aerodynamics and forewing-hindwing interaction of a model dragonfly in forward flight. *J. Exp. Biol.* **2005**, *208*, 3785–3804. [CrossRef] [PubMed]
11. Hu, Z.; Deng, X.Y. Aerodynamic interaction between forewing and hindwing of a hovering dragonfly. *Acta Mech. Sin.* **2014**, *30*, 787–799. [CrossRef]
12. Usherwood, J.R.; Lehmann, F.O. Phasing of dragonfly wings can Improver aerodynamic efficiency by removing swirl. *J. R. Soc. Interface* **2008**, *5*, 1303–1307. [CrossRef] [PubMed]
13. Wang, Z.J. Dissecting insect flight. *Annu. Rev. Fluid Mech.* **2005**, *37*, 183–210. [CrossRef]
14. Wang, Z.J. The role of drag in insect hovering. *J. Exp. Biol.* **2004**, *207*, 4147–4155. [CrossRef] [PubMed]
15. Vogel, S. *Lift in Moving Fluids*; Princeton University: Princeton, NJ, USA, 1996.
16. Ellington, C.P. The aerodynamics of Hovering Insect Flight. III. Kinematics. *Philos. Trans. R. Soc. Lond. B* **1984**, *305*, 41–78. [CrossRef]
17. Ellington, C.P. The Novel Aerodynamics of Insect Flight: Applications to Micro-air Vehicles. *J. Exp. Biol.* **1999**, *202*, 3439–3448. [PubMed]
18. San, S.P.; Dickinson, M.H. The Control of Flight Force by a Flapping Wing: Lift and Drag Production. *J. Exp. Biol.* **2001**, *204*, 2607–2626.
19. Dickinson, M.H. The Effects of Wing Rotation on Unsteady Aerodynamic Performance at Low Reynolds Numbers. *J. Exp. Biol.* **1994**, *192*, 179–206. [PubMed]
20. Dickinson, M.H. Wing rotation and the Aerodynamic Basis of Insect Flight. *Science* **1999**, *284*, 1954–1960. [CrossRef] [PubMed]
21. Bergou, A.J.; Xu, S.; Wang, J. Passive wing pitch reversal in insect flight. *J. Fluid Mech.* **2007**, *591*, 321–337. [CrossRef]
22. Whitney, J.P.; Wood, R.J. Aerodynamics of passive rotation in flapping flight. *J. Fluid Mech.* **2010**, *60*, 197–220. [CrossRef]
23. McIntosh, S.H.; Agrawal, S.K.; Khan, Z. Design of a Mechanism for Biaxial Rotation of a Wing for a Hovering Vehicle. *IEEE/ASME Trans. Mechatron.* **2006**, *11*, 145–153. [CrossRef]
24. Fenelon, M.A.A.; Furukawa, T. Design of an active flapping wing mechanism and a micro aerial vehicle using a rotary actuator. *Mech. Mach. Theory* **2010**, *45*, 137–146. [CrossRef]

25. Karásek, M.; Hua, A.; Nan, Y.; Lalami, M.; Preumont, A. Pitch and Roll Control Mechanism for a Hovering Flapping Wing MAV. *Int. J. Micro Air Veh.* **2014**, *6*, 253–264. [CrossRef]
26. Nguyen, Q.V.; Park, H.C.; Goo, N.S.; Byun, D. Characteristics of a Beetle's Free Flight and a Flapping-Wing system that Mimics Beetle Flight. *J. Bionic Eng.* **2010**, *7*, 77086. [CrossRef]
27. Karásek, M. Robotic Hummingbird: Design of a Control Mechanism for a Hovering Flapping Wing Micro Air Vehicle. Ph.D. Thesis, Department of Mechanical, Engineering and Robotics, Delft University, Delft, The Netherlands, 2014.
28. Wang, H.; Zeng, L.; Liu, H.; Yin, C. Measuring wing kinematics, flight trajectory and body attitude during forward flight and turning maneuvers in dragonflies. *J. Exp. Biol.* **2003**, *206*, 745–757. [CrossRef] [PubMed]
29. Azuma, A.; Azuma, S.; Watanabe, I.; Furuta, T. Flight Mechanics of a Dragonfly. *J. Exp. Biol.* **1985**, *116*, 79–107.
30. Okamoto, M.; Yasuda, K.; Azuma, A. Aerodynamic Characteristics of the Wings and Body of a Dragonfly. *J. Exp. Biol.* **1996**, *199*, 281–294. [PubMed]
31. Hu, Z.; McCauley, R.; Schaeffer, S.; Deng, X. Aerodynamics of Dragonfly Flight and Robotic Design. In Proceedings of the 2009 IEEE International Conference on Robotics and Automation (ICRA), Kobe, Japan, May 12–17 May 2009; pp. 3061–3066.
32. Koehler, C.; Lisang, Z.; Gaston, Z.; Wan, H.; Dong, H. 3D reconstruction and analysis of wing deformation in free-flying dragonflies. *J. Exp. Biol.* **2012**, *215*, 3018–3027. [CrossRef] [PubMed]
33. Agrawal, A.; Agrawal, S.K. Design of Bio-inspired Flexible Wings for Flapping-Wing Micro-sized Air Vehicle Applications. *Adv. Robot.* **2009**, *23*, 979–1002. [CrossRef]
34. Du, G.; Sun, M. Effects of wing deformation on aerodynamic forces in hovering hoverflies. *J. Exp. Biol.* **2010**, *213*, 2273–2283. [CrossRef] [PubMed]
35. Pelletier, A.; Mueller, T.J. Low Reynolds Number Aerodynamics of Low-Aspect_Ratio, Thin/Flat/Cambered-Plat Wings. *J. Aircr.* **2000**, *37*, 825–832. [CrossRef]
36. Caetano, J.V.; Percin, M.; van Oudheusden, B.W.; Remes, B.; de Wagter, C.; de Croon, G.C.H.E.; de Visser, C.C. Error Analysis and Assessment of Unsteady Forces Acting on a Flapping Wing Micro Air Vehicle: Free-Flight versus Wind Tunnel Experimental Methods. *Bioinspir. Biomim.* **2015**, *10*, 056004. [CrossRef] [PubMed]

14

Pick and Place Operations in Logistics using a Mobile Manipulator Controlled with Deep Reinforcement Learning

Ander Iriondo [1,*], Elena Lazkano [2], Loreto Susperregi [1], Julen Urain [1], Ane Fernandez [1] and Jorge Molina [1]

1 Department of Autonomous and Intelligent Systems, Fundación Tekniker, Iñaki Goenaga, 5-20600 Eibar, Spain; loreto.susperregi@tekniker.es (L.S.); julen.urain@tekniker.es (J.U.); ane.fernandez@tekniker.es (A.F.); jorge.molina@tekniker.es (J.M.)

2 Faculty of Computer Science, P° Manuel Lardizabal, 1-20018 Donostia-San Sebastián, Spain; e.lazkano@ehu.es

* Correspondence: ander.iriondo@tekniker.es

Abstract: Programming robots to perform complex tasks is a very expensive job. Traditional path planning and control are able to generate point to point collision free trajectories, but when the tasks to be performed are complex, traditional planning and control become complex tasks. This study focused on robotic operations in logistics, specifically, on picking objects in unstructured areas using a mobile manipulator configuration. The mobile manipulator has to be able to place its base in a correct place so the arm is able to plan a trajectory up to an object in a table. A deep reinforcement learning (DRL) approach was selected to solve this type of complex control tasks. Using the arm planner's feedback, a controller for the robot base is learned, which guides the platform to such a place where the arm is able to plan a trajectory up to the object. In addition the performance of two DRL algorithms ((Deep Deterministic Policy Gradient (DDPG)) and (Proximal Policy Optimisation (PPO)) is compared within the context of a concrete robotic task.

Keywords: deep reinforcement learning; mobile manipulation; robot learning

---

1. Introduction

Logistics applications demand the development of flexible, safe and dependable robotic solutions for part-handling including efficient pick-and-place solutions.

Pick and place are basic operations in most robotic applications, whether in industrial setups (e.g., machine tending, assembling or bin picking) or in service robotics domains (e.g., agriculture or home). In some structured scenarios, picking and placing is a mature process. However, that is not the case when it comes to manipulating parts with high variability or in less structured environments. In this case, picking systems only exist at laboratory level, and have not reached the market due to factors such as lack of efficiency, robustness and flexibility of currently available manipulation and perception technologies. In fact, the manipulation of goods is still a potential bottleneck to achieve efficiency in the industry and the logistic market.

At the same time, the market demands more flexible systems that allow for a reduction of costs in the supply chain, increasing the competitiveness for manufacturers and bringing a cost reduction for consumers. The introduction of robotic solutions for picking in unstructured environments requires the development of flexible robotic configurations, robust environment perception, methods for trajectory planning, flexible grasping strategies and human–robot collaboration.

This study focused specifically on the development of adaptive trajectory planning strategies for pick and place operations using mobile manipulators. A mobile manipulator is a mobile platform carrying one or more manipulators. The typical configuration of mobile manipulators in industry is an anthropomorphic manipulator mounted on a mobile platform complemented with sensors (laser, vision, and ultrasonic) and tools to perform operations. The mobility of the base extends the work-space of the manipulator, which increases the operational capacity.

Currently, mobile manipulators implementation in industry is limited. One of the main challenges is to establish coordinated movements between the base and the arm in a unstructured environment depending on the application.

Although navigation and manipulation are fields where much work has been done, mobile manipulation is a less known area because it suffers from the difficulties and uncertainties of both previously mentioned fields. We propose to solve the path planning problem using a reinforcement learning (RL) strategy. The aim is to avoid explicit programming of hard-to-engineer behaviours, or at least to reduce it, taking into account the difficulty of foreseeing situations in unstructured environments. RL-based algorithms have been successfully applied to robotic applications [1] and enable learning complex behaviours only driven by a reward signal.

Specifically, this study focused on learning to navigate to such a place that the mobile manipulator's arm is able to pick an object from a table. Due to the limited scope of the arm, not all positions near the table are feasible to pick the object and to calculate those positions analytically is not trivial. The goal of our work was to evaluate the performance of deep reinforcement learning (DRL) [2] to acquire complex control policies such as mobile manipulator positioning by learning only through the interaction with the environment. Specifically, we compared the performance of two model-free DRL algorithms, namely Deep Deterministic Policy Gradient (DDPG) and Proximal Policy Optimisation (PPO) algorithms, in two simulation tests.

The simulated robot we used is the mobile manipulator miiwa, depicted in Figure 1, which has been used in industrial mobile manipulation tasks (e.g., [3]).

Figure 1. Kuka miiwa.

2. Literature Review

There are different approaches to establishing coordinated movements between the base and the arm depending on the application. Nassal et al. [4] specified three types of cooperation, which differ in the degree of cooperation, the associated complexity and the potential for manipulation capabilities: (1) Loose cooperation is when the mobile base and the manipulator are considered two separate systems, and the base serves as transport. There are several very well-known methods for navigation and trajectory planning [5]. (2) Full cooperation is when the two systems are seen as one (with nine degrees of freedom), and both the base and the arm move simultaneously to position the tool center. There are various approaches to solve the path planning problem [6,7], however, in these cases, the computational cost is very high. (3) Transparent cooperation is the combination of the previous two: the manipulator control compensates for the base and the base moves accordingly to maximise a cost function related to the positioning of the manipulator to perform the task. In [4], an approach to this type of control is proposed.

In general, current robotic solutions follow the loose cooperation approach. The coordination of the two subsystems for mobile manipulation depends on the task that needs to be solved. Berntorp et al. [8] addressed the issue of pick and place where the robotic system must take a can of a known position and place it in another combining the movement of the arm and the base. In [9], the problem of opening doors is considered where the movement of the arm and base are coupled by sensing the forces generated between the arm and the door, and coordinating the forward movement of the base.

One of the principal goals of artificial intelligence (AI) is to enable agents to learn from their experience to become fully autonomous. This experience is obtained by interacting with the environment. Thus, the agent should continue improving its behaviour through trial and error until it behaves optimally. Deep learning enables reinforcement learning to face decision-making problems that were previously infeasible. The union of both of them, namely DRL, makes it possible to learn complex policies guided by a reward signal and has been applied to learn multiple complex decision making tasks in a wide variety of areas. For instance, in [10], the authors learn a chess player agent by self playing that is able to defeat a world champion. In [11], DRL is used to learn an intelligent dialogue generator agent. It also has been successfully applied to computer vision, for example to train a 2D object tracker agent that outperforms state-of-the-art real-time trackers [12].

The field of DRL applied to robotics is becoming more and more popular, and the number of published papers related to that topic is increasing quickly. Applications extend from manipulation [13–16] to autonomous navigation [17,18] and locomotion [19,20].

DRL has been successfully applied in some robotic path planning and control problems. For example, Gu et al. [16] used DRL to learn a low-level controller that is able to control a robotic arm to perform door opening. On the one hand, 25 variables are used to represent the environment's state. More specifically, those variables correspond to the joint angles and their time derivatives, the goal position and the door angle together with the handle angle. On the other hand, the action is represented by the torque for each joint. Robot navigation is another area where DRL also has been applied successfully; for example, in [18], DRL is applied to learn a controller to perform mapless navigation with a mobile robot. In this work, a low dimensionality state representation is also used, specifically 10-dimensional laser range findings, the previous action and the relative target position. The action representation includes linear and angular velocities of the differential robot.

Alternatively, several methods use 2D or 3D images instead of low-dimensional state representations. In [15], for example, 2D images are used with the goal of learning grasping actions. The authors tried to map images directly with low-level control actions and, in addition, the developed system is able to generalise and pick objects that the model has not been trained with. Moreover, other applications (e.g., [21]) use 3D depth images as state representation, but instead of learning directly from pixels, they encode the images to a lower dimensional space before training the model. This encoding enables to accelerate the training process.

Although there are multiple works about learning to control either robotic arms and mobile bases, to the best of our knowledge, references about DRL applied to mobile manipulation are few. The goal of this paper is to show that it is possible to use the arm path planner's feedback to learn a low-level controller for the base. The learned controller is able to guide the robot to a feasible pose in the environment to do the picking trial. To do that, a low-dimensional state representation is used.

In [22], the authors proposed to learn, through interaction with the environment, which place is the best to locate the base of a mobile manipulator, in such a way that the arm is able to pick an object from a table. In that way, the robot takes into account its physical conditions and also the environment's conditions to learn which is the optimal place to perform the grasping action. In this case, they acquire experience off-line, and, after applying some classifiers, such as support vector machines, they are able to learn a model that maps the state of the robot with some feasible places in the environment to do the picking trial. Then, the trained model is used on-line.

In comparison with the previous approach, the goal of the RL-based algorithms is to learn on-line, while the agent interacts with the environment, improving its policy until it reaches to the optimal behaviour. In addition, the goal of our work is to learn a controller, which gives a low-level control signal in each state to drive the mobile robot, while, in [22], instead of learning a controller to drive the robot up to the optimal pose, they tried to find directly which is the optimal pose.

3. Methodological Approach

According to Sutton and Barto [23], a reinforcement learning solution to a control problem is defined as a finite horizon Markov Decision Process (MDP). At each discrete time-step t, the agent observes the current state of the environment $s_t \in S$, takes an action $a_t \in A(s_t)$, receives a reward $r{:}S \times A \rightarrow \mathbb{R}$ and observes the new state of the environment s_{t+1}. At each episode of T time-steps, the environment and the agent are reset to their initial poses. The goal of the agent is to find a policy, deterministic $\pi_\theta(s)$ or stochastic $\pi_\theta(a|s)$, parameterised by θ under which the expected reward is maximised.

Traditional reinforcement learning algorithms are able to deal with discrete state and action spaces but, in robotic tasks, where both state and action spaces are continuous, discretising those spaces does not work well. Nevertheless, most state-of-the-art algorithms use deep neural networks to map observations with low-level control actions to be able to deal with continuous spaces. In our approach, a DRL algorithm is used to learn where to place the mobile manipulator to make a correct picking trial through the interaction with the environment. Those DRL algorithms need a huge amount of experience to be able to learn complex robotic behaviours and, thus, it is infeasible to train them acquiring experience in real world. In addition, the actions taken by the robot in the initial learning iterations are nearly random and both the robot and the environment might end up damaged as a result. Therefore, DRL algorithms are usually trained in simulation. Learning in simulation enables a faster experience acquisition and avoids material costs. Our study used a simulation-based approach, and we based it on open source robotic tools such as Gazebo simulator, Robot Operating System (ROS) middleware [24], OpenAI Baselines DRL library [25] and Gym/Gym-gazebo toolboxes [26,27].

DRL algorithms follow value-based, policy search or actor–critic architectures [28]. Value based-algorithms estimate the expected reward of an state or state–action pair and are able to deal with discrete action spaces, typically using greedy policies. To be able to deal with continuous action spaces, policy search methods use a parameterised policy and do not need a value function. Usually, those methods have the difficulty of not easily being able to evaluate the policy and they have high variance. Actor–critic architecture is the most used one in the state-of-the-art algorithms and combines the benefits of the two previous algorithm types. On the one hand, actor–critic-based methods use a parameterised policy (actor), and, on the other hand, use a value or action–value function that evaluates the quality of the policy (critic).

The proposed approach follows the actor–critic architecture. On the one hand, a parameterised policy π_θ is used to be able to deal with both continuous state and action spaces in stochastic environments, encoded by the parameter vector θ. On the other hand, a parameterised value function is used to estimate the expected reward at each state or state–action pair, where w is the parameter vector that encodes the critic. The state value function $V_w(s)$ estimates the expected average reward of all actions in the state s. The action–value function $Q_w(s, a)$, instead, estimates the expected reward of executing action a in state s. Then, the critic's information is used to update both actor and critic. In DRL algorithms, both actor and critic are parameterised by deep neural networks, and the goal is to optimise those parameters to get the agent's optimal behaviour.

In addition, DRL algorithms are divided into two groups, on-policy and off-policy, depending on how they are able to acquire experience. On-policy algorithms expect that the experience used to optimise their behaviour policy is generated by the same policy. Off-policy methods, instead, can use experience generated by another policy to optimise its behaviour policy. Those methods are said to be able to better explore the environment than on-policy methods because they use a more exploratory policy to get experience. In this study, we compared an on-policy algorithm and an off-policy algorithm, to see which type of methods adjusts better to our mobile manipulation behaviour learning. Specifically, PPO and DDPG were used, being those on-policy and off-policy, respectively. The first one learns stochastic policies, which maps states with actions represented by Gaussian probability distributions. DDPG, instead, is able to deterministically map states with actions. Both algorithms follow actor–critic architecture.

3.1. *Algorithms*

In this section, we describe the theoretical basis of PPO and DDPG.

3.1.1. PPO

PPO [29] follows the actor–critic architecture and it is based on the trust-region policy optimisation (TRPO) [30] algorithm. This algorithm aims to learn a stochastic policy $\pi_\theta(a_t|s_t)$ that maps states with Gaussian distributions over actions. In addition, the critic is a value function $V_w(s_t)$ that outputs the mean expected reward in state s_t. This algorithm has the benefits of TRPO and in general of trust region based methods but it is much simpler to implement it. The intuition behind trust-region based algorithms is that, at each parameter update of the policy, the output distribution cannot diverge too much from the original distribution.

To update the actor's parameters, a clipped surrogate objective is used. Although another loss function is also proposed, using a Kullback–Leibler (KL) divergence [31] penalty on the loss function instead of the clipped surrogate objective, the experimental results obtained are not as good as with the clipped one. Let $r_t(\theta)$ denote the probability ratio defined in Equation (1), so that $r_t(\theta_{old}) = 1$.

$$r_t(\theta) = \frac{\pi_\theta(a_t|s_t)}{\pi_{\theta_{old}}(a_t|s_t)} \tag{1}$$

θ_{old} is the actor's parameter vector before the update. The objective of TRPO is to maximise the objective function $L(\theta)$ defined in Equation (2). Here, $\mathbb{E}[...]$ indicates the average over a finite batch of samples. The usage of the advantage $\hat{A}_t$ in policy gradient algorithms was popularised by Schulman et al. [32] and indicates how good the performed action is with respect to the average actions performed in each state. To compute the advantages, the algorithm executes a trajectory of T actions and computes them as defined in Equation (3). Here, t denotes the time index $[0, T]$ in the trajectory of length T and γ is the discount factor.

$$L(\theta) = \mathbb{E}\left[\frac{\pi_\theta(a_t|s_t)}{\pi_{\theta_{old}}(a_t|s_t)}\hat{A}_t\right] \tag{2}$$

$$\hat{A}_t = -V_w(s_t) + r_t + \gamma r_{t+1} + ... + \gamma^{T-t+1} r_{T-1} + \gamma^{T-t} V_w(s_T) \tag{3}$$

At each policy update, if the advantage has a positive value, the policy gradient is pushed in that direction because it means that the action performed is better than the average. Otherwise, the gradient is pushed in the opposite direction.

Without any constraint, the maximisation of the loss function $L(\theta)$ would lead to big changes in the policy at each training step. PPO modifies the objective function so that penalises big changes in the policy that move $r_t(\theta)$ away from 1. Maintaining $r_t(\theta)$ near to 1 ensures that, at each policy update, the new distribution does not diverge to much from the old one. The objective function is defined in Equation (4).

$$L^{CLIP}(\theta) = \mathbb{E}[min(r_t(\theta)\hat{A}_t, clip(r_t(\theta), 1-\epsilon, 1+\epsilon))\hat{A}_t] \quad (4)$$

Here, ϵ is a hyper-parameter that changes the clip range.

To update the value function $V_w(s)$ (the critic), the squared-error loss function is used (Equation (5)) between the current state value and a target value. The target value is defined in Equation (6).

$$J(w) = (V_w(s_t) - V_t^{targ})^2 \quad (5)$$

$$V_t^{targ} = \hat{A}_t + V_w(s_t) \quad (6)$$

The PPO algorithm is detailed in Algorithm 1. Although this algorithm is designed to be able to have multiple parallel actors getting experience, only one actor is being used.

Algorithm 1 Proximal Policy Optimisation (PPO).

for $e \in episodes$ **do**
    for $a \in actors$ **do**
        Run policy $\pi_{\theta_{old}}$ in environment for T time-steps
        Compute advantage estimates $\hat{A}_1 ... \hat{A}_T$
    end for
    Optimise actor's loss function L^{CLIP} with regard to θ, with K epochs and minibatch size N $\leq T \cdot actors$
    $\theta_{old} \leftarrow \theta$
end for

3.1.2. DDPG

DDPG [33] combines elements of value function based and policy gradient based algorithms, following the actor–critic architecture. This algorithm aims to learn a deterministic policy $\pi_\theta(s) = a$ and it is derived from the deterministic policy gradient theorem [34].

Following the actor–critic architecture, DDPG uses an action–value function $Q_w(s,a)$ as critic to guide the learning process of the policy and it is based on the deep Q-network (DQN) [35]. Prior to DQN, it was believed that learning value functions with large and nonlinear function approximators was difficult. DQN is able to learn robust value functions due to two innovations: First, the network is trained off-policy getting experience samples from a replay buffer to eliminate temporal correlations. In addition, target networks are used, which are updated more slowly, and this gives consistent targets during temporal difference learning.

The critic is updated according to the gradient of the objective defined in Equation (7).

$$L(w) = \mathbb{E}[(Q_w(s_t, a_t) - y_t)^2] \quad (7)$$

where

$$y_t = r(s_t, a_t) + \gamma Q_w(s_{t+1}, a_{t+1})|_{a_{t+1} = \pi_\theta(s_{t+1})} \quad (8)$$

The actor is updated following the deterministic policy gradient theorem, defined in Equation (9). The intuition is to update the policy in the direction that improves $Q_w(s, a)$ most .

$$\nabla J(\theta) = \mathbb{E}[\nabla_\theta \pi_\theta(s_t) \nabla_a Q_w(s_t, a_t)|_{a=\pi_\theta(s_t)}] \tag{9}$$

As mentioned before, the target value defined in Equation (8) is calculated using the target networks $\pi'_{\theta'}$ and $Q'_{w'}$, which are updated more slowly and this gives more consistent targets when learning the action–value function.

As DDPG is an off-policy algorithm, the policy used to get experience is different from the behaviour policy. Despite the behaviour policy being deterministic, typically, a stochastic policy is used to get experience, being able to better explore the environment. This exploratory policy is usually achieved adding noise to the behaviour policy. Although there are common noises such as normal noise or Ornstein–Uhlenbeck noise [36], which are added directly to the action generated by the policy, in [37], Plapperta et al. proposed adding noise to the neural network's parameter space to improve the exploration and to reduce the training time. The DDPG algorithm is described in Algorithm 2.

Algorithm 2 Deep Deterministic Policy Gradient (DDPG).

Initialise the actor $\pi_\theta(s)$ and the critic $Q_w(s, a)$ networks.
Initialise the target networks Q' y π' with the weights $\theta' \leftarrow \theta$, $w' \leftarrow w$
Initialise the replay buffer
for $e \in episodes$ **do**
    Initialise noise generation process
    Get the first observation
    for $t \in steps$ **do**
        Select the action $a_t = \pi_\theta(s_t) + N$
        Execute the action a_t, get the reward r_t and the observation s_{t+1}
        Store the transition $< s_t, a_t, r_t, s_{t+1} >$ in replay buffer
        Get M experience samples $< s_i, a_i, r_i, s_{i+1} >$ from the replay buffer
        $y_i = r_i + \gamma Q'_{w'}(s_{i+1}, \pi'_{\theta'}(s_{i+1}))$
        Update the critic minimising the loss :

$$L = \frac{1}{M}\sum_i (y_i - Q_w(s_i, a_i))^2$$

        Update the policy of the actor:

$$\nabla_\theta J \approx \frac{1}{M}\sum_i \nabla_a Q_w(s, a)|_{s=s_i, a=\pi_\theta(s_i)} \nabla_\theta \pi_\theta(s)|_{s=s_i}$$

        Update target networks:

$$\theta' \leftarrow \tau\theta + (1 - \tau)\theta'$$

$$w' \leftarrow \tau w + (1 - \tau)w'$$

    end for
end for

3.2. *Simulated Layout*

To model our world, we used the Gazebo model based simulator. The elements that are placed in this simulated world are the robot miiwa, the table and an object, which is on top of the table, as depicted in Figure 2.

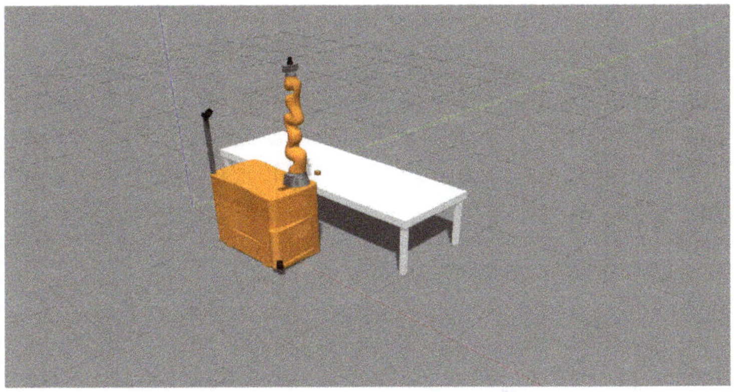

Figure 2. Simulated world in Gazebo.

The mobile manipulator miiwa is composed of a 7 DoF arm and a 3 DoF omnidirectional mobile base. To be able to control the mobile base, the *gazebo planar move* plugin was used.

To test if our environment was modelled correctly and to know if the algorithms could learn low-level control tasks in that environment, the learning process was divided into two simulation tests. In both, the algorithm must learn to control the mobile base with velocity commands, such that, at each discrete time-step, the algorithm gets the state of the environment and publishes a velocity command. The objective of those tests was to learn a low-level controller to drive the robot to a place where the arm can plan a trajectory up to the object on the table. To learn those controllers, the feedback of the arm's planner was used. The summary of the tests is listed in Table 1.

Table 1. Test setup summary.

Test	Robot Initial Pose	Box Initial Pose	Objective
Test 1	Variable	Variable	The arm to be able to plan a trajectory up to the box
Test 2	Variable	Constant	To plan a trajectory with an obstacle in the table

4. Implementation

The application was implemented in a modular way using ROS for several reasons. On the one hand, it enabled us to modularise the application and to parallelise processes. On the other hand, Gazebo is perfectly integrated with ROS and offers facilities to control the simulation using topics, services and actions.

OpenAI Gym is a toolkit to do research on DRL algorithms and to do benchmarks between DRL algorithms. This library includes some simulated environments that are used to test the quality of new algorithms and its use is widespread in the DRL community. Gym offers simple functions to interact with the environments, which are mostly modelled in the Mujoco [38] simulator. Due to the simplicity of the interface that Gym offers, it has become a standard way to interact with environments in DRL algorithms. Gym-gazebo is an extension of Gym that enables the user to create robotic environments in Gazebo and offers the same simple interface to the algorithm to be able to interact with the environment. All the environments we modelled were integrated with Gym-gazebo, which enabled us to straightforwardly use OpenAI Baselines DRL library, which is designed to work with Gym.

Gym-gazebo wraps the used DRL algorithms in ROS nodes and enables interaction with the environment. Nevertheless, another ROS node has been developed that is in charge with controlling all the elements of the simulation and works as bridge between the Gym-gazebo node and the simulator. To be able to control the simulation physic updates and to compute them as fast as possible, we developed a Gazebo plugin, which in turn is a ROS node. Thus, using ROS communication methods, we could control each simulated discrete time-step and we simulated those steps faster than real time.

Specifically, this node takes care of all time-steps being of the same length, executing a fixed number of physic updates at each step.

The *tf broadcaster* node uses the *tf* tool that ROS offers to keep track of all the transformations between frames. We used this node to publish some transformations such as the transformation between the *object* and *world* frames. Consequently, the robot is always aware of where the object is in order to be able to navigate up to it. The implemented architecture is shown in Figure 3.

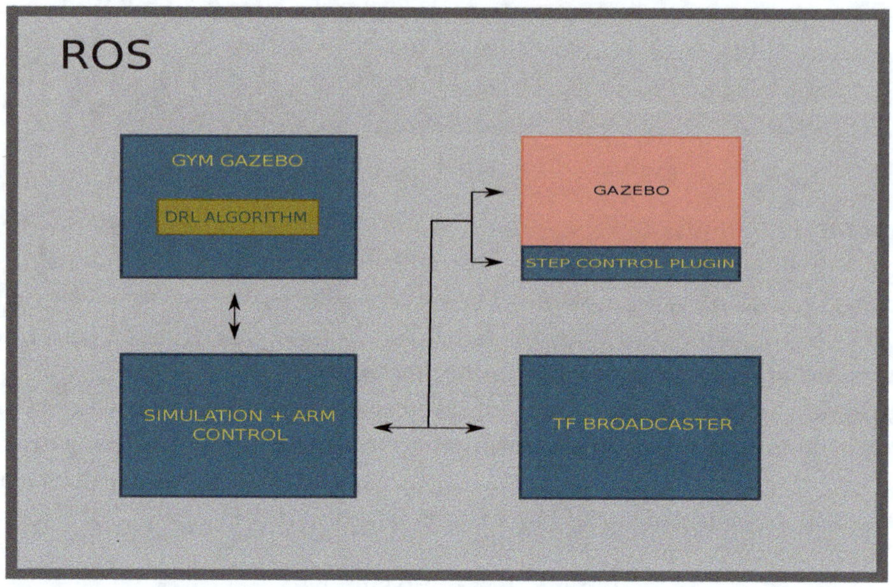

Figure 3. Implemented architecture. ROS: Robot Operating System.

4.1. Simulation

The learning process was carried out during a fixed number of time-steps, which were divided into episodes of 512 time-steps. Gazebo's *max step size* T_m is an important parameter that indicates the time-step at which Gazebo computes successive states of the model. The default value is 1 ms, but, in this case, $T_m = 2$ ms was used, which gave enough stability and enabled a faster simulation. Thus, each iteration of the physic engine meant 2 ms of simulated time and those iterations could be computed as quickly as possible to be able to accelerate the simulation. Besides, the *real time update rate* U_r parameter indicates how many physic iterations will be tried in one second. The *real time factor* is defined in Equation (10) and indicates how much faster the simulation goes in comparison with the real time. Thus, $rtf = 1$ indicates that the simulation is running on real time.

$$rtf = T_m \cdot U_r \tag{10}$$

In addition, a *frame-skip* N_{T_m} was defined to be able to get a reasonable control rate. In this application, $N_{T_m} = 4$ was used so the discrete time-step size is defined in Equation (11) and doing that we achieve a control rate of 125 Hz.

$$T_s = T_m \cdot N_{T_m} = 8 \text{ ms} \tag{11}$$

Thus, the length of each discrete time-step is 8 ms and those steps are computed faster than real time. Gazebo does not allow the control of the physics engine iterations, so a Gazebo plugin was developed to be able to execute the simulation for some iterations and to be able to compute those iterations as fast as possible. When the plugin was told to run a time-step, it ran N_{T_m} physics iterations and the *real time update rate* was set to 0 to compute those iterations as quickly as possible. Doing this, the *real time factor* increased and enabled us to run the simulation about 5–10 times faster than in real time.

At each step, the learning algorithm generates an action, specifically a velocity command, which is sent to the mobile base. After executing it and at the end of the step, a reward signal is given evaluating the quality of the action performed by the base. The action is the same for all the experiments and is defined in Equation (12).

$$a = [v_x \;\; v_y \;\; \omega_z] \tag{12}$$

To be able to control the mobile base, the *gazebo planar move* plugin was used, which enabled the sending of velocity commands to the omnidirectional base. In addition, it published the odometry information, which was used to know where the robot is respect to a parent coordinate frame. The path planning and control of the arm was made using MoveIt! [39] and enabled us to plan collision free trajectories. Besides that, MoveIt! uses an internal planning scene where objects can be added to be taken into account when the trajectory is planned.

4.2. Network Architectures

The library of DRL algorithms used is OpenAI Baselines. This library offers high quality implementations of many state-of-the-art DRL algorithms implemented in Tensorflow [40]. Although the implementation offers complex networks such as convolutional neural networks (CNN) or recurrent neural networks (RNN), we used fully connected multi-layer perceptrons (MLP) to parameterise both policies and value functions.

4.2.1. PPO

The network architecture used in this algorithm is the one proposed by Schulman et al. [29] in the original paper. To parameterise both the policy $\pi_\theta(a_t|s_t)$ and the value function $V_w(s_t)$, a four layered MLP was used. In both actor and critic, the first layer's size depends on the size of the state encoding, which was different in each test. The actor's input layer is followed by two hidden layers of 64 neurons each. The output layer's size depends on the action space's size. In this case, the action was a vector of three elements representing the velocity command, which has to be sent to the mobile base. Specifically, the velocity command is composed of the linear velocities in x and y axes, and the angular velocity in z axis. As mentioned before, this algorithm uses a stochastic policy and, thus, each action is defined by a Gaussian distribution. Each distribution is characterised by a mean and a standard deviation. Hence, the neurons of the last hidden layer are fully connected with the mean of each action. In addition, three variables are used to store and update the standard deviation.

The activation function applied to the output of each neuron of the hidden layers is *Tanh*. Instead, no activation function is applied to the output of the neurons of the last layer.

Regarding the value function $V_w(s_t)$, the first three layers have the same architecture as the actor's first three layers. The output layer, instead, is composed of a single neuron, which is fully connected with each neuron of the last hidden layer and outputs the expected average reward of a state. The activation function used in the hidden layers is *Tanh* as well.

4.2.2. DDPG

The network architecture used is the default implementation that the OpenAI Baselines library offers, which is very similar to the architecture proposed by Lillicrap et al. [33]. As in PPO, to parameterise both the policy $\pi_\theta(s_t)$ and the action–value function $Q(s_t, a_t)$, a four layered MLP was used. This algorithm aims to learn a deterministic policy so each state will be mapped with a concrete action. The actor's input layer also depends on the state encoding and it is followed by two hidden layers, composed of 64 neurons each. Concerning the output layer, it is composed of three neurons, one per action, and each neuron is fully connected with each neuron of the last hidden layer.

With respect to the activation functions, the *Rectified Linear Unit* (ReLU) function is used in each neuron of the hidden layers. In the output layer, instead, *Tanh* is used to bound the actions between -1 and 1. In addition, a process called *layer normalisation* [41] is applied to the output of each hidden layer to simplify the learning process and to reduce the training time. DDPG is an off-policy algorithm, hence it uses a more exploratory policy to get experience. As explained in the algorithm's description (Section 3.1.2), there are multiple types of noise but, in this application, the noise is added in the neural network's parameter space.

The action–value function $Q_w(s,a)$ receives as input the state and the action and outputs the expected future reward of executing action a_t in state s_t. The input layer depends on the size of the state codification. The input layer is followed by two hidden layers of 64 neurons each and actions are not included until the second hidden layer. The output layer only has one neuron and outputs the q-value. The activation function used in the critic network is ReLU for each neuron of the hidden layers. The last layer's neurons do not have any activation function, because the action–value does not have to be bounded.

To minimise the complexity of the critic and to avoid over-fitting, a $L2$ regularisation term is added to the critic's loss function [42]. This penalises the model for having parameters with high values.

4.3. Test Setup

Here, we describe the simulation setup in addition to the state codification and the reward function we used in each test. For both PPO and DDPG, the setup and reward functions were the same in all tests.

The robot's work-space is limited and a new episode starts when it goes out of limits. Those limits are defined in Equation (13).

$$x_r^w = [-1.0, 3.0] \qquad y_r^w = [-1.5, 0.0] \tag{13}$$

In addition, although the robot was penalised by navigating with high speed, the environment bounds the velocities applied to the mobile base. The maximum linear velocity in x and y axes was 1 m/s and the maximum angular velocity in z axis was 1 rad/s. Thus, if the algorithm predicted velocities higher than those, the limit velocities were applied and the agent was penalised proportionally. Besides, high accelerations were also penalised proportionally, but in this case the environment did not bound them.

The learning process was divided into episodes where the robot had T time-steps to complete the task. The episode length was $T = 512$ time-steps, which is about 4 s, and a discrete time-step t was terminal if:

- The robot collides with the table.
- Robot poses out of limits.
- $t = T$.

The tuning hyper-parameters used in each algorithm are described in Table 2 and were not changed across tests.

Table 2. Hyper-parameters.

PPO Setup Hyper-Parameters	
Actor/Critic learning rate	$f(step) = step \cdot 3 \times 10^{-4}$
Clip-range	0.2
Discount factor γ	0.99
Batch size	512
Mini-batch size	64
Updates	$\frac{training\ time\_steps}{batch\_size}$
Training epochs per update	10
DDPG Setup Hyper-Parameters	
Actor's learning rate	1×10^{-4}
Critic's learning rate	1×10^{-3}
Batch size	128
Discount factor γ	0.99
Critic l2 regularisation	1×10^{-2}
Running epochs	500
Cycles per epoch	10
Rollouts per cycle	512
Updates	$epochs \cdot cycles$
Training iterations per update	100

4.3.1. Test 1

Here, the objective was to learn a controller that guides the robot to a place near the table so that the arm could plan a trajectory up to the object. The robot and object initial poses were variable, which are defined in Equations (14) and (15). The robot's initial y_r^w coordinate was constant so that the robot always began at the same distance from the table. The state codification is defined in Equation (16) and it is composed of the following 10 variables:

Robot's position in world coordinate system: x_r^w, x_r^w
Robot's rotation on z axis in world coordinate system: yaw_r^w
Robot's linear velocities on x and y axes: v_x, v_y
Robot's angular velocity in z axis: ω_z
Object's position in world coordinate system: x_{obj}^w, y_{obj}^w
Distance between the robot and the object: $d(p_r, p_{obj})$
Remaining time steps to end the episode: t

$$x_r^w = [0.0, 3.0] \quad y_r^w = -1.5 \quad yaw_r^w = [-\pi, \pi] \tag{14}$$

$$x_{obj}^w = [0.5, 2.5] \quad y_{obj}^w = [0.5, 0.75] \quad z_{obj}^w = 1.2 \tag{15}$$

$$s = [x_r^w \ y_r^w \ yaw_r^w \ v_x \ v_y \ \omega_z \ x_{obj}^w \ y_{obj}^w \ d(p_r, p_{obj}) \ t] \in \mathbb{R}^{10} \tag{16}$$

The linear and the angular velocities included in the state were the velocities sent in the previous time-step to the robot (previous action). The reward function used is defined in Equation (17). A nonlinear function was used to give higher rewards when the robot was close to the object and high velocities and accelerations were penalised to encourage smooth driving. Here, Δv means the velocity difference (acceleration) between current and previous action. Instead, *v_high* penalised each linear or angular velocity being higher than a threshold. As mentioned before, the maximum linear and angular velocities were 1 m/s and 1 rad/s, respectively. In addition, in the last time-step of the episode, an additional reward was given if the arm could plan a trajectory up to the object. The number of remaining time-steps was included in the state for robot to be aware when this last step was coming. The *collision* variable was equal to 1 when a collision occurred, and 0 otherwise.

$$r(s_t, a_t) = \frac{1}{d(p_r, p_{obj})} \cdot (1 - collision) - 0.5 \cdot \Delta v - 0.5 \cdot v\_high(a_t) + 100 \cdot success \tag{17}$$

$$v\_high(a_t) = \sum_{i=1}^{3} a_{t_i}, \text{ if } a_{t_i} > k \tag{18}$$

4.3.2. Test 2

In this test, as in the previous one, the robot had to navigate to a place near the table such that the arm could plan a trajectory up to the object in the table. In this case, a wall was placed near the object to see if the algorithm could discard poses that were behind the wall. The robot's initial pose was variable (Equation (19)) and the object's initial pose was constant (Equation (20)). As in the first test, the robot's initial y_r^w coordinate was constant so that the robot always began at the same distance from the table. The state codification defined in Equation (21) is composed of the following 8 variables:

Robot's position in object's coordinate system: x_r^{obj}, x_r^{obj}
Robot's rotation on z axis in object's coordinate system: yaw_r^{obj}
Robot's linear velocities on x and y axes: v_x, v_y
Robot's angular velocity in z axis: ω_z
Distance between the robot and object's coordinate system origin: $d(p_r, \mathbb{O})$
Remaining time steps to end the episode: t

$$x_r^w = [0.0, 3.0] \quad y_r^w = -1.5 \quad yaw_r^w = [-\pi, \pi] \tag{19}$$

$$x_{obj}^w = 1.7 \quad y_{obj}^w = 0.75 \quad z_{obj}^w = 1.005 \tag{20}$$

$$s = [x_r^{obj} \;\; y_r^{obj} \;\; yaw_r^{obj} \;\; v_x \;\; v_y \;\; \omega_z \;\; d(p_r, \mathbb{O}) \;\; t] \in \mathbb{R}^8 \tag{21}$$

The used reward function is defined in Equation (22). In this case, the distance was computed between the robot's position p_r and object's coordinate system origin $\mathbb{O}$.

$$r(s_t, a_t) = \frac{1}{d(p_r, \mathbb{O})} \cdot (1 - collision) - 0.5 \cdot \Delta v - 0.5 \cdot v\_high(a_t) + 100 \cdot success \tag{22}$$

The environments used in the first and second tests are depicted in Figure 4a,b, respectively.

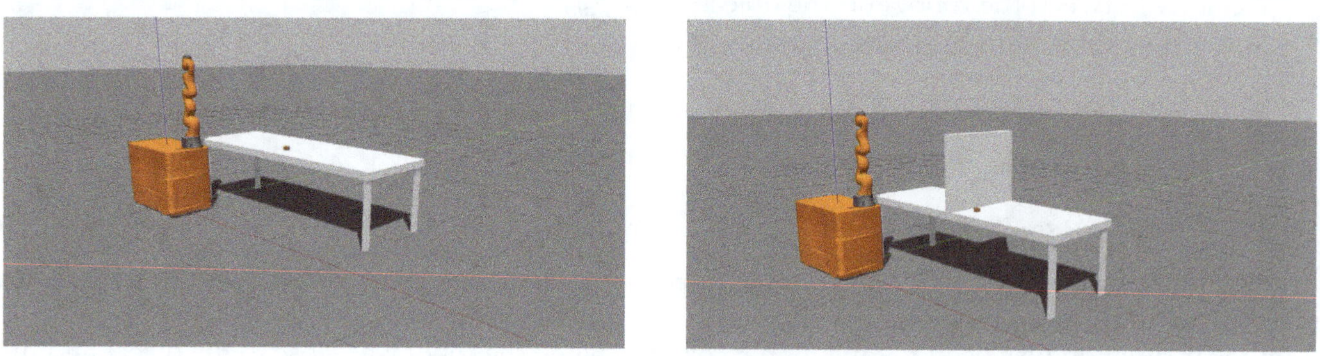

(**a**) Environment 1. (**b**) Environment 2.

Figure 4. Simulated environments.

5. Results

The learning process was carried out during a fixed number of time-steps and, to be able to evaluate the performance of the algorithms, several evaluation periods of 10 episodes each were made. Specifically, 500 evaluation periods were made uniformly distributed over the learning process and the metrics used to evaluate the performance were the mean accumulated reward and the success rate, which are the most used ones in the DRL community.

5.1. Test 1

The goal of this test was to learn a low-level controller that could drive the mobile manipulator's base close to the table, so that the arm could plan a trajectory up to the object. The learning process was carried out during 5M time-steps approximately, and an evaluation period was performed every 20 episodes. The results obtained with both PPO and DDPG algorithms are depicted in Figure 5. Figure 5a shows the mean accumulated rewards obtained in each of the 500 evaluation periods. The obtained success rates are depicted in Figure 5b. Due to the unstable nature of DRL algorithms, the obtained success rates vary considerably. Thus, to better understand the results, the mean value and the maximum/minimum values are shown per 10 test periods.

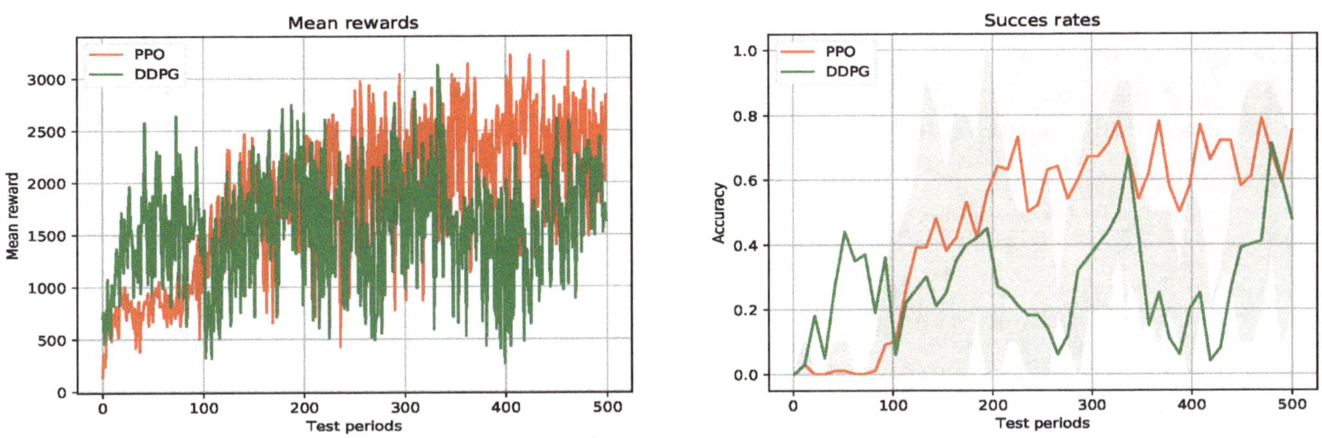

(**a**) Accumulated mean rewards. (**b**) Success rates.

Figure 5. Test 1 results.

Concerning the accumulated mean rewards, DDPG converged faster than PPO but, when the learning process moved along, PPO obtained higher rewards. The fact that DDPG obtained higher initial rewards is explained by two main reasons: (1) The policy bounds the actions between -1 and 1 so that it is not penalised for high velocities. (2) From the initial learning steps, this algorithm predicts smooth velocity changes across consecutive time-steps and, thus, is not penalised for high accelerations. Instead, PPO does not bound the actions and that is why it is penalised for high velocities and accelerations in the initial learning steps. Nevertheless, this algorithm is able to understand that lower velocities/accelerations are not penalised and, once learned, it obtains higher rewards than DDPG.

Regarding the success rates, it can be seen clearly that DDPG approximated the goal faster than PPO. Although, in some evaluation periods, it succeeded 100% of the time, it presented an unstable behaviour. As explained before, PPO takes more time to approximate the goal but, when it learns to drive smoothly, it shows a more stable behaviour. Besides, it could get 100% success rate considerably more times than DDPG. Off-policy algorithms are said to be able to better explore the environment than on-policy algorithms. In our environment, due to the initial random poses and the limited work-space of the robot, the exploration problem decreased considerably. PPO is an on-policy algorithm and takes more time to explore the environment than DDPG and that is another reason DDPG approaches the goal faster than PPO. Nevertheless, PPO relatively quickly improves enough for its policy to be able to succeed.

DRL has been applied to goal reaching applications either in manipulation or in navigation. Typically, those goals are not surrounded by obstacles and this makes the learning process easier. In this case, the goal (the object) was on top of the table and the robot had to learn to approximate to it without colliding with the obstacle. The reward function defined in this test encouraged the robot approximating to the object as much as possible and the robot had to use the contact information to learn where the table was to not collide with it. Although an additional reward was given when the arm's planning succeeded, it first tried to get as close as possible to the table, sometimes colliding with it and that is one of the reasons that caused the unstable behaviour of both algorithms.

Due to the variable pose of the object, in some episodes, if the object was near to table's edge, it had the possibility to get closer, scoring higher rewards. In addition, the initial poses of both the robot and the object in every test period were totally random and thus the accumulated rewards and/or success rates varied considerably.

5.2. Test 2

The goal of this test was to learn a low-level controller that could drive the mobile manipulator's base close enough to the table for the arm to be able to plan a trajectory up to the object. In this case, a wall was placed near the object with the aim of making the decision making problem more difficult. The learning process was carried out during 2.5M time-steps and an evaluation period aws performed every 10 episodes. In Figure 6, the results obtained with both PPO and DDPG algorithms are depicted.

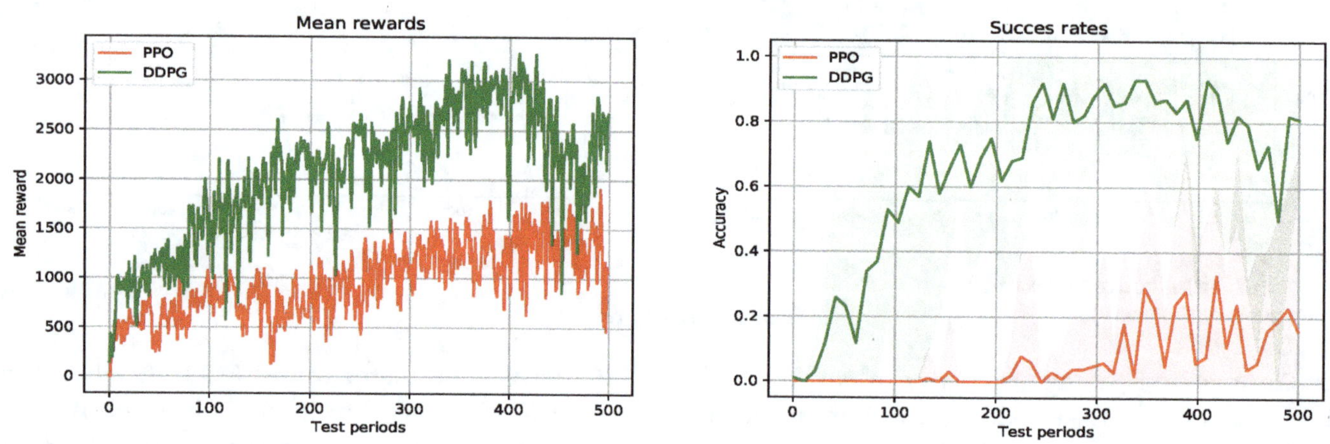

(**a**) Accumulated mean rewards. (**b**) Success rates.

Figure 6. Test 2 results.

Concerning the accumulated mean rewards (Figure 6a), DDPG's performance was better than PPO's for the overall learning process. Due to the bounded actions and the smooth output of DDPG, it could get higher rewards in the initial learning steps. Thanks to the fact that this algorithm is off-policy, it is able to better explore the environment and, thus, it learns to locate the robot's base much closer to the object than PPO. In addition, the constant pose of the box simplified the exploration problem, since the base always had to navigate to the same area of the environment and, consequently, the mean rewards were not as irregular as in the first test.

The success rates depicted in Figure 6b indicate that DDPG's performance was much better than PPO's, being the former able to score 100% success rate multiple times. DDPG could learn relatively quickly where the grasping zone is, discarding the poses over the wall. After navigating to the grasping zone, the learned policy stoped the robot sending near to zero velocities to the base. In addition, the deterministic behaviour of the algorithm enabled the robot to learn a more stable behaviour near the table, avoiding collisions. Even though PPO could learn to drive the robot near the correct grasping zone, it could not learn to stop it and continued sending velocities high enough to get the robot out of the grasping zone. Besides, due to the stochastic policy, the robot's performance near the table was not as robust as DDPG's, sometimes colliding with it.

6. Discussion and Future Work

In this work, we successfully implemented several DRL algorithms for learning picking operations using a mobile manipulator. It is shown that it is possible to use the arm's feedback to learn a low-level controller that drives the base to such a place that the arm is able to plan a trajectory up to the object.

Two state-of-the-art DRL algorithms were applied and compared to learn a mobile manipulation task. Specifically, the arm planner's feedback was used to learn to locate the mobile manipulator's base. To the best of our knowledge, this is the first approach that uses the arm's feedback to acquire a controller for the base.

Although DRL enables the learning of complex policies driven only by a reward signal, the unstable nature of those algorithms makes it difficult to obtain a robust behaviour. In addition, the sensitivity of DRL algorithms to hyper-parameters hinders finding the best parameter combination to get a robust and stable behaviour. Even though an optimisation over those hyper-parameters could be made, the large training times makes this process very expensive and commonly the default values proposed in the literature are used. Concerning the algorithms tested in this work, the results obtained show that the behaviour of the algorithms is dependant on several properties of the environment such as the state/action codification and the reward function definition. In addition, the network architecture used to encode either policies and the value-function has a large effect in the learning process. Even though the same network architecture and hyper-parameters were used in both tests, the results obtained are very different. Therefore, each environment needs the algorithm to be tuned in the best way to solve the problem, which is why the learned policies are not reproducible.

The reward function definition is another key point of the learning process, since it is entirely guided by this signal. A logical approach could be to give a reward to the robot in the last step only if the arm plans a trajectory, but, unfortunately, using only sparse rewards does not work well. In several goal reaching applications, either with arms or mobile bases, a distance dependant reward is proposed. In our application, this encourages the robot to navigate close to the object so the arm can plan a trajectory. After some tests, we saw that a nonlinear distance function accelerates the learning process.

Although DRL algorithms are typically trained in simulation, the experience acquisition is still the bottleneck of DRL based applications applied to robotics. Even though we accelerated the simulation, the entire learning process took several hours. Nevertheless, to be able to transfer the learned policy to a real robot and reduce the reality gap, the robot must be simulated accurately. Thus, a balance between simulation accuracy and training time acceleration should be found.

Moreover, it is complex to tune the algorithms to get a robust performance. We intend to increase the perception capabilities of the robot to be able to navigate in a more secure way and to be aware of the dynamical obstacles placed in the environment, using 2D/3D vision for example. Most of the applications in the literature map observations directly with low-level control actions and this black-box approach is not scalable. To be able to learn multiple behaviours and to combine them, hierarchical DRL proposes to learn a hierarchy of behaviours in different levels. In that vein, our goal is to learn a hierarchy of behaviours and, after training them in simulation, test those behaviours in a real robot.

Author Contributions: Conceptualisation, A.I. and L.S.; methodology, A.I., L.S., and E.L.; software, A.I.; formal analysis, A.I., J.U., A.F., and J.M.; data curation, A.I.; writing—original draft preparation, A.I.; writing—review and editing, E.L. and L.S.; and supervision, E.L., L.S.

Abbreviations

The following abbreviations are used in this manuscript:

RL	Reinforcement learning
DRL	Deep reinforcement learning
MDP	Markov Decision Process
ROS	Robot Operating System
CNN	Convolutional Neural Network
RNN	Recurrent Neural Network
MLP	Multi Layer Perceptron
DDPG	Deep Deterministic Policy Gradient
PPO	Proximal Policy Optimisation

References

1. Kober, J.; Bagnell, J.A.; Peters, J. Reinforcement learning in robotics: A survey. *Int. J. Robot. Res.* **2013**, *32*, 1238–1274. [CrossRef]
2. Li, Y. Deep reinforcement learning: An overview. *arXiv* **2017**, arXiv:1701.07274.
3. Dömel, A.; Kriegel, S.; Kaßecker, M.; Brucker, M.; Bodenmüller, T.; Suppa, M. Toward fully autonomous mobile manipulation for industrial environments. *Int. J. Adv. Robot. Syst.* **2017**, *14*. [CrossRef]
4. Nassal, U.; Damm, M.; Lüth, T. A mobile platform supporting a manipulator system for an autonomous robot. In Proceedings of the 5th World Conference on Robotics Research, Cambridge, MA, USA, 27–29 Spetember 1994.
5. Siciliano, B.; Khatib, O. *Springer Handbook of Robotics*; Springer: Berlin, Germany, 2016.
6. Padois, V.; Fourquet, J.Y.; Chiron, P. From robotic arms to mobile manipulation: On coordinated motion schemes. In *Intelligent Production Machines and Systems*; Elsevier: Amsterdam, The Netherlands, 2006; pp. 572–577.
7. Tan, J.; Xi, N.; Wang, Y. Integrated task planning and control for mobile manipulators. *Int. J. Robot. Res.* **2003**, *22*, 337–354. [CrossRef]
8. Berntorp, K.; Arzén, K.E.; Robertsson, A. Mobile manipulation with a kinematically redundant manipulator for a pick-and-place scenario. In Proceedings of the 2012 IEEE International Conference on Control Applications (CCA), Dubrovnik, Croatia, 3–5 October 2012; pp. 1596–1602.
9. Meeussen, W.; Wise, M.; Glaser, S.; Chitta, S.; McGann, C.; Mihelich, P.; Marder-Eppstein, E.; Muja, M.; Eruhimov, V.; Foote, T.; et al. Autonomous door opening and plugging in with a personal robot. In Proceedings of the 2010 IEEE International Conference on Robotics and Automation (ICRA), Anchorage, AK, USA, 3–8 May 2010; pp. 729–736.
10. Silver, D.; Hubert, T.; Schrittwieser, J.; Antonoglou, I.; Lai, M.; Guez, A.; Lanctot, M.; Sifre, L.; Kumaran, D.; Graepel, T.; et al. Mastering chess and shogi by self-play with a general reinforcement learning algorithm. *arXiv* **2017**, arXiv:1712.01815.
11. Li, J.; Monroe, W.; Ritter, A.; Galley, M.; Gao, J.; Jurafsky, D. Deep reinforcement learning for dialogue generation. *arXiv* **2016**, arXiv:1606.01541.
12. Yoo, S.; Yun, K.; Choi, J.Y.; Yun, K.; Choi, J. Action-Decision Networks for Visual Tracking with Deep Reinforcement Learning. In Proceedings of the 2017 IEEE Conference on Computer Vision and Pattern Recognition (CVPR), Honolulu, HI, USA, 21–26 July 2017.
13. Levine, S.; Pastor, P.; Krizhevsky, A.; Ibarz, J.; Quillen, D. Learning hand-eye coordination for robotic grasping with deep learning and large-scale data collection. *Int. J. Robot. Res.* **2018**, *37*, 421–436. [CrossRef]
14. Popov, I.; Heess, N.; Lillicrap, T.; Hafner, R.; Barth-Maron, G.; Vecerik, M.; Lampe, T.; Tassa, Y.; Erez, T.; Riedmiller, M. Data-efficient deep reinforcement learning for dexterous manipulation. *arXiv* **2017**, arXiv:1704.03073.
15. Quillen, D.; Jang, E.; Nachum, O.; Finn, C.; Ibarz, J.; Levine, S. Deep Reinforcement Learning for Vision-Based Robotic Grasping: A Simulated Comparative Evaluation of Off-Policy Methods. In Proceedings of the 2018 IEEE International Conference on Robotics and Automation (ICRA), Brisbane, QLD, Australia, 21–25 May 2018.

16. Gu, S.; Holly, E.; Lillicrap, T.; Levine, S. Deep reinforcement learning for robotic manipulation with asynchronous off-policy updates. In Proceedings of the 2017 IEEE International Conference on Robotics and Automation (ICRA), Singapore, 29 May–3 June 2017; pp. 3389–3396, doi:10.1109/ICRA.2017.7989385. [CrossRef]
17. Chen, Y.F.; Everett, M.; Liu, M.; How, J.P. Socially aware motion planning with deep reinforcement learning. In Proceedings of the 2017 IEEE/RSJ International Conference on Intelligent Robots and Systems (IROS), Vancouver, BC, Canada, 24–28 September 2017; pp. 1343–1350.
18. Tai, L.; Paolo, G.; Liu, M. Virtual-to-real deep reinforcement learning: Continuous control of mobile robots for mapless navigation. In Proceedings of the 2017 IEEE/RSJ International Conference on Intelligent Robots and Systems (IROS), Vancouver, BC, Canada, 24–28 September 2017; pp. 31–36.
19. Peng, X.B.; Berseth, G.; Yin, K.; Van De Panne, M. Deeploco: Dynamic locomotion skills using hierarchical deep reinforcement learning. *ACM Trans. Graph. (TOG)* **2017**, *36*, 41. [CrossRef]
20. Heess, N.; Sriram, S.; Lemmon, J.; Merel, J.; Wayne, G.; Tassa, Y.; Erez, T.; Wang, Z.; Eslami, A.; Riedmiller, M.; et al. Emergence of locomotion behaviours in rich environments. *arXiv* **2017**, arXiv:1707.02286.
21. Breyer, M.; Furrer, F.; Novkovic, T.; Siegwart, R.; Nieto, J. Flexible Robotic Grasping with Sim-to-Real Transfer Based Reinforcement Learning. *arXiv* **2018**, arXiv:1803.04996.
22. Stulp, F.; Fedrizzi, A.; Beetz, M.; Autonomous, I.; Group, S. Learning and Performing Place-Based Mobile Manipulation. In Proceedings of the 2009 IEEE 8th International Conference on Development and Learning, Shanghai, China, 5–7 June 2009; pp. 1–7.
23. Sutton, R.S.; Barto, A.G. *Reinforcement Learning: An introduction*; MIT press: Cambridge, UK, 1998; Volume 1.
24. Quigley, M.; Conley, K.; Gerkey, B.; Faust, J.; Foote, T.; Leibs, J.; Wheeler, R.; Ng, A.Y. ROS: An open-source Robot Operating System. In Proceedings of the ICRA Workshop on Open Source Software, Kobe, Japan, May 17 2009; VoLume 3, p. 5.
25. Dhariwal, P.; Hesse, C.; Klimov, O.; Nichol, A.; Plappert, M.; Radford, A.; Schulman, J.; Sidor, S.; Wu, Y. OpenAI Baselines. 2017. Available online: https://github.com/openai/baselines (accessed on 18 January 2019).
26. Brockman, G.; Cheung, V.; Pettersson, L.; Schneider, J.; Schulman, J.; Tang, J.; Openai, W.Z. OpenAI Gym. *arXiv* **2016**, arXiv:1606.01540.
27. Zamora, I.; Gonzalez Lopez, N.; Vilches, V.M.; Hernández Cordero, A.; Robotics, E. Extending the OpenAI Gym for Robotics: A Toolkit for Reinforcement Learning Using ROS and Gazebo. *arXiv* **2017**, arXiv:1608.05742v2.
28. Arulkumaran, K.; Deisenroth, M.P.; Brundage, M.; Bharath, A.A. A brief survey of deep reinforcement learning. *arXiv* **2017**, arXiv:1708.05866.
29. Schulman, J.; Wolski, F.; Dhariwal, P.; Radford, A.; Klimov, O. Proximal policy optimization algorithms. *arXiv* **2017**, arXiv:1707.06347.
30. Schulman, J.; Levine, S.; Moritz, P.; Jordan, M.I.; Abbeel, P. Trust Region Policy Optimization. *arXiv* **2015**, arXiv:1502.05477.
31. Kullback, S.; Leibler, R.A. On information and sufficiency. *Ann. Math. Stat.* **1951**, *22*, 79–86. [CrossRef]
32. Schulman, J.; Moritz, P.; Levine, S.; Jordan, M.; Abbeel, P. High-dimensional continuous control using generalized advantage estimation. *arXiv* **2015**, arXiv:1506.02438.
33. Lillicrap, T.P.; Hunt, J.J.; Pritzel, A.; Heess, N.; Erez, T.; Tassa, Y.; Silver, D.; Wierstra, D. Continuous Control With Deep Reinforcement Learning. *arXiv* **2016**, arXiv:1509.02971v5.
34. Silver, D.; Lever, G.; Heess, N.; Degris, T.; Wierstra, D.; Riedmiller, M. Deterministic Policy Gradient Algorithms. In Proceedings of the 31st International Conference on Machine Learning (ICML 2014), Beijing, China, 21–26 June 2014.
35. Mnih, V.; Kavukcuoglu, K.; Silver, D.; Rusu, A.A.; Veness, J.; Bellemare, M.G.; Graves, A.; Riedmiller, M.; Fidjeland, A.K.; Ostrovski, G.; et al. Human-level control through deep reinforcement learning. *Nature* **2015**, *518*, 529–533. [CrossRef]
36. Uhlenbeck, G.E.; Ornstein, L.S. On the theory of the Brownian motion. *Phys. Rev.* **1930**, *36*, 823. [CrossRef]
37. Plappert, M.; Houthooft, R.; Dhariwal, P.; Sidor, S.; Chen, R.Y.; Chen, X.; Asfour, T.; Abbeel, P.; Openai, M.A. Parameter Space Noise for Exploration. *arXiv* **2018**, arXiv:1706.01905v2.

38. Todorov, E.; Erez, T.; Tassa, Y. Mujoco: A physics engine for model-based control. In Proceedings of the 2012 IEEE/RSJ International Conference on Intelligent Robots and Systems (IROS), Vilamoura, Portugal, 7–12 October 2012; pp. 5026–5033.
39. Chitta, S.; Sucan, I.; Cousins, S. Moveit![ROS topics]. *IEEE Robot. Autom. Mag.* **2012**, *19*, 18–19. [CrossRef]
40. Abadi, M.; Barham, P.; Chen, J.; Chen, Z.; Davis, A.; Dean, J.; Devin, M.; Ghemawat, S.; Irving, G.; Isard, M.; et al. TensorFlow: A System for Large-Scale Machine Learning. In Proceedings of the 12th USENIX Symposium on Operating Systems Design and Implementation (OSDI), Savannah, GA, USA, 2–4 November 2016; Volume 16, pp. 265–283.
41. Ba, J.L.; Kiros, J.R.; Hinton, G.E. Layer Normalization. *arXiv* **2016**, arXiv:1607.06450.
42. Ng, A.Y. Feature selection, L 1 vs. L 2 regularization, and rotational invariance. In Proceedings of the Twenty-First International Conference on Machine Learning, Banff, AB, Canada, 4–8 July 2004; ACM: New York, NY, USA, 2004; p. 78.

Permissions

All chapters in this book were first published in MDPI; hereby published with permission under the Creative Commons Attribution License or equivalent. Every chapter published in this book has been scrutinized by our experts. Their significance has been extensively debated. The topics covered herein carry significant findings which will fuel the growth of the discipline. They may even be implemented as practical applications or may be referred to as a beginning point for another development.

The contributors of this book come from diverse backgrounds, making this book a truly international effort. This book will bring forth new frontiers with its revolutionizing research information and detailed analysis of the nascent developments around the world.

We would like to thank all the contributing authors for lending their expertise to make the book truly unique. They have played a crucial role in the development of this book. Without their invaluable contributions this book wouldn't have been possible. They have made vital efforts to compile up to date information on the varied aspects of this subject to make this book a valuable addition to the collection of many professionals and students.

This book was conceptualized with the vision of imparting up-to-date information and advanced data in this field. To ensure the same, a matchless editorial board was set up. Every individual on the board went through rigorous rounds of assessment to prove their worth. After which they invested a large part of their time researching and compiling the most relevant data for our readers.

The editorial board has been involved in producing this book since its inception. They have spent rigorous hours researching and exploring the diverse topics which have resulted in the successful publishing of this book. They have passed on their knowledge of decades through this book. To expedite this challenging task, the publisher supported the team at every step. A small team of assistant editors was also appointed to further simplify the editing procedure and attain best results for the readers.

Apart from the editorial board, the designing team has also invested a significant amount of their time in understanding the subject and creating the most relevant covers. They scrutinized every image to scout for the most suitable representation of the subject and create an appropriate cover for the book.

The publishing team has been an ardent support to the editorial, designing and production team. Their endless efforts to recruit the best for this project, has resulted in the accomplishment of this book. They are a veteran in the field of academics and their pool of knowledge is as vast as their experience in printing. Their expertise and guidance has proved useful at every step. Their uncompromising quality standards have made this book an exceptional effort. Their encouragement from time to time has been an inspiration for everyone.

The publisher and the editorial board hope that this book will prove to be a valuable piece of knowledge for researchers, students, practitioners and scholars across the globe.

List of Contributors

Uwe Jahn, Daniel Heß, Merlin Stampa, Christof Röhrig and Carsten Wolff
IDiAL Institute, Dortmund University of Applied Science and Arts, Otto-Hahn-Str. 23, 44227 Dortmund, Germany

Andreas Sutorma
Faculty of Information Technology, Dortmund University of Applied Science and Arts, Sonnenstraße 96, 44139 Dortmund, Germany

Peter Schulz
Faculty of Engineering and Computer Science, Hamburg University of Applied Sciences, 20099 Hamburg, Germany

Noé Pérez-Higueras, Alberto Jardón, Ángel Rodríguez and Carlos Balaguer
Robotics Lab, Department of Systems Engineering and Automation, University Carlos III of Madrid, Avda. Universidad, 30, Leganés, 28911 Madrid, Spain

Sendren Sheng-Dong Xu and Tai-Chun Chiu
Graduate Institute of Automation and Control, National Taiwan University of Science and Technology, Taipei City 10607, Taiwan

Hsu-Chih Huang and Shao-Kang Lin
Department of Electrical Engineering, National Ilan University, Yilan City 26047, Taiwan

Dong Zhao and Hao Guo
Key Laboratory of Mechanism Theory and Equipment Design of The State Ministry of Education, Tianjin University, Tianjin 300072, China

Carlos Villaseñor, Nancy Arana-Daniel, Alma Y. Alanis, Carlos Lopez-Franco and Javier Gomez-Avila
Centro Universitario de Ciencias Exactas e Ingenierías, Universidad de Guadalajara, Blvd Marcelino García Barragán 1421, Guadalajara 44430, Mexico

Youdong Chen and Ling Li
School of Mechanical Engineering and Automation, Beihang University, Beijing 100191, China

Yunsheng Fan, Hongyun Huang and Yuanyuan Tan
School of Marine Electrical Engineering, Dalian Maritime University, Dalian 116026, China

Yiqing Huang, Zhikun Li, Yan Jiang and Lu Cheng
College of Electrical Engineering, Anhui Polytechnic University, Wuhu 241000, China

Jingchuan Wang, Ming Zhao and Weidong Chen
Department of Automation, Shanghai Jiao Tong University, Shanghai 200240, China
Laboratory of System Control and Information Processing, Ministry of Education of China, Shanghai 200240, China

Sheng Wen
Engineering Foundation Teaching and Training Center, South China Agricultural University, Guangzhou 510642, China
National Center for International Collaboration Research on Precision Agriculture Aviation Pesticides Spraying Technology, Guangzhou 510642, China

Quanyong Zhang, Jizhong Deng, Yubin Lan, Xuanchun Yin and Jian Shan
National Center for International Collaboration Research on Precision Agriculture Aviation Pesticides Spraying Technology, Guangzhou 510642, China
Engineering College, South China Agricultural University, Guangzhou 510642, China

Jong-Chih Chien
Degree Program of Digital Space and Product Design, Kainan University, Taoyuan 33857, Taiwan

Zih-Yang Dang
Department of Electrical Engineering, Chang Gung University, Taoyuan 33302, Taiwan

Jiann-Der Lee
Department of Electrical Engineering, Chang Gung University, Taoyuan 33302, Taiwan
Department of Neurosurgery, Chang Gung Memorial Hospital at LinKou, Taoyuan 33302, Taiwan
Department of Electrical Engineering, Ming Chi University of Technology, New Taipei City 24301, Taiwan

Guo Li and Songhao Piao
Department of Computer Science and Technology, Harbin Institute of Technology, Harbin 150001, China

Rui Lin, Maohai Li and Rongchuan Sun
School of Mechanical and Electrical Engineering, Soochow University, Suzhou 215000, China

Jae Hyung Jang and Gi-Hun Yang
Robotics Group, Korea Institute of Industrial Technology, Ansan-si 15588, Gyeonggi-do, Korea

Ander Iriondo, Loreto Susperregi, Julen Urain, Ane Fernandez and Jorge Molina
Department of Autonomous and Intelligent Systems, Fundación Tekniker, Iñaki Goenaga, 5-20600 Eibar, Spain

Elena Lazkano
Faculty of Computer Science, Po Manuel Lardizabal, 1-20018 Donostia-San Sebastián, Spain

Index

www.ingramcontent.com/pod-product-compliance
Lightning Source LLC
LaVergne TN
LVHW081038210926
848485LV00004B/1053